T0211399

Handbook
of
Culture of Abalone
and
Other Marine
Gastropods

Editor

Kirk O. Hahn

Bodega Marine Laboratory
University of California
Bodega Bay, California

CRC Press
Taylor & Francis Group
Boca Raton London New York

CRC Press is an imprint of the
Taylor & Francis Group, an **informa** business

CRC Press
Taylor & Francis Group
6000 Broken Sound Parkway NW, Suite 300
Boca Raton, FL 33487-2742

Reissued 2019 by CRC Press

A Library of Congress record exists under LC control number:

Publisher's Note
The publisher has gone to great lengths to ensure the quality of this reprint but points out that some imperfections in the original copies may be apparent.

Disclaimer
The publisher has made every effort to trace copyright holders and welcomes correspondence from those they have been unable to contact.

ISBN 13: 978-0-367-21078-6 (hbk)
ISBN 13: 978-0-367-21081-6 (pbk)
ISBN 13: 978-0-429-26526-6 (ebk)

Visit the Taylor & Francis Web site at http://www.taylorandfrancis.com and the
CRC Press Web site at http://www.crcpress.com

THE EDITOR

Kirk O. Hahn is currently studying for his Ph.D. in Endocrinology at the University of California at Davis with an emphasis in molluscan aquaculture. He received his B.S. in Biology from the University of California at Irvine in 1975 and attended Tohoku University in Sendai, Japan in 1981.

While attending school in Japan, Mr. Hahn was awarded the Monbusho Scholarship which is given to foreigners by the Japanese government. He conducted research on abalone aquaculture and also learned the techniques of aquaculture (abalone, oysters, pearl oysters, etc.). These projects allowed Mr. Hahn to conduct work on *Trochus* aquaculture and visit facilities throughout the South Pacific, New Zealand, Australia, and Hawaii. As a result of this experience, he has formed a consulting firm, Pacific Aquaculture, to develop aquaculture techniques and transfer technology to countries and private companies around the world.

In addition to his duties as consultant, Mr. Hahn has also been a visiting lecturer at the University of California at Berkley and a teaching assistant at the Bodega Marine Laboratory. He is a member of the World Aquaculture Society and is the recipient of numerous grants. His doctoral research concentrates on the endocrine regulation of reproduction in abalone and *Trochus niloticus*.

ACKNOWLEDGMENTS

There were many people who helped produce this book. First and foremost, I want to thank my major professor, Ernest S. Chang, for his support (both financial and intellectual) and encouragement. Truly, this book would never have been done without his help.

The National Sea Grant College Program and James Sullivan, Director of the California Sea Grant College Program, are thanked for supporting my graduate research with a traineeship. The traineeship was invaluable and allowed me to devote my full effort to research.

Mitsue Doi translated the many Japanese research papers into English. She did an excellent job in the difficult task of translating the highly technical papers. Her friendship and help over the years has been greatly appreciated. Teva Siu translated the French literature on abalone aquaculture at Argenton, France into English. His work was very important to the writing of the section on French abalone aquaculture. George Trevelyan and Nancy Coultrup assisted me in collecting data for my research on the reproduction of the tropical top shell reported in this book.

The wide scope of this book is partially due to the efforts of Eleanor Uhlinger, research librarian at the Bodega Marine Laboratory. She was able to locate obscure and difficult to find papers published throughout the world. I thank Eleanor Uhlinger, George Trevelyan, Michael Brody, and Fred Sly for reviewing my chapters. Additionally, George Trevelyan drew all the figures in my chapters and Michael Brody redrew the ink drawings of the abalone larvae.

I also thank the Japan Ministry of Education, Tohoku University (Sendai, Japan), and the University of California for giving me the Monbusho Scholarship to spend 1 year in Japan to study abalone aquaculture and reproduction. Tetsuo Seki (Oyster Research Institute) and Nagahisa Uki (Tohoku Regional Fisheries Research Laboratory) spent many hours with me discussing their aquaculture facilities. Their help and assistance made my stay in Japan very educational.

I thank the following people for supplying information on their research and/or allowing me to visit their facilities: John Grant (Australia), Len Tong (New Zealand), Jean-Pierre Flassch (France), Jacques Clavier (France), Nagahisa Uki (Japan), Tetsuo Seki (Japan), Terepai Maoate Jr. (Cook Islands), Karen Norman-Boudreau (U. S.), George Trevelyan (U. S.), Lisa Holsinger (U. S.), and Keith Scott (U. S.).

I thank NOAA, National Sea Grant College Program, and the Department of Commerce for partial support of my research on abalone and tropical top shell (Grant NA80AA-D-00120, Project R/A-61 to Ernest S. Chang). I also received partial funding from the University of California Jastro-Shields Research Scholarship, the Lerner-Gray Fund for Marine Research, Sigma Xi, and the University of California Research Expeditions Program.

I thank all the professors, students, and staff at the Bodega Marine Laboratory, University of California, Davis for their help and friendship during my stay at the laboratory.

I thank the contributing authors: Mia Tegner, Robert Butler, John McMullen, Tim Thompson, Earl Ebert, James Houk, Sung Kyoo Yoo, and Hon-Cheng Chen for making this book possible. Each author took valuable time away from their research or business to write their chapter. The chapters written by the contributing authors were invaluable to the scope and usefulness of this book.

This book is dedicated to my parents, Joy and Otto. They encouraged me to learn and had the patience to allow me.

CONTRIBUTORS

Robert A. Butler
Staff Research Associate
Ocean Research Division
Scripps Institution of Oceanography
LaJolla, California

Hon-Cheng Chen, Ph.D.
Professor
Department of Zoology
Institute of Fishery Science
National Taiwan University
Taipai, Taiwan

Earl E. Ebert
Department of Fish and Game
Marine Culture Laboratory
State of California
Monterey, California

Kirk O. Hahn
Ph.D. Candidate
Bodega Marine Laboratory
University of California
Bodega Bay, California

James L. Houk
Procemar, LDTA
Arica, Chile

John D. McMullen
President
AbLab
Port Nueneme, California

Mia J. Tegner, Ph.D.
Associate Research Marine Biologist
Scripps Institution of Oceanography
LaJolla, California

Tim Thompson
Vice President
AbLab
Port Nueneme, California

Sung Kyoo Yoo, Ph.D.
Department of Aquaculture
National Fisheries University of Pusan
Pusan, Korea

TABLE OF CONTENTS

SECTION I: BIOLOGY OF ABALONE

SECTION II: CULTURE TECHNIQUES

Biology of Abalone

SURVEY OF THE COMMERCIALLY IMPORTANT ABALONE SPECIES IN THE WORLD

Kirk O. Hahn

INTRODUCTION

Abalone are large herbivorous marine gastropods with all species in one genus, *Haliotis* (Table 1).[1] (Older scientific papers also use the genus name, *Notohaliotis, Euhaliotis,* or *Sanhaliotis.*[2]) Aristotle (ca. 347), in *Historia Animalium,* called abalone *"Agria lepas"* (wild limpet) and *"Thalattion us"* (marine ear).[3] Linnaeus in *Systema Naturae,* Ed. II (1740) named the genus "Haliotis", which means sea ear.[4] Table 2 gives some of the common generic names for Haliotid.[3-5]

The abalone is one of the most primitive gastropods in form and structure.[6] The round or ear-shaped shell has a row of respiratory pores located along the left margin of the shell. As the animal grows, older pores are successively filled in and closed. The number of open pores varies among different abalone species.[7] All species of *Haliotis* like well-oxygenated sea water with a stable salinity.[8]

DISTRIBUTION

There are about 100 species of *Haliotis* world wide. They are found in both hemispheres with the larger species found in temperate regions and smaller species in tropical and arctic regions.[9] The greatest number of species is found in the central and south Pacific, and parts of the Indian Ocean, but none of these species are large in size.[6,7]

Abalone are found along the coasts of most temperate and tropical regions except South America and eastern North America. Abalone are found along the west coast of North America (Baja California to Alaska); along the eastern and southern coasts of Asia (U.S.S.R., Korea, Japan, China, Taiwan, Indonesia, Borneo, Malaysia, Cambodia, Thailand, India, and Sri Lanka); islands in the Pacific Ocean as far east as the Toumotus; Australia; New Zealand; Africa (Egypt, Tanzania, Mozambique, Natal, Madagascar, Cape of Good Hope, Gold Coast, and the islands of Madeira and Azores); and Europe (France, Spain, Italy, Yugoslavia, and Greece).[7]

COMMERCIALLY IMPORTANT SPECIES

At the present time, abalone aquaculture consists of culturing local species. Satisfactory results have been obtained by several culture facilities but there have not been any great advances in production. Future emphasis will be placed in selecting abalone species or producing hybrids that have improved qualities (e.g., meat quality, growth rate) over local species. For example, a temperate species may have a higher growth rate when reared in semitropical regions with constant water temperatures, or a species from a different region of the world may have more tender or better tasting meat than local species. Abalone are relatively easy to hybridize and several species could be crossed to select for superior qualities. This chapter will survey the characteristics (e.g., physical appearance, habitat, distribution) of the commercially important abalone species in the world.

Abalone species are identified by specific characteristics of the animal and not by the shell alone. Care must be taken during identification since some species living in the same region appear identical, especially when small. Do not be misled by common names which usually refer to shell color. The shell color may not be visible or the name may not accurately describe the shell color.[1]

Table 1
SYSTEMATICS OF
ABALONE

Kingdom Animalia
Phylum Mollusca
Class Gastropoda
Subclass Prosobranchia
Order Archeogastropoda
Suborder Zygobranchia
Superfamily Pleurotomariacea
Family Haliotidae
Genus Haliotis

Table 2
COMMON NAMES OF WORLD HALIOTIDS

Abalone	U.S.
Awabi	Japan
Cholburi	Thailand
Holley	Amboina (Molluccas), Ceram
Telinga Maloli	Malaysia
Ria Scatsjo	Malaysia
Mutton fish	Australia
Paua	New Zealand, Tasmania (Maori)
Karariwha	New Zealand, Tasmania (Maori)
Aulon, Aulone	Spanish American
Aulone	Mexican
Ormer, Ormier, Omar	England
Venus ear	England
Norman shell	Old English
Orielle de Mer	France
Si-ieu	France
Ohrsnecke	Germany
Meerohren	Germany
Lapa Burra	Portugal
Senorinas	Spain
Orecchiale	Italy
Patella Reale	Sicily
Venus ear	Greece
Orechio de San Pietro	Adriatic, Dalmatia (Yugoslavia)
Perlemoen	South Africa

(From Ebert, E.E., Abalone, in *The Encyclopedia of Marine Resources*, Firth, F. E., Ed., Van Nostrand Reinhold, New York, 1969, 1. With permission).

North America
Haliotis rufescens

The red abalone, *H. rufescens,* is the largest in the world, often reaching lengths greater than 27.5 cm and weighing over 1.7 kg.[10] Red abalone has traditionally been the most popular and commercially important species in California. Shell exterior is dull brick red with a red shell edge, and the respiratory pores (three to four open holes) are raised slightly and oval in shape. The shell interior is iridescent green, pink, and copper, and has a large muscle scar that usually has dark green markings. The side of the body is black but may appear to be slightly striped with dark and light bands in some individuals. The epipodium is scalloped with tentacles extending beyond the edge of the shell.[6]

Red abalone are found along the west coast of the U.S. and Mexico, with most animals found from intertidal to 20 m.[6,11] The depth range reflects the preferred environmental conditions. Red abalone prefer water in the temperature range of 7 to 16°C and areas of active wave action along rocky headlands and promontories.[1,6] They are rarely found in sheltered bays.[6] Red abalone usually eat *Macrocystis* spp. in their southern range and *Nerocystis* spp. in their northern range, but will eat most brown macroalgae.[1]

Haliotis fulgens

H. fulgens is known by the common names; green, southern green, or blue abalone.[6,12] Individuals can reach 25 cm in length but most individuals are between 12.5 and 20 cm. The shell is oval, thick, and the surface has numerous broad, flat, topped ribs separated by narrow straight-sided grooves parallel to the respiratory pores. The shell edge is very thin and reddish brown in color. The exterior surface of the shell is reddish brown or olive green, usually with a red band along the respiratory pores. The respiratory pores are small, circular, slightly elevated, and five to seven pores are open. The shell interior is extremely beautiful with green, blue, and copper patterns, and a brilliant muscle scar.[1] The shell is considered to be one of the most beautiful in the world. Green abalone have short, thick, light green or gray tentacles that extend a short distance from the shell edge. The epipodium has brown and light green patches and is scalloped along the edge with small protuberances which give it a rough frilled surface. These abalone are found from Santa Barbara to Baja California and on the Channel Islands. They are found from low tide to between 17 and 20 m, but most are in 3 to 7 m.[6]

Haliotis corrugata

H. corrugata, pink, corrugated or yellow abalone, can grow to 25 cm, but most are 15 to 17.5 cm, averaging 1 kg in weight.[6,10,12] The shell exterior is a dull green or reddish color, but may be entirely turquoise, and has a corrugated dark red to turquoise shell edge. The shell is thick, circular, and highly arched with rough corrugations on the surface. There are two to four large, highly elevated open respiratory pores. The shell interior is iridescent blue, green, pink, and copper with a large muscle scar. They have long, slender black tentacles which extend beyond the shell edge. The black body has a lace-like epipodium with black and white spots.[1] This species is found along the coasts of southern California below Point Conception, the Channel Islands, and Baja California. Individuals are found from intertidal to 60 m with most in 7 to 26 m depth.[6]

Haliotis sorenseni

H.sorenseni is commonly called white, sorensen, or Chinese abalone. Shell length can reach 25 cm but most are 12.5 to 20 cm.[6] The shell is oval, highly arched, and very thin; reddish in color; and has a bright orange apex. The shell edge is very thin and red. Roughly 25 to 50% of the individuals have an orange band parallel to the respiratory pores. This band fades to a pale red after the shell is 2 inches in length. Hybrids with white abalone have an orange band the entire length of the shell. The shell has regular spiral lines and three to five elevated open respiratory pores. Shell interior is pearly white with tints of pink, and no muscle scar.[1] The epipodium is scalloped and lacy, and mottled with yellowish green or brownish beige. Tentacles are yellow and may have a greenish tint. The mantle over the head region is purple while the body is typically yellow or orange, and the foot is bright orange. The maroon-colored digestive gland in the larval white abalone is distinctive of this species.[13] White abalone occur from Point Conception, California to Baja California in 12 m to over 35 m of water.[1] They are mainly found on the Channel Islands and Mexico.[6] White abalone is deep water species and found from 5 to 50 m with the greatest concentration between 27 to 35 m. *H. sorenseni* is the most highly valued abalone species in North America due to its tender white meat.[10,13]

Haliotis assimilis

H. assimilis, threaded abalone, is sometimes called *H. kamtschatkana assimilis*.[1] Most individuals are smaller than 10 cm but some can grow to 17.5 cm. The shell is oval, moderately thin, and arched. The shell exterior has prominent, broad spiral ribs interspaced by several low, narrow ribs or ridges. There are four to six small, tubular, raised open respiratory pores. Shell color varies more than in other species in California and ranges from white, pink, pale orange, red, dark brown, light green, turquoise, to dark green. The most common color is dark turquoise. The shell edge is reddish brown and green, and the shell interior is pearly white with no muscle scar.[1] The body is yellowish cream with brown blotches, and the foot, mantle, and eye stalks are orange. The light brown or cream epipodium has numerous small rounded protuberances on the surface and a fringe-like upper edge. Tentacles are short and brownish yellow or tan. Threaded abalone are found from central California to Baja California, Mexico. They are found in greatest densities from 23 to 35 m.[6]

Haliotis cracherodii

The black abalone, *H. cracherodii*, can reach 20 cm in length but most are 7.5 to 12.5 cm. The shell is circular with a smooth exterior with little marine growth, and is typically dark blue or greenish black, but sometimes orange. Respiratory pores are small in diameter, flush with the surface, and five to nine are open. The shell interior is silvery with green and pink reflections, and no muscle scar. The smooth, black body is scalloped along the upper edge of the epipodium, and has scattered short, slender, black tentacles.[6] Black abalone are found from northern California to Baja California in water from 7 to 24°C.[1] Most individuals are found intertidally but some are found to depths of 6 m. They are usually found crowded together and sometimes stacked two to three animals high. Black abalone graze on algae from the shell of other individuals since the intertidal zone sometimes lacks seaweeds. This causes the shells to lack encrusting marine growth. This species is not a preferred food due the dark meat color and inferior flavor.[6]

Haliotis walallensis

Most adult *H. walallensis*, flat or northern green abalone, are 7.5 to 12.5 cm in length but can reach 17.5 cm. The shell is oval, very flat, and has ridges parallel to the respiratory pores with raised growth lines perpendicular to these ridges. There are usually five to six open respiratory pores. Shell color is dark brick red with occasional spots of greenish blue and white. The shell edge is dark red or green.[1] The shell interior is pale pink with green reflections and no muscle scar. The body is mottled yellow and brown with tinges of green, and the epipodium is lacy along the upper edge, colored a yellowish-green with large brown and yellow splotches. Tentacles are dark green or brown, and slender.[6] Flat abalone range from British Columbia to the Mexican border. They are found from intertidal to 17 m in northern California, 7 to 20 m in central California, and at depths greater than 27 m in southern California.[1]

Haliotis kamtschatkana

H. kamtschatkana, pinto abalone, is sometimes called *H. kamtschatkana kamtschatkana*.[1] The shell length is usually 10 cm but sometimes reaches 15 cm. The shell exterior is covered with the same marine growth found on the substrata, making distinguishing them from their surroundings difficult. There are four to six open respiratory pores and shell form is highly variable.[1,6] In its northern range (Alaska to Point Sur, California), the shell is long, narrow, highly arched, has a rough irregular surface, and a prominent spire. In its southern range (Point Sur to Point Conception, California), the shell is oval, not as highly arched, surface is more regular and smooth, and the spire is not as high.[6] Shell color is reddish brown, pale orange or greenish-brown, with blue, green, or white streaks. The shell edge is usually pale red or green, and the interior is an iridescent pearly white with no muscle scar.[1] The raised lumpy areas on the exterior are mirrored by hollows in the interior. The body is mottled tan and greenish brown with tinges of orange. Tentacles are slender, light yellowish brown or green, and the tips extend from under the shell edge when the animal is moving or feeding. The epipodium is scalloped, lacelike along the upper edge and colored a mottled greenish brown, pale yellow or dark brown, with orange ends.[6]

The pinto abalone is found from the Aleutians to Point Conception, California. They are found intertidally in the northern range and 12 to 17 m in the southern range. Individuals live on top of boulders and rocks, in contrast to other abalones which live in crevices or under rocks. Food consists of smaller algae growing on the substratum rather than larger sea weeds. This gives the shell the varied colors.[6] The pinto abalone has a lower meat weight in proportion to shell length than other North America species.[14]

Japan
Haliotis discus hannai

Ezo awabi, *H. discus hannai*, is probably the most thoroughly studied abalone species in the world.[7],[9] *H. discus hannai* has at times been considered a variety of *H. gigantea*, *H. kamtschatkana*, and *H. discus*.[2,15-17] There has been a lot of confusion over the classification of this species. *H. discus hannai* was once thought to be the same species as *H. kamtschatkana* of North America. *H. discus hannai* was then considered a subspecies of *H. discus* even though ezo awabi is smaller, the shell is thinner and "bumpy", and individuals are restricted to shallow water. However, *H. discus hannai* gradually changed to look like *H. discus* when it was transplanted to southern regions. It is now known that *H. discus hannai* is morphologically, ecologically, and genetically different from *H. discus*.[7]

Ezo awabi grows to between 18 and 20 cm, but is usually harvested at 9 cm.[7] The shell is thin and elliptical, and there are three to five open respiratory pores. The yellow foot sole has a black edge.[9] *H. discuss hannai* are found from Siberia to China.[7] This species is not found along the eastern coast of Hokkaido which always has a cold current. Natural distribution is along the Pacific coast north of Ibaraki Prefecture and the Japan Sea coast of Hokkaido.[9] Ezo awabi is the only species in Japan tolerant to cold water temperatures. The water temperature is below 5°C for 3 months of the year in many regions inhabited by ezo awabi.[18-20] *H. discus hannai* is restricted to shallow water from 1 to 5 m of depth.[9] Ezo awabi is the most commercially valuable abalone in Japan and makes up 46% of the catch.[21] Also, the majority of cultured abalone seed in Japan are *H. discus hannai*.

Haliotis discus

H. discus is commonly called kuro awabi (black abalone), oni, or onigai.[7,12] It is also called *H. gigantea discus* in older scientific papers.[16] Kuro awabi grows to 20 cm in length. The shell is elliptical and the exterior surface is smoother than other species in Japan. The respiratory pore size is intermediate between *H. sieboldii* and *H. gigantea*.[9] The black foot sole has a blue edge. Kuro awabi is found along central and southern Japan and is most abundant along the Pacific coast of Honshu. Kuro awabi, a shallow water species, is most abundant at 2 m with the abundance dropping rapidly at depths greater than 8 m.[7] Distribution is about the same as *H. gigantea* and *H. sieboldii* but the vertical distribution is even shallower. *H. discus* makes up 26% of the Japanese catch although the meat is slightly hard.[9]

Haliotis diversicolor supertexta

Adult tokobushi, *H. diversicolor supertexta*, are about 5 cm in length.[22] Tokobushi have six to nine open respiratory pores and are smaller than other Japanese abalone. Distributon of the species is restricted to warm water current regions. *H. diversicolor supertexta* has a smaller distribution than *H. discus* in regions of warm water temperatures. Tokobushi is of minor importance to the commercial fishery.[9]

Haliotis gigantea

Madaka, *H. gigantea*, is the largest Japanese species and reaches a length of 25 cm.[7,9] The shell is convex with four to five open respiratory pores which protrude prominently. The foot sole is mostly light brown. Madaka are found in deep water down to 50 m with the majority in water greater than 10 m. Individuals are found south of Ibaraki Prefecture and along islands in the Sea of Japan southwest of Hokkaido.[9] All Japanese abalone species except *H. diversicolor supertexta* were once thought to be varieties of this species.[12,16] This species is commercially valuable (14% of the total catch) and the meat is tender.[9,21]

Haliotis sieboldii

Megai, *H. sieboldii*, has a round shell and grows to about 17 cm.[7] The shell surface has regular radiating lines with low respiratory pores. The foot sole is light brown. *H. sieboldii* has the same geographical distribution as *H. gigantea* except it is found in shallower water.[9] Megai is harvested in central and southern Japan at depths of 2 to 28 m and makes up 14% of the total catch.[21] The meat is soft and tender.[9]

Haliotis asinina

H. asinina, mimigai (ear shell), has a very unusual, thin, long, and narrow shell. The shell is very small and only covers a small portion of the body. *H. asinina* are found in 7 m of water in the subtropical and tropical regions of Japan (Kyushu and Ryukyu).

Korea

There are two recognized regions (northern and southern) along the Korean coast. The regions are separated by the 12°C isothermal line that runs at a depth of 25 m in February east to west just north of Saishu-to Island (Cheju-do) in the Korean Strait. *H. discus hannai* is restricted north of this line, and *H. discus*, *H. gigantea*, and *H. sieboldii* are found south of this line.[7]

Australia

Australia has three commercially valuable large abalone species. They are found along the southern coast and Tasmania between 30 to 50° south with water temperatures from 11 to 23° C.[7,9,23] Shepherd[24] believes *H. ruber* and *H. laevigata* show great promise for aquaculture because their growth is "faster than that recorded for any other species of abalone".

Haliotis ruber

H. ruber, black-lip abalone or red-ear shell, is the most important species in Australia.[9] Individuals are usually 12 to 14 cm in length but some grow to 20 cm. Black-lip abalone are generally found in 1 to 10 m of water. Large animals will produce 0.6 kg edible meat.[25]

Haliotis laevigata

H. laevigata, green-lip abalone or mutton fish, commonly grows to between 13 and 14 cm, and has white flesh.[25] Individuals occur at depths of 5 to 40 m in moderate to rough water.[9,25]

Haliotis roei

H. roei, Roe's abalone, grows to between 7 and 8 cm in length, and is the smallest of the commercial Australian species.[26] Individuals are found in less than 5 m of water.[24]

New Zealand
Haliotis iris

H. iris, paua or black paua, grow to 17 cm in length. The shell of *H. iris* is exceptional for its iridescence.[27] Individuals are found from just below the low tide line to a depth of 9 m. *H. iris* has a clear distinction between the juvenile and adult habitats. Individuals less than 7 cm occur in crevices and under stones in the intertidal zone along rocky shores, while adults are subtidal. Migration of animals toward deeper water probably occurs as the animal grows. Field observations suggest that *H. iris* moves very little during a 24-hr period, but moves slightly more at night than during the day.[28] *H. iris* is found discontinuously along the coasts of the two main islands of New Zealand, and Chatham, Stewart, and Snares Islands.[28,29] Although the shell is highly prized and very valuable, the meat is of poor quality. The meat is generally dark, has inferior flavor and is covered with black pigment.[30]

Haliotis australis

H. australis, silver, queen, or yellow-foot paua, has a maximum shell length of 12.5 cm. Silver paua has a wide distribution along the rocky coasts of New Zealand. The sole of the foot is golden yellow and ringed with a black epipodium. Juveniles and adults use the same habitat (narrow subtidal crevices or rarely under stones), so migration of juveniles is unlikely. *H. australis* is very active and seen moving around during the day.[28]

Haliotis virginea

H. virginea is commonly known as virgin paua. There are four subspecies: *H. virginea crispata* occurs from Cape Maria van Diemen around the North Island of New Zeland to the Hauraki Gulf; *H. virginea virginea* occurs around the rest of the North Island of New Zealand, the South Island of New Zealand, and Stewart Island; *H. virginea morioria* occurs only in the Chatham Islands; and *H. virginea huttoni* occurs in the sub-Antarctic islands.[28,31] The maximum shell length is 7 cm. Virgin paua lives beneath rocks in the subtidal zone and, although it is rare, has the widest distribution of all the species in New Zealand.[31]

France
Haliotis tuberculata

H. tuberculata is commonly called ormer (contraction of "oreille de mer").[4] The maximum shell length is 12 cm.[32] *H. tuberculata* is the only commercial species in Europe and is taken in small quantities off Spain, France, and Channel Islands.[12] Natural distribution of ormer is along the northwest coast of France, from the west coast of Normandy to the south coast of Brittany until Belle-Ile, as well as around the British Channel Islands.[8,32] *H. tuberculata* is not found on the British and Irish coasts. There are some rare cases of mosaic hermaphroditic animals.[33] Adult size (reproductive age) is small and varies depending on the area.[32]

South Africa
Haliotis midae

The only commercial population of abalone (*H. midae*) in Africa is found near Cape Town, South Africa. The other species in Africa are small in size and are harvested in low quantities.[7] *H. midae* is a medium-sized abalone which reaches 9 cm in 6 years. Juveniles less than 7 cm in length live in shallow protected bays under loose boulders. Individuals 7 to 12 cm in length live in shallow water 0.3 to 1.8 m below the mean low-tide mark. Large individuals occur over a wide range of depths (0 to 23 m) and are concentratred in beds where there may be 18 to 24 abalone per square meter.[34,35]

REFERENCES

1. **Howorth, P.C.,** *The Abalone Book,* Naturegraph, Happy Camp, Calif. 1978, 1.
2. **Ino, T.,** Biological studies on the propagation of Japanese abalone (genus *Haliotis*), *Bull. Tokai Reg. Fish. Res. Lab.,* 5, 1, 1952.
3. **Olley, J. and Thrower, S. J.,** Abalone — an esoteric food, in *Advances in Food Research,* Vol. 23, Chichester, C. O., Mrak, I. M., and Stewart, G. F., Eds., Academic Press, New York, 1977, 144.
4. **Crofts, D. R.,** Haliotis, *Liverpool Mar. Biol. Comm. Mem. Typ. Br. Mar. Plants Anim.,* 29, 1, 1929.
5. **Ebert, E. E.,** Abalone, in *The Encyclopedia of Marine Resources,* Firth, F. E., Ed., Van Nostrand Reinhold Co., New York, 1969, 1.
6. **Cox, K. W.,** Review of the abalone in California, *Calif. Fish Game,* 46 (4), 381, 1960.
7. **Ino, T.,** *Fisheries in Japan, Abalone and Oyster,* Japan Marine Products Photo Materials Assoc., Tokyo, 1980, 165.
8. **Flassch, J. and Aveline, C.,** Production de jeunes ormeaux a la station experimentale d'Argenton, *Pub. C.N.E.X.O., Rap. Sci. Tech.,* 50, 1, 1984.
9. **Imai, T.,** *Aquaculture in shallow seas: progress in shallow sea culture,* Amerind Pub., New Delhi, 1977, Part IV.
10. **McAllister, R.,** California marine fish landings for 1974, *Calif. Fish Game Fish. Bull.,* 166, 1, 1976.
11. **Cuthbertson, A.,** *The Abalone Culture Handbook,* Tasmanian Fisheries Development Authority, Hobart, Australia, 1985, 1.
12. **Mottet, M. G.,** A review of the fishery biology of abalones, *Washington Dept. Fisheries, Tech. Rep.,* 37, 1, 1978.
13. **Leighton, D.L.,** Laboratory observations on the early growth of the abalone, *Haliotis sorenseni,* and the effect of temperature on larval development and settling success, *Fish. Bull.,* 70(2), 373, 1972.
14. **Livingstone, R.,** Preliminary investigation of the southeastern Alaska abalone *(Haliotis kamtschatkana), Commer. Fish. Rev.,* 14(9), 8, 1952.
15. **Tanikawa, E. and Yamashita, J.,** Chemical studies on the meat of abalone *(Haliotis discus hannai).* I, *Bull. Fac. Fish. Hokkaido Univ.,* 12(3), 210, 1961.
16. **Yamamoto, T.,** Biometric study on abalone shells, *Bull. Jpn. Soc.Sci. Fish.,* 15(5), 209, 1949.
17. **Ino, T.,** American *Haliotis kamtschatkana* and the far-eastern *Haliotis kamtschatkana, Bull. Biogeogr. Soc. Jpn.,* 15(1), 39, 1951.
18. **Tomita, K. and Saito, K.,** The growth of the abalone, *Haliotis discus hannai,* at Rebun Island, Hokkaido, *J. Hokkaido Fish. Sci. Inst.,* 23(11), 555, 1966.
19. **Kikuchi, S. and Uki, N.,** Technical study on artificial spawning of abalone, genus *Haliotis.* I. Relation between water temperature and advancing sexual maturity of *Haliotis discus hannai* Ino, *Bull. Tohoku Reg. Fish. Res. Lab.,* 33, 69, 1974.
20. **Uki, N. and Kikuchi, S.,** Regulation of maturation and spawning of an abalone, *Haliotis* (Gastropoda) by external environmental factors, *Aquaculture,* 39, 247, 1984.
21. **Inoue, M.,** Abalone section, in *Marine Aquaculture Data Book,* Oikawa, K., Ed., Suisan Shuppan, Tokyo, 1976, 19.
22. **Oba, T.,** Studies on the propagation of the abalone *Haliotis diversicolor supertexta* Lischke. II. On the development, *Bull. Jpn. Soc. Sci. Fish.,* 30, 809, 1964.
23. **Shepherd, S. A.,** Distribution, habitat and feeding habits of abalone, *Aust. Fish.,* 34(1), 12, 1975.
24. **Shepherd, S. A.,** Breeding, larval development and culture of abalone, *Aust. Fish.,* 35(4), 7, 1976.
25. **Harrison, A. J.,** The Australian abalone industry, *Aust. Fish.,* 28(9), 1, 1969.
26. **Shepherd, S. A. and Laws, H. M.,** Studies on southern Australian abalone (genus *Haliotis).* II. Reproduction of five species, *Aust. J. Mar. Freshwat. Res.,* 25, 49, 1974.
27. **Wright, R. R.,** Export value of paua shell to New Zealand, in *Proceedings of the Paua Fish. Workshop,* Fish. Res. Div. Pub. No. 41, Akroyd, J. M., Murray, T. E., and Tayler, J. L., Eds., New Zealand Min. Agric. Fish., Wellington, 1982, 24.
28. **Dutton, S. and Tong, L.,** New Zealand paua species, *Catch 81,* June, 15, 1981.
29. **Sainsbury, K. J.,** Population dynamics and fishery management of the paua, *Haliotis iris.* I. Population structure, growth, reproduction, and mortality, *N. Z. J. Mar. Freshwater Res.,* 16, 147, 1982.
30. **Wast, G. H.,** Paua — the potential for alternative product development, in *Proceedings of the Paua Fisheries Workshop,* Fish. Res. Div. Pub. No. 41, Akroyd, J. M., Murray, T. E., and Tayler, J. L., Eds., New Zealand Min. Agric. Fish., Wellington, 1982, 22.
31. **Murray, T.,** Could the virgin paua appeal?, *Catch '83,* June, 13, 1983.
32. **Koike, Y.,** Biological and ecological studies on the propagation of the ormer, *Haliotis tuberculata* Linnaeus. I. Larval development and growth of juveniles, *La Mer,* 16(3), 124, 1978.
33. **Cochard, J.,** Recherches sur les Facteurs Determinant la Sexualite et la Reproduction chez *Haliotis tuberculata* L., Ph.D. thesis, Universite de Bretagne Occidentale, 1980.

34. **Newman, G. G.,** Distribution of the abalone *(Haliotis midae)* and the effect of temperature on productivity, *Div. Sea Fish. Invest. Rep.,* 74, 1, 1969.
35. **Newman, G. G.,** Movements of the South African abalone, *Div. Sea Fish. Invest. Rep.,* 56, 1, 1966.

GONAD REPRODUCTIVE CYCLES

Kirk O. Hahn

INTRODUCTION

Having detailed information on the reproduction of cultured species is very important to the culturist. This information can be used to design efficient and effective techniques for conditioning, spawning induction, larval development, and settlement. Despite a great need, it is difficult to obtain the required data since studies on reproduction are usually time consuming, labor intensive, and expensive. Field studies on reproduction usually consist of collecting specimens at regular intervals and simultaneously measuring environmental factors (e.g., temperature, photoperiod, lunar phase) that may be controlling the reproductive cycle. In addition, laboratory analyses on levels of biochemical components (e.g., hormones, vitellogenin) can be conducted on the gonad and other tissues.[1]

The reproductive cycle is defined as the time interval between successive spawnings in a population. This time interval can be determined by two methods: (1) monitoring changes in gonad histology and (2) measuring changes in gonad indices.[2] Marine invertebrates that release their gametes into the environment at spawning (broadcast fertilizers) generally have a large gonad relative to their total body mass and well-defined annual reproductive cycles that can be divided into four phases: initiation, control of gametogenesis, maturation, and spawning.[3] Each stage takes a finite length of time and is probably controlled by a different environmental or physiological factor.

This chapter describes the methods most commonly used to analyze data on reproduction and evaluates the efficiency of each method. The best information comes from using as many different techniques as possible, since this provides more data and may help reveal important subtle factors in the reproductive cycle.[1]

GONAD INDICES

Sampling frequency during a study is dependent upon the length of the reproductive cycle and time scale of the study. Monthly sampling of gonads is probably adequate, with more frequent sampling during the spawning period, for animals (e.g., *Haliotis* spp.) with an annual reproductive cycle. Animals (e.g., *Trochus niloticus*) with a shorter reproductive cycle, such as one synchronized by the lunar cycle, must be sampled more frequently.[1] The number of specimens collected in a reproduction study is usuallly relatively small (10 to 15 animals per sampling period) due to the large amount of work required to analyze samples. Since there is variability between samples, the largest sample size possible is preferred to determine the stages in the reproductive cycle.[3]

The reproductive cycles of individuals in a population are usually relatively in phase. The gonad increases in size, until the gonad is fully ripe, as the layer of gonad tissue surrounding the digestive gland increases in size as the developing gametes increase in quantity or size. A ripe gonad may make up 15 to 20% of the body weight in a fully mature animal and remains this size until spawning rapidly decreases the gonad size.[4]

After spending so much time and effort, it is important the maximum amount of information be obtained from the data and the results accurately reflect the information contained in the data. To quantify the spawning season, several researchers have developed methods to standardize the gonad development by comparing the gonad size to a measurement (i.e., shell length, size of digestive gland, or body weight) of the animal that is independent of the gonad size.[2,5-12] The conical appendage, gonad, or digestive gland area can be measured

Table 1
GONAD INDEXES

GI^6 =	(Gonad area / shell length) × 100
GI^{51} =	Conical appendage diameter / digestive gland diameter
GI^7 =	[(Conical appendage diameter − digestive gland diameter) / digestive gland diameter] × 100
GI^4 =	Weight of gonad / weight of soft body
GBI^8 =	(Gonad area / conical appendage area) × 100

by using graph paper, a planimeter, or weighing a tracing cut out of paper. Table 1 shows several commonly used gonad indices. Statistical analysis of the data ensures an objective study of the reproductive cycle.[1]

Gonad Index

The Gonad Index (GI) is the simplest and most frequently used method to represent the reproductive cycle. This index can be calculated on the basis of dry weight, wet weight, or caloric content. Calculation of GI by dry weight avoids problems which may arise if the water content of the animal is variable.[1]

GI assumes that a spent or immature gonad is small and a ripe gonad is large, and the absolute gonad size is proportional to the animal's size. The latter assumption is not valid in all species of abalone, especially large or small individuals of a species. This problem can be solved by sampling only adult animals of uniform size.[10] GI also assumes that the digestive gland size (the gonad surrounds the digestive gland to form the conical appendage) is correlated with shell length. This assumption has been found to be incorrect in several abalone species.[5,12] Since GI only relates gonad area to a constant (e.g., shell length) for the animal, it does not take variations in digestive gland size into account. This index cannot distinguish whether the gonad area is a thin layer around a large digestive gland (spent) or if the gonad area is a thick layer around a small digestive gland (ripe).[12] GI is not a good criterion for monitoring the reproductive cycle in *H. rufescens*, *H. kamtschatkana*, and *H. walallensis*.[5,12]

Gonad Bulk Index

The Gonad Bulk Index (GBI) is simply the percentage of gonad tissue in the conical appendage. This index is better than GI at detecting the reproductive cycle and spawning. There is also a correlation between GBI and gonad histology.[12] GBI accounts for the fluctuations in digestive gland size and allows a clear picture of the abalone reproductive cycle. GBI and GI are independent indexes. They do not correlate to one another due to high variability between the total area of gonad sections and shell length.[5]

Modified Gonad Bulk Index

Tutschulte and Connell[13] developed a modification of the Gonad Bulk Index, (GBI_{tc}). They found it impractical to routinely determine gonad volume by dissecting the gonad from the digestive gland of each animal as was done by Webber and Giese.[4] The GBI_{tc} calculates an estimate of the gonad volume (EGV) by assuming that almost all the gonad is contained in the conical appendage, and the conical appendage and digestive gland are concentric cones with parallel sides. Fresh or frozen samples of the entire conical appendage can be used to obtain the linear dimensions necessary to calculate the lower EGV.

EGV is calculated from a histological section at the midpoint of the conical appendage length. This section is assumed to have been taken at the midpoint (h/2) of a right circular cone where r = radius of conical appendage at midpoint, r′ = radius of digestive gland at midpoint, h′ = height of digestive gland, and h = height of conical appendage (Figure 1).

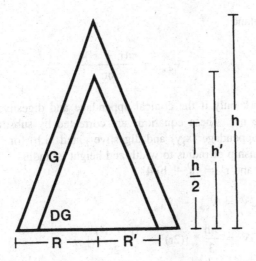

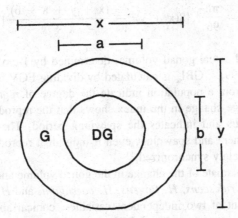

FIGURE 1. Top — Schematic drawing of conical appendage showing gonad (G) and digestive gland (DG); h = height of conical appendage; h' = height of digestive gland; h/2 = midpoint of conical appendage; R = radius of conical appendage; R' = radius of digestive gland. Bottom — Schematic drawing of transverse section taken at midpoint of conical appendage; a = width of digestive gland; b = height of digestive gland; x = width of conical appendage; y = height of conical appendage. (From Tutschulte, T. and Connell, J. H., *The Veliger*, 23(3), 195, 1981. With permission.)

The gonad volume is calculated by estimating the conical appendage (R) and digestive gland (R') radii at the base, and digestive gland height (h'). Value of "h" is measured from the intact gonad before sectioning.[2]

In a right circular cone, R = 2r and (r — r') = (R — R'), therefore R' = r + r'. Also, h/R = h'/R', thus h' = h (r + r')/2r.

Volume of the conical appendage

$$V_{c.a.} = \frac{\pi(2r)^2 h}{3}$$

Volume of digestive gland

$$V_{d.g.} = \frac{\pi(r + r')^3}{6r}$$

These equations work only if the conical appendage and digestive gland are perfectly round, which they are not. These equations are corrected by substituting the width and height of the conical appendage (x,y) and digestive gland (a,b) for r and r', respectively (Figure 1). The relationship of radius to width and height is: radius = (width + height)/4. Thus, $r = (x + y)/4$, and $r' = (a + b)/4$

$$EGV = V_{c.a.} - V_{d.g.}$$

$$EGV = \frac{\pi h}{3} * [(2r)^2 - \frac{(r + r')^3}{2r}]$$

$$EGV = \frac{\pi h}{3} * [\frac{(x + y)^2}{4} - \frac{(x + y + a + b)^3}{32(x + y)}]$$

$$EGV = \frac{\pi h}{96} * [8(x + y)^2 - \frac{(x + y + a + b)^3}{x + y}]$$

Correlation of EGV and actual gonad volume, determined by 1-cm serial sections, was highly significant $(t > 0.999)$.[13] GBI_{tc} is calculated by dividing EGV by body weight. The GBI_{tc} means and ranges from a population indicate the degree of reproduction synchrony within a population. A large change in the index shows that the reproductive cycles of the individuals are synchronous and indicates the spawning period. The mean GBI_{tc} shows periods of gonad development and spawning, even if individual reproductive cycles in the population are only moderately synchronized.[13]

GBI_{tc} gives a sensitive estimate of the change in the gonad volume and is a good indicator of gonad maturation in *H. rufescens, H. fulgens, H. corrugata*, and *H. sorenseni*.[2,13] This index uses the measurement of two independent variables, conical appendage radius and length, while GI and GBI use only radius.[2] GBI_{tc} is very informative when studying the reproductive cycle, but due to variability in the values at each sampling period, GBI_{tc} should not replace histological analysis for determining the exact reproductive stage of a given animal. The variability of GBI_{tc} is too large to allow determination of reproductive stage.[2]

Fluctuation in Gonad and Digestive Gland Section Area

The digestive gland section area usually remains fairly constant throughout the year as compared to the gonad section area which fluctuates from month to month. However, the digestive gland will increase in size just before or immediately after spawning.[11,12] Boolootian[6] found an inverse relationship between gonad size and digestive gland size, which indicated that material might be transported from the digestive gland to the gonad. This increase can be dramatic, with the digestive gland area doubling in size. The GBI sometimes does not decrease to levels found prior to spawning periods, but the digestive gland will show a distinctive increase in size — either surpassing or equaling the gonad in area. A drop in the gonad area with a corresponding increase in digestive gland area will pinpoint the spawning period. This phenomenon has been noted in California for *H. rufescens, H. cracherodii, H. kamtschatkana*, and *H. walallensis*.[11,12,14]

SYNCHRONIZATION OF THE REPRODUCTIVE CYCLE

Determining the mechanisms that regulate gametogenesis is one of the most important aspects of controlling reproductive synchrony in the laboratory. Spawning synchronization is very important to animals with broadcast fertilization because a large amount of metabolic energy is invested into the gametes during gametogenesis.[4]

While it has been postulated that synchronization of the reproductive cycle is controlled by exogenous factors, usually water temperature or photoperiod, the mechanism is probably not this simple. There are two separate aspects of the reproductive cycle, rhythmicity of gonad development and gametogenesis in the population, and synchronicity of spawning between individuals. These two phenomena are probably controlled by separate mechanisms. Gonad development may be controlled by a long-term endogenous rhythm with an annual or lunar phase and an additional zeitgeber (time giver) by the exogenous environment. Once the gonads are fully ripe, the synchronization of spawning is probably determined by a combination of water temperature, lunar cycle, or photoperiod.[15]

The female reproductive cycle can be separated into initiation of gametogenesis, vitellogenesis, and oocyte growth (6 to 8 months), and spawning. Gametogenesis is probably initiated by the release of the ripe eggs during spawning. Webber and Giese[4] found no resting period in *H. cracherodii*, with gametogenesis beginning immediately after spawning. Vitellogenesis and oocyte growth is probably initiated by a decrease in water temperature, but there are several cases where a rise in water temperature might control vitellogenesis. The effect of a rising water temperature is combined with an increasing food supply. Spawning in nature may be induced by a sudden temperature change caused by the tides or internal waves at the interface between two layers of different water temperatures. All individuals in a local population of abalone would sense this sudden temperature change almost simultaneously and this could provide a trigger for a mass spawning.[15] A synchronous trigger is required for successful external fertilization.[16] *H. corrugata* show a high degree of synchronization, even though there are different temperature patterns over their depth range. This contradicts the theory that water temperature levels and fluctuations are the main exogenous factors controlling the reproductive cycles of invertebrates.[3,13]

The importance of local environmental parameters in regulating the reproductive rhythmicity is emphasized by *H. ruber* and *H. cracherodii* which have different spawning periods at different locations.[4,10] The spawning period of *H. ruber* at different locations in Australia is highly variable; West Island in spring and autumn, Tipara Reef during fall and winter, Tasmania in spring and fall, and New South Wales coast during late winter and spring.[16] In California, two populations of *H. cracherodii* only 7 miles apart had different spawning periods.[4]

Both photoperiod and water temperature might induce spawning in *H. rufescens*. The major spawning period (April) in southern California correlates with a decrease in average water temperature from the previous month, and this spawning period and a minor spawning period in September correspond to rapid changes in the natural photoperiod.[11]

Although some abalone species will spawn in the lab when the water temperature is rapidly changed, this phenomenon does not occur very often in nature.[17-22] It is possible, however, for ambient water temperatures to change 2.5°C (a quarter of the annual change) in one day. Owen[23] observed a spontaneous spawning of *H. corrugata* which followed an abrupt midday increase in water temperature. Laboratory animals have also spawned following a natural increase in water temperature.[23]

Tutschulte and Connell[13] found that *H. sorenseni* in shallow water (18 m) spawn in February and March, when the water temperature at this depth is fairly constant and the rapid decrease in water temperature occurred 4 to 6 months previously. However, animals at a deeper depth (30 m) spawn in March and April during a period of rapidly declining water temperature.[13] The high degree of synchronization in both populations, even though there were significant differences in water temperature patterns, raises questions concerning the significance of water temperature fluctuation as an exogenous factor controlling the reproductive cycle phases in this species.[3,13]

Fretter and Graham[24] stated that gonad maturation is controlled by annual temperature fluctuations, while a combination of other factors may trigger spawning. Giese[3] also noted there was a difference between the stimulus to induce spawning and maturation. The natural inducer of spawning may not be unique to each abalone species. Five abalone species, which were being maintained in running sea water, were observed to begin spawning simultaneously at Pacific Mariculture in Cayucos.[23]

Periods of rapid gametogenesis and vitellogenesis, and the precise timing of spawning in *H. laevigata, H. cyclobates,* and *H. rufescens* are correlated to the abundance of kelp.[10] In regions where food is abundant, the spawning season is extended from March to the end of May, as compared to regions where the food is scarce.[16] Gametogenic changes correlate more closely with changes in the amount of food eaten than with sea temperature changes alone.[10] The production of gametes requires a large amount of nutrients for metabolic requirements and synthesis of vitellogenin, which fuels larval development. Kelp provides ample nutrition for both growth and gamete production.[25]

The source of the metabolic energy for gametogenesis is not clearly known, but both the foot and digestive gland are indicated as sources. During gonad development, the absolute size of the foot decreases and levels of glycogen decrease in correlation to the decrease in size.[14] Foot muscle also decreases as the gonad develops.[26] Relative to shell weight, the foot is 120% in January and 82% in August just prior to spawning.[26] Glycogen may be converted to lipids and transferred to the ovary. Large amounts of lipids are packaged with the vitellogenin. Lipid content in the abalone varies throughout the reproductive cycle, lipids increase to high levels in the digestive gland tissue, and then concentrate in the ovaries as the eggs ripen.[14] Lipids appear to be stored in the digestive gland, reaching about 30% of dry weight during the period when the gonad is spent. Lipid levels in the digestive gland steadily decrease in correlation with increasing gonad development, and presumably with vitellogenesis. Lipid levels reach the lowest level (10% of dry weight) a month before spawning occurs. This evidence supports the hypothesis that the digestive gland plays an important role in supplying lipids to developing oocytes. There is a reciprocal relationship between lipid levels in the gonad and digestive gland. The digestive gland lipid level is lowest when the ovary lipid content is highest. It is not surprising that the ovary lipid levels are well correlated with the level of egg maturation in the gonad. This indicates that lipids are rapidly being deposited in the ovary while eggs are maturing.[14] Fatty-acid composition in the ovary is very similar in different abalone species and there is a high level of arachidonic acid present, which is unusual for a marine species.[26] Arachidonic acid is a precursor to prostaglandin, the hormone implicated in inducing spawning in abalone.[27]

HISTOLOGY

Examination of gonad histological sections provides more detailed information on reproduction than is possible from gonad indices alone. Subjectively assigning a gonad development stage is the simplest analysis of histological sections. This method is relatively rapid and can be used on both males and females. However, it is subjective and different people probably will not analyze the samples the same. Grant and Tyler[1] have proposed using the term "maturity index" for these methods and restricting the use of "gonad index" to the quantitative analysis of relative gonad weight or size.

Gonad Structure

Gametogenesis is the same in all abalone species studied. The structure of the gonad is so simple and uniform that any vertical section through the gonad is relatively the same.[5,8,12,25,28-36] The gonad is located between the outer epithelial layer of the conical appendage and the digestive gland. The outer gonad wall consists of an epidermis which is a simple, glandular columnar epithelium covered with a thin cuticle.[5] A fibromuscular capsule composed of connective tissue and muscle fibers lies beneath the epithelial layer. The coelomic epithelium, together with finer strands of underlying connective tissue, grows outward in thin folds that form trabeculae between the overlying connective tissue of the digestive gland and integument. Primordial germ cells are formed on the lobular epithelia originating from the fibromuscular capsule.[35] Nutritive cells (stain positive with eosin and PAS) are located between the gametes, and these cells decrease in quantity with advancing gonad development.[28,35]

Testis

Spermatogenesis in abalone can be monitored, although it is difficult since the spermatocytes are very small.[28] Branching tubes of connective tissue, surrounded by cuboidal germinal epithelium, produce spermatogonia. The nuclei of spermatogonia are 5 to 7 μm in diameter, and have chromomeres of condensed heterochromatin pressed against the nuclear envelope. The later stages of prophase I and the first meiotic division are not seen, which suggests these stages are of short duration. Nuclei of both the primary spermatocytes and spherical secondary spermatocytes are approximately 3 to 5 μm in diameter. Secondary spermatocytes frequently have visible condensed chromatin: however, spermatids do not have condensed chromatin. Spermiogenesis begins with the formation of an eosinophilic acrosome which undergoes rapid condensation, followed by elongation of the nucleus. Mature spermatozoa have a head with an anterior eosinophilic acrosome, a short clear space (2.5 μm), and a nucleus (3.5 μm), for a total length of 6 μm, and width of 1 to 1.5 μm. There is also a short eosinophilic mass at the posterior end of the spermatozoa. The flagellum is just visible at 1000 times.[5]

Ovary

Polygonally arranged trabeculae (connective tissue and maybe some muscle fibers), possibly serving as supports, extend from the outer ovarian wall and may stop in the lumen or attach to the inner ovarian wall. Oocyte development starts at the distal end of the trabeculae. As the oogonia develop into primary oocytes, the oocytes migrate from the germinal epithelium toward the digestive gland. The teardrop-shaped oocytes remain attached to the trabeculae by a delicate stalk during the growth phase.[5,28] The oocyte cytoplasm becomes strongly basophilic, which might indicate protein synthesis. Small spherical vacuoles (yolk) appear and increase in number as the primary oocyte grows.[5]

Chromosomes appear as fuzzy threads uniformly distributed along the periphery of the nucleus during leptotene of prophase I. The nucleolus is small, indistinct, and oval, and is next to the nuclear membrane. The chromosomes shorten into thick strands which radiate from one side of the nucleus during zygotene and pachytene. The nucleolus becomes spherical and a dense aggregate of chromosomes may form a synizetic knot when they attach to the nucleolus.[5]

A network of flattened basophilic organelles, maybe a Golgi apparatus, appears on the vegetal side of the nucleus of primary oocytes. As the oocyte grows, the organelle network gradually thickens, forms basophilic strands parallel to the oocyte polar axis, and develops one or more large lobes on the side adjacent to the nucleus. The lobes flatten and the organelles thinly spread out around the nucleus vegetal pole as the primary oocyte reaches its maximum volume and moves toward the digestive gland. These organelles are not seen in detached primary oocytes.[5]

As the primary oocyte grows, a thin eosinophilic membrane forms around the plasma membrane, except where the oocyte is attached to the follicular wall. It can not be determined if this membrane is a vitelline membrane or a chorion, however two eosinophilic layers quickly become clear. Young and DeMartini[5] called the inner layer the vitelline membrane and the outer layer the chorion. The vitelline membrane quickly becomes about 1 μm thick at its widest point. The chorion of a fully mature oocyte (about 180 to 270 μm in diameter depending on the species) is surrounded by a thick gelatinous layer. Mature oocytes have numerous yolk and lipid granules in the cytoplasm, and pigment granules in the cortical layer prior to spawning.[5,28,35] The breakdown of the trabeculae near the digestive gland allows mature gametes to move out of the gonad during spawning.[8]

Necrosis of unspawned oocytes has been found in several species. Oocyte necrosis first becomes evident in the nucleus. The chromatin becomes evenly distributed, less granular, and strongly eosinophilic. The nuclear membrane becomes convoluted and breaks down, followed by a shrinking and slow disappearance of the nucleus. Numerous small vacuoles in the cytoplasm fuse and increase in size, and large eosinophilic bodies (up to 40 μm in diameter) and basophilic inclusions form in the cytoplasm. The cell surface becomes highly irregular, the vitelline membrane disappears, the chorion ruptures, and disperses cellular contents into the ovarian lumen. During advanced stages of necrosis, amoebocytes are found in the ovarian lumen. Amoebocytes, acting as phagocytes, might be released by the breakdown of trabeculae. In fresh ovarian tissue, necrotic areas are characteristically small flaccid patches, while viable areas are firm. During advanced stages of necrosis, proliferation of oogonia in the distal portions of the trabeculae is common.[5]

Reproductive Stages

The reproductive cycle of a species can be determined by monitoring changes in the gonad histology.[2] It is important to make a distinction between terms referring to gametogenesis, and gonad maturation or development. Gametogenesis refers to the stages an individual gamete undergoes during development, and gonad maturation refers to the stages the gonad undergoes during the reproductive cycle. The number of stages assigned to gametogenesis vary depending on the researcher and sex of the animal. Tomita[31] classified spermatogenesis into three stages: spermatogonium, spermatocyte, and spermatid; and oogenesis into seven stages: (1) oogonium (3 to 5 μm in diameter), (2) chromatin-nucleolus (10 μm), (3) yolkless (50 μm), (4) oil drop (oil drops — about 7 μm, eggs — 50 to 90 μm), (5) primary yolk globule (150 μm), (6) secondary yolk globule (150 to 180 μm), and (7) mature (150 to 200 μm).

The number of subjective gonad maturation stages used can range from 5 to 10, depending on how rapidly and completely the animal spawns.[1] The names of these stages differ depending on the researcher but usually refer to the same phenomena. Generally, the reproductive cycle of an organism is characterized by 5 to 6 distinct stages: Giese[3] and Boolootian[37] called them (1) activation (initiation of gametogenesis), (2) gametogenesis, (3) increase in gonad size due to increase in the number or size of gametes, (4) spawning, (5) resorption of unspawned gametes, and (6) resting; Tomita[32] called them pre-mature, mature, spawning, spent, and recovery; Lee[35] called them multiplication, growing, mature, spent, and recovery (the multiplication and growing stages are the longest); Ault[2] called them preproliferative, proliferative, new stalk, old stalk, and free; and Giorgi and DeMartini[25] called them active, ripe, partially spawned, spent, and necrotic.

The stages assigned by each researcher are not totally equivalent and, in some cases, represent phenomenon which only occur in one species or one sex. Descriptions of the stages used by Ault[2], and Giorgi and DeMartini[25] are given to represent the process of gonad maturation.

Ault developed his ovarian stages to follow gonad maturation in female abalone under laboratory conditions. The individuals were artificially spawned and gonad maturation was followed at precise regular intervals.[2]

Preproliferative — Gonad is immature, and has little or no germinal epithelium between the outer epidermis and digestive gland.[2]

Proliferative — Germinal epithelial cells become cuboidal and develop into oogonia. Oogonia range in size from 10 to 25 μm and form clusters on the trabeculae walls. Gametogenesis occurs during this stage.[2]

New stalk — Oogonia develop into primary oocytes and become stalked at about 25 μm. Vitellogenesis begins during this stage.[2]

Old stalk — Primary oocytes are greater than 50 μm in diameter during this stage and vitellogenesis is intense. Oocytes become teardrop shaped and extend from the germinal epithelium, but remain attached by a stalk.[2]

Free — Oocytes (170 to 190 μm in diameter) detach from the trabeculae and are completely surrounded by a chorion. Free oocytes are found between the trabeculae and in the ovarian lumen next to the digestive gland.[2]

Ault[2] found the reproductive cycle of *H. rufescens* larger than 100 mm had a regular cycle. Spawning individuals did not immediately show a large reduction in gonad volume. The trabeculae probably prevent complete collapse of the gonad and act as temporary reinforcement.[2] Ovarian growth and proliferation is extensive within 15 to 30 days postspawning. Old stalk and free oocytes predominate the ovary but proliferation is still extensive by day 30. Even though gametogenesis is initiated soon after spawning and there are numerous developing gametes, the gonad does not show quantitative growth in size until approximately 45 days postspawning, when vitellogenesis begins. Old stalk oocytes continue to grow and more than 85% of the gonad volume is filled with late old stalk and free stage oocytes by the day 75.[2]

Giorgi and DeMartini[25] developed their gonad maturation stages for a natural population of *H. rufescens*. These stages qualitatively measure gonad development and can be used for both sexes.

Active — This stage is characterized by intense gametogenic activity. At the beginning, the ovary contains principally small oocytes (<50 μm in diameter) and the testis contains spermatogonia and primary spermatocytes. The gametes continue to increase in quantity and develop during this stage, until the ovary contains oocytes from 10 to 250 μm in diameter (in almost equal proportions), and the testis contains almost exclusively spermatozoa.[25]

Ripe — An ovary is classified as ripe when almost all of the primary oocytes are larger than 160 μm (maximum size — 250 μm). A ripe ovary may still have slight proliferation of oocytes, less than 50μm, along the peripheral wall of the gonad. A ripe testis principally contains spermatozoa, but early gametogenic stages may be present, but restricted to areas surrounding the trabeculae.[25]

Partially spawned — The gonads contain lower densities of gametes compared to ripe gonads. This classification is a qualitative decision and the quantity of remaining gametes can range from almost complete spawning to almost nonspawning. Care must be taken when classifying individuals as partial spawners. Individuals collected during the "spawning period", which have released only a portion of the viable gametes, may have completed their spawning if left in the population, or may have already finished spawning and have residual, ripe gametes.[25]

Spent — The gonad is greatly reduced in size, lacks any ripe gametes, and has very little gametogenic activity. The rarity of spent individuals in populations of some species may indicate that gametogenesis is initiated immediately after spawning.[4,25]

Necrotic — The quantity of necrotic oocytes in the ovary is variable, but necrosis is usually limited to oocytes larger than 150 μm. Individuals with necrotic oocytes are found throughout the year in some populations. Necrosis is believed to be autolysis and degeneration of oocytes that have remained after the spawning period.

The ovary characteristically has yellowish granular substances (when the gonad is stained with hematoxylin and eosin) and associated cells surrounding the advanced necrotic ova, along the wall between the digestive gland and gonad, and in the digestive gland. The associated cells are believed to be resorptive. The granular substances and associated cells are rare in ovaries with viable oocytes; however, the quantity of associated cells increases in proportion to the quantity of necrotic oocytes. Granular substances and associated cells are also found in testes but not in the quantity found in ovaries.[25]

Giorgi and DeMartini[25] used the quantity of necrotic oocytes present in the ovary to classify the individuals to a spawning type for the previous spawning period; none (Type I — complete spawning), increasing quantities (Type II — partial spawning), or almost the entire gonad (Type III — nonspawning).[25]

Oocyte Size Frequency

Data on oocyte size frequencies are very informative and should be collected if histological sections are used to calculate gonad indices. Oocyte sizes are measured directly from a microscope slide with an eyepiece caliper calibrated against a stage micrometer. It is important to measure only those oocytes sectioned through the nucleus. Developing eggs are first stalked, become polygonal as they mature, and finally become round.[10] The data collected for oocyte size frequency are standardized by measuring the lesser diameter of stalked oocytes, or the mean diameter of polygonal and round ova. The oocyte size range is divided into several size classes, usually 10 to 15, depending on the size range and a convenient size increment while using the eyepiece caliper. Between 100 to 200 oocytes in each ovary are measured and grouped into size classes. As a precaution against heterogeneity of development within the ovary, it is best to measure oocytes from several sections and along perpendicular axes on each section.[1] If the individaul has spawned completely with only a few oocytes present and is classified as "spent", it is scored as "0" without attempting to measure the remaining oocytes.[1]

ANALYSIS OF GONAD DATA

Gonad Indexes

Grant and Tyler[1,38] described the analysis of reproductive cycles in marine invertebrates. The statistical methods described in these papers are very useful for following abalone reproduction and helping to determine the endogenous and exogenous factors affecting gonad maturation and spawning. In addition, they discuss possible sources of error while evaluating reproductive cycles in marine invertebrates.

All data (gonad weight, histological examination, biochemical and physiological levels, etc.) should be collected from each animal. If this is not possible, the mean values of each parameter for the entire sample must be used to evaluate relationships between two variables. Relationships between variables could be masked by using the sample means, even though the variables are closely linked, if there is large within sample variation of the development stage (which is common in environments which are not strongly seasonal). Examining only a subsample of the gonads at each collection period and calculating dry weights or gonad indexes on the remaining gonads should be avoided. Gonad indexes using wet weight may be necessary to save the gonads for microscopic examination, or a large gonad can be divided after weighing to obtain a dry-weight gonad index and a biochemical analysis.[1]

There are two possible sources of errror when using gonad indexes. First, the gonad size may not remain constant in proportion to the body as the animal grows. The goal of a gonad index is to standardize gonad size, independent of animal size, to produce an indicator of gonad development. If this assumption is not true, differences in mean gonad index may only reflect differences in animal size rather than differences in gonad size. The second potential problem with gonad indexes is the inability to accurately measure somatic body weights in some animals (e.g. damaged during collection). These two problems are solved by using an analysis of covariance (ANCOVA) on gonad weights.[1]

The relationship of the GI to body length is determined by including body length as a covariant in an analysis of variance. If the covariant accounts for a significant amount of the variance, the gonad index is not independent of body size. The ANCOVA reduces the error mean square and increases the precision of estimating the sample mean.[1]

Even if the gonad index appears to be independent of body size, it may be helpful to analyze gonad weight at each sampling period using an ANCOVA with body weight as covariant, or log [gonad weight] with log [body weight] or log [body length] as a covariant. It is important to be sure the assumptions of ANCOVA are satisfied before using either method. The analysis with log [gonad weight] has the added advantage that the relationship between body weight (or length) is represented by: gonad weight = a (body weight)^b. This formula gives the rate at which reproductive effort increases with size.[1]

Minimum size of sexual maturity is determined during histological examination of the gonad. Scatter diagrams of female gonad index and mean oocyte size vs. body length or weight are useful to determine the minimum size of sexual maturity. Animals below a certain size will have smaller and less variable gonad indexes, and smaller mean oocyte sizes. Immature animals should never be included in the analysis of gonad indexes or oocyte size frequencies since they have smaller gonads to body size than mature animals, and their gonad weights and oocyte sizes do not follow the adult reproductive cycle.[1]

The mean, standard deviation, and standard error of the mean are calculated for each sampling period. Standard deviation and standard error represent different aspects of the data and should be used only when appropriate, especially in graphs. The standard deviation of a sample represents the variation of the values in the sample about the mean. Large standard deviations indicate a lack of synchrony in reproduction or variation in the level of reproductive effort. The standard error represents the accuracy of estimating the true value of the population mean from the sample mean. A large standard error indicates the true population mean has not been estimated accurately. When graphing gonad indexes standard deviation bars are plotted to display the degree of intrasample synchrony, and 95% confidence intervals are plotted to show differences between sample means and whether or not those differences are statistically significant.[1]

The data should be first analyzed with one-way analysis of variance (ANOVA) or the nonparametric Kruskall-Wallis test to determine if the sample means are statistically significantly different. ANOVA is a relatively robust and powerful test. If an ANOVA determines the means to be significantly different, then they are probably different. However, the differences between means can be hidden if a few individuals in the sample have extreme values. The data should be transformed to symmetry if it is skewed (the mode is to the left or right of the mean) or if the sample variances are greatly unequal. A log transformation is helpful if the means and variances of the samples are correlated. The nonparametric Kruskall-Wallis test is less sensitive to the skewness of the data and may be preferred for maturity indexes. These indexes only have a few values and probably will not be normally distributed.[1]

Data from males and females should be analyzed separately. However, it is informative to know if the gonads develop the same way in both sexes. A two-way ANOVA should be calculated with sex and sampling date as the two factors. If the two-way ANOVA is significant, the pattern of development in each sex is different. Differences can usually be seen from graphs of the data, which may show one sex with a larger gonad after spawning or having an extended spawning period. If there is no significant difference between the sexes, sampling periods should be tested for significance. There will probably be differences in the sample means, even in species that are continuous spawners. These calculations should not be performed on maturity indexes, since there is no guarantee the data are comparable between sexes.[1]

If sample means are different, the form of these differences is seen with a regression analysis. A periodic regression is used to fit a sine curve to cyclic phenomena. A periodic regression shows how much of the difference in sample means is attributed to an annual cycle. A cubic polynomial is used to study the annual cycle if the samples were collected for only 1 year.[1]

It is fairly common for researchers to study the reproductive cycle in several subpopulations of the same species. If comparisons between populations are made at each sample period, the probability of drawing a wrong conclusion is much greater than the probability level chosen for statistical analysis.[1]

Gonad indexes can be analyzed with a two-way analysis of variance with sample period and location as the factors, or a three-way analysis of variance with sample period, location, and sex as the factors. Data on oocyte sizes also can be analyzed with sample period and location as factors.[38]

Oocyte Size Frequency

Oocyte size frequency is the best method to use for studying the reproductive cycle. If histological examination of the gonads is conducted, then calculating the oocyte size frequencies in the females is invaluable. Data from oocyte size frequencies are more complex than data from gonad indexes or subjective development stages. The data from each animal consist of 100 or more oocyte measurements and thus are treated differently than gonad indexes[1]

The mean and standard deviation of oocyte sizes for each individual, and mean and standard error of the mean are calculated for each sample. In addition, calculating the coefficient of variation (SD $\times$ 100)/mean is an useful method to determine the oocyte variation within an animal. A size-frequency polygon should be graphed for each individual and sample.[38]

The percentage of spent individuals plotted against the mean oocyte size for remaining individuals in the sample is helpful in determining how rapidly the population spawns. The mean oocyte size of the population does not show a decrease until all individuals in the population have spawned. However, mean maturity index decreases as soon as spawning begins and the standard deviation increases.[38]

A one-way analysis of variance or Kruskall-Wallis test is used to analyze mean oocyte size. The Kruskall-Wallis test can be calculated in two ways, using only the mean oocyte sizes of individuals not spent, or including spent individuals and assigning the value of 0 for oocyte size. If a computer is used, calculating the test both ways is probably most informative.[38]

The data from oocyte size frequency are more complex and have additional variation. Variation in oocyte sizes in the population is explained by the sample means, the individual means, and the residual variance within individuals. The data can be transformed to satisfy the assumptions of ANOVA without changing the values of the variance components and F ratios. If most of the variation is from the sample means, the developmental stage of the individuals in each sample period is synchronized. If most of the variation is from the individual means, the oocytes within an individual ripen at the same time but individuals within the sample are not synchronized. Variation in either of these two levels may or may not be significant, which causes four possible patterns with biological implications (Table 2).[38]

Contingency table analysis can determine if the observed differences in the oocyte size frequency polygons of individuals within a sample or sample averages are statistically significantly different. Oocyte size-frequency data for several individuals or sample periods can be thought of as an (R $\times$ C) contingency table where "R" is the number of individuals or sample periods and "C" is the number of size classes.[38]

Table 2
BIOLOGICAL IMPLICATIONS OF RESULTS OF NESTED DESIGN ANALYSIS OF VARIANCE

	Between individuals	
	Not significant	**Significant**
Significant	Unlikely. Would correspond to seasonal breeding with perfect synchrony between animals. May be caused by measuring an insufficient number of oocytes per animal.	Typical pattern in seasonally breeding species
Between samples		
Not significant	Breeding continous. No synchronization of development either within or between individuals.	No distinct season but some animals ripe at any time. Could be caused by sampling too few animals in each sample.

(From Grant, A. and Tyler, P. A., *Int. J. Invert. Reprod.*, 6, 271, 1983. With permission).

If the individuals are assumed to be at the same stage of development, the expected frequency at each position, e_{ij}, is calculated:

$$e_{ij} = \frac{R_i \times C_j}{n}$$

where R_i is the total number of oocytes in the ith size class summed over all individuals, C_j is the total number of oocytes measured in the jth animal and n is the total number of oocytes measured. The observed frequencies (o_{ij}) are compared to the expected frequencies (e_{ij}) by the formula:

$$G = \sum \frac{(o_{ij} - e_{ij})^2}{e_{ij}}$$

where the summation is over all (R × C) positions in the matrix. If the computed value G is greater than the value of χ^2 with $(R - 1)(C - 1)$ degrees of freedom for the significance level chosen, the individuals are not at the same stage of development and there is heterogeneity within the sample. The same analysis can be done on summed size frequencies for each sample period to determine if there is heterogeneity between samples.

The standardized residuals of the (R × C) contingency analysis shows which size classes, or which individuals or sample periods contribute to the value of χ^2. A (R × C) contingency table of standardized residuals allows analysis of the growth of oocytes, and differences in development stage (within and between samples). The residual is calculated as:

$$r_{ij} = \frac{o_{ij} - e_{ij}}{(e_{ij})^{0.5}}$$

and standardized by dividing by the expected variance, v_{ij}:

$$V_{ij} = \left[1 - \frac{R_i}{n} \right] \times \left[1 - \frac{C_j}{n} \right]$$

Residuals in each individual or sample period. vs. size class is tabulated to show the range of variation in each size class. The standardized residuals show the same information as the size-frequency polygons, but are easier to interpret. A positive residual indicates the frequency of oocytes in that size class is greater than expected, and a negative residual indicates a frequency lower than expected. The contingency table of standardized residuals allows the rapid determination of the size classes in a particular individual or sample period which contain more than the expected number of oocytes. For example, a seasonal spawner will have a cohort of oocytes which can be easily followed by the positions of the positive residuals (Table 3) and a continuous spawner will have a fairly constant ratio of oocytes size at all times (Table 4). Routine photographs of all examined females is helpful when analyzing the contingency tables for individuals.[38]

Relationships of Gonad Maturation with Exogenous and Endogenous Factors

If biochemical factors have been assayed, the mean values are drawn with either 95% confidence intervals or ± 1 SD bars, and plotted vs. the mean oocyte size, gonad index, or maturity index. The correlation coefficient between the biochemical factor and gonad development shows if there is any relationship.[38]

The relationship betwen exogenous factors (e.g., water temperature, photoperiod, lunar cycle) and gonad development are frequently investigated. This relationship is calculated by using Spearmann correlation coefficients between the exogenous factor and the mean of the gonad value (e.g., oocyte size, gonad index) and repeating the calculation for time lags of one, two, or more sampling intervals. A relationship is suggested when the correlatior. coefficient reaches a peak value for a particular time lag; however, cause and effect cannot be drawn from this method.[38]

Summary

Gonad indexes have been used for many years and will probably be used in the future; however, greater knowledge and details about the reproductive cycle can be obtained by analyzing oocyte size frequency. This method requires more work but the benefits of the method exceed any negative aspects. The data are independent of adult size and give detailed information on gonad histology and development. Histological examination allows observation of important phenomenon (e.g., vitellogenesis, necrosis). Growth of oocytes, and differences in developmental stage between individuals and samples are easily studied with contingency table analysis. Also, oocyte size-frequency analysis is the most suitable for evaluating exogenous factors and is more powerful than gonad indexes and maturity indexes.[38]

REPRODUCTIVE CYCLES OF THE MAJOR ABALONE SPECIES

Before attempting to culture a new species, it is important to know the natural reproductive cycle of the species and possible exogenous factors controlling the reproductive cycle. A careful examination of all the factors, both exogenous and endogenous, affecting the reproductive cycle may give a clue to the natural control of gonad maturation and help researchers develop methods to artificially induce gonad maturation throughout the year. During the initial phases of culture, before conditioning methods are developed, the broodstock must be obtained from wild populations and spawning induction is limited by the length of time

Table 3
EXAMPLE OF RESIDUALS IN A CONTINGENCY TABLE FOR SEASONALLY BREEDING SPECIES

Sample	Oocyte size class (μm)									
	0—15	15—25	25—35	35—45	45—55	55—65	65—75	75—85	85—95	95—115
Date										
I	19.33	27.67	-14.24	-12.00	-7.21	-4.04	-3.95	-5.64	-6.95	-11.04
II	9.75	28.66	-9.84	-10.58	-6.35	-4.05	-4.30	-5.07	-6.89	-9.84
III	14.53	39.83	-14.60	-14.58	-8.76	-5.58	-6.09	-7.62	-9.35	-13.58
IV	-3.80	21.03	4.55	-8.31	-4.99	-3.18	-3.60	-4.34	-5.42	-7.74
V	-6.69	-29.86	22.83	42.63	18.44	-2.98	-6.13	-8.90	-11.95	-16.86
VI	-7.55	-11.45	11.02	3.76	4.01	11.29	5.60	3.07	-0.91	-5.72
VII	-14.57	-35.15	-7.41	-5.02	8.25	8.51	13.34	16.97	20.36	34.17
VIII	-13.51	-32.44	-3.51	-1.28	-2.77	3.71	6.88	13.35	24.15	33.92
IX	-4.49	-10.91	10.50	7.16	-0.36	1.47	2.87	1.64	-1.14	-2.22
X	6.53	8.03	-2.02	-4.26	-2.88	-1.83	-1.55	-1.77	-2.76	-4.46
XI	0.10	7.57	8.24	-4.80	-3.55	-2.26	-2.56	-3.09	-3.86	-5.27
XII	5.73	8.48	3.77	-4.68	-4.02	-2.56	-2.89	-3.49	-4.08	-6.22

(From Grant, A. and Tyler, P.A., *Int. J. Invertebr. Reprod.*, 6, 271, 1983. With permission).

Table 4
EXAMPLE OF RESIDUALS IN A CONTINGENCY TABLE FOR CONTINUOUS BREEDING SPECIES

Sample	Oocyte size class (µm)									
	0—15	15—25	25—35	35—45	45—55	55—65	65—75	75—85	85—95	95—115
Date										
I	2.72	-2.64	-1.35	-1.69	1.63	0.56	0.79	0.74	0.90	1.07
II	-3.63	-1.79	-3.52	-0.70	2.75	2.95	3.09	3.55	3.18	3.28
III	0.83	-8.07	-2.62	3.10	5.81	5.57	3.30	3.30	-1.75	-1.73
IV	1.75	-3.07	0.24	-0.03	-0.03	0.22	0.00	1.20	-1.49	1.82
V	5.12	17.31	3.28	-7.61	-8.04	-8.05	-4.58	-5.50	-3.42	-3.37
VI	0.62	1.31	4.52	-0.27	0.43	-2.09	-2.98	-2.73	-2.57	-2.77
VII	4.21	5.61	2.41	-1.34	-5.66	-4.28	-3.68	-3.58	-1.39	-0.99
VIII	-10.27	-1.33	2.63	-0.25	0.30	1.20	3.32	0.37	6.97	0.63
IX	-2.73	-6.61	-3.35	7.03	2.96	3.18	0.81	1.98	0.37	1.55

(From Grant, A. and Tyler, P.A., Int. J. Invertebr. Reprod., 6, 271, 1983. With permission).

Table 5
SPAWNING SEASONS OF ABALONE (NORTHERN HEMISPHERE)

Species	Winter			Spring			Summer			Fall			Ref.
	Jan	Feb	Mar	Apr	May	Jun	Jul	Aug	Sep	Oct	Nov	Dec	
Halioti: corrugata					X	X	X	X	X				52
H. cracherodii									X	X	X	X	4
						X	X	X	X	X			6
				X	X	X	X	X	X				51
H. discus										X	X	X	19
	X									X	X	X	35,36
											X	X	53
H. discus hannai									X	X			6
	X									X	X	X	7
					X	X				X	X	X	7
										X	X	X	19
								X	X	X	X		19
								X	X	X	X	X	19
								X	X	X	X	X	31,32
							X	X	X	X	X	X	35,36
									X	X			54
								X	X	X			55
									X	X			56
							X	X	X				57
H. diversicolor supertexta							X	X	X	X	X		58
H. fulgens				X	X	X	X	X					52
H. gigantea											X	X	6
										X	X	X	19
	X										X	X	35,36
											X	X	53
											X	X	54
H. kamtschatkana							X	X	X				9
					X								12
				X			X	X					46
H. lamellosa							X	X	X	X	X	X	29,30
H. pustulata							X	X	X				10
H. rufescens	X	X	X	X	X	X	X	X	X	X	X	X	5
	X	X	X	X	X	X	X	X	X	X	X	X	6
	X												11
							X	X	X				18
					X	X	X						25
	X	X	X										59
	X	X	X										60
							X	X	X	X			61
					X	X							62
							X	X	X	X			63
H. sieboldii										X	X	X	19
	X										X	X	35,36
											X	X	53
H. tuberculata							X	X					6
							X	X	X	X	X	X	64
							X	X	X	X			65
H. walallensis				X	X	X							12

ripe individuals are present in the wild. Tables 5 and 6 show the spawning seasons for the important abalone species in the world.

When discussing spawning periods, the time interval refers to the spawning of the population and not to the spawning of an individual within a population. An individual in the

Table 6
SPAWNING SEASONS OF ABALONE (SOUTHERN HEMISPHERE)

Species	Summer			Fall			Winter			Spring			Ref
	Jan	Feb	Mar	Apr	May	Jun	Jul	Aug	Sep	Oct	Nov	Dec	
Haliotis australis				X					X	X			9
H. cyclobates										X	X	X	10
H. iris			X	X									9
		X											9
H. laevigata										X	X	X	10
	X	X	X	X						X	X	X	10
	X									X	X	X	66
H. midae				X						X	X	X	8
							X	X	X				8
		X	X	X					X	X			8
H. roei				X	X	X	X	X	X	X	X	X	10
H. ruber				X	X	X	X	X	X	X			10
				X	X	X	X	X	X				10
										X	X	X	10
	X	X							X	X			47
H. scalaris	X	X	X	X	X	X	X	X			X	X	10
	X	X	X										67

population may only spawn once during a fairly narrow time period; however, the population may have an annual reproductive cycle where gametes are released over a longer period.

Japan
Haliotis discus hannai
The spawning period of *H. discus hannai* is determined by water temperature and photoperiod. Spawning begins 1 hr after sunset on the day the water temperature drops below 20°C. Generally, northern populations spawn earlier than southern populations; however, the spawning periods vary depending on the water temperature.[39] *H. discus hannai* tends to aggregate during the spawning season.[40]

Spermatogonia are found in the testis during all seasons of the year. Secondary spermatocytes are found less often than other gametogenic stages, which indicates that the maturation division is very rapid. In the testis, the spent stage is from late August to December, recovery stage is from mid October to the following May, premature stage is from May to mid June, and mature stage is from mid June to late August.[31]

All gametogenic stages, from oogonium to oil drop, are found in the ovary throughout the year. In the ovary, the spent stage is from late August to late October, recovery stage is from early October to the following late April, premature stage is from late March to late June, and mature stage is from June to late August. Oogenesis is very slow during the winter due to the cold water temperature (~5°C). The oocytes begin to develop as soon as the water temperature increases above 7.6°C (approximately March) and development becomes very rapid as the water temperature increases in summer to 24°C.[32]

Southern Japan Species
Although there are differences in the exact timing of spawning between *H. discus hannai* (a cold water species), and *H. gigantea, H. sieboldii, H. discus,* and *H. diversicolor supertexta* (warm water species), each species initiates spawning after sunset on the day the water temperature decreases to 20°C.[19] However, the exact spawning period of each species varies depending on the water temperature. Each abalone species in southern Japan aggregates during the spawning season.[40]

North America
Haliotis rufescens

The spawning period of *H. rufescens* has been reported as either one time per year, two times per year, or the entire year.[5,11,25] The last finding is substantiated by the observation that ripe individuals capable of spawning can be collected from the wild during most of the year.[41]

Even when the same population is studied, different conclusions have been made about the reproductive cycle. Young and Demartini[5] found *H. rufescens* in northern California capable of spawning throughout the year. Ripe oocytes and necrotic vegetative oocytes were present in each monthly sample, and the GI showed no significant change during the year.[5] Giorgi and DeMartini[25] studied the same *H. rufescens* population, and found the spawning period occurred from April to July with a peak in May.[2] Spontaneous spawnings were seen on several occasions from March to July but only a portion of the population spawned.[25] This observation was confirmed by histological examination of the gonads which proved only a portion of the population spawned during this period. Partially spawned individuals were found during the spring with the active phase occurring during the summer. Most large oocytes (~90%) were necrotic by the end of fall. Gametogenesis continued throughout winter and maximum ripeness was reached in February.[25] The GBI_{tc} indicated there were three stages in the reproductive cycle. There was an initial flat section during the early active phase, the curve became steeper during the late active stage and vitellogenesis, and then flattened out during the ripe phase.[2]

Ault[2] reexamined the histological slides of the specimens collected by Giorgi and DeMartini.[25] He believed some of the individuals in the population which were classified as nonspawners (i.e., retaining mature eggs) during the previous spawning period could have become ripe by early August, if they had spawned in April (since it takes approximately 120 days to complete gametogenesis in the wild). However, the presence of necrotic oocytes along with mature eggs in an ovary proved the oocytes were residual from the previous spawning period.[2]

Most of the studies on *H. rufescens* have been conducted in northern California. One exception was the study by Price[11] on a population of *H. rufescens* from Point Loma, San Diego. Price determined the spawning period by following fluctuations in the digestive gland area and found the reproductive cycle of *H. rufescens* in this area was different from that found in northern California populations. The digestive gland cross-section area remained fairly constant throughout the year but almost doubled in size between March 2 and April 30 and was larger than the gonad cross-section area at the end of this period. There was a large decline in the GBI in both sexes at the end of April and minor decreases at the end of January and September.[11]

Haliotis fulgens

H. fulgens individuals spawn once per year but the population has a prolonged spawning period with an intermediate degree of syncrhony over the natural shallow depth range. The spawning period along the California Channel Islands is the same as the spawning periods near Ensenada, Mexico, and Cedros Island, Mexico.[13,42,43] *H. fulgens* collected from the field can be induced to spawn from April to October.[41]

Haliotis corrugata

H. corrugata spawn several times during the year. These spawnings are not well synchronized among the individuals in the population.[13] *H. corrugata* can be induced to spawn in the laboratory, between late winter and early summer, with the peak spawning period from April to November.[44]

Haliotis sorenseni

Individual *H. sorenseni* have a high degree of rhythmicity in their annual reproductive cycle and synchrony of spawning in the population. There is a single short spawning period during winter, when the entire population spawns.[41] Gonad development begins in August-September. Spawning can be induced from January to April.[44]

Haliotis cracherodii

The gonad size of *H. cracherodii* remains relatively constant during the spring but increases rapidly in early summer until it makes up 20% of the body weight. When the ovary is fully ripe, the eggs reach a maximum size of 160 μm. Spawning occurs in late summer, however, there can be as much as 6 weeks difference in the time of spawning between populations only 7 mi. apart. Gametogenesis is initiated immediately (no resting phase) after spawning and again during the following spring. A pool of early gametogenic stages present in the gonad throughout the year allows rapid production of later gametogenic stages right before spawning.[4] Spawning can be induced in the laboratory from early spring to early fall.[41]

Haliotis kamtschatkana

Spring is the major spawning period for *H. kamtschatkana* in Canada, although individuals in the population may spawn until July in some areas and ripe individuals are found throughout the year.[45,46] Breen[45] observed a natural spawning of about half the individuals in a population. Spawning began at 3 p.m. and continued until 6:30 p.m. Spawning individuals formed aggregations (up to six abalone) with one individual attached to a rock and the others stacked on top of the bottom one. Sometimes the top abalone was 20 to 25 cm above the substratum of the bottom abalone. Most of the spawning aggregations and many single spawners were on the highest points available (rock ridge or boulder top) in the surrounding area. These aggregations might be caused by a tendency for abalone to be as high as possible during spawning or might act as a mechanism for ensuring a high fertilization rate by exposing the released eggs to more sperm as they fall through the water column before settling on the bottom.[45]

New Zealand

Haliotis iris

Two size clases of oocytes are found in the ovary of *H. iris* during the year. The oocytes of the small size class are round with a clearly visible nucleus and pale cytoplasm, and become stalked as they grow to 130 μm. The larger size class contains large stalked yolky oocytes with hazy nuclei, which develop into round mature oocytes (250 μm). The two classes are hard to distinguish in post-spawned individuals due to the presence of a wide range of gametogenic stages.[9]

Poore[9] found the reproductive cycles of populations from Kaikoura and Taylors Mistake were similar but there were some major differences. At Taylors Mistake spawning occurred about 1 month before the population at Kaikoura, the gonads required more time to recover after spawning and were "spent" a longer time. The ovary in Kaikoura individuals recovered rapidly after spawning, with many ripe oocytes and younger stalked stages present in June. There was also a partial spawning in the Taylors Mistake population in April and some of the individuals retained ripe eggs through winter.[9]

Haliotis australis

H. australis spawns twice during the year and has a more complex reproductive cycle than *H. iris*. Variability of the spawning period from year to year makes it impossible to determine the precise spawning period. Poore[9] found two size classes of oocytes were found in *H. australis* ovaries and the oocytes had the same characteristics as *H. iris*. Young stalked oocytes were most numerous in September and October, and again in April. There was rapid recovery after both spawnings and maximum gonad size was attained several months before the fall spawning. The spawning period in fall was a shorter duration than in spring but both spawnings could be important to the species.[9]

Australia
Haliotis ruber

A H. ruber population at West Island had partial spawnings in spring and early summer, with complete spawning by fall. Spawning is poorly synchronized as a population, with peaks of spawning in spring and fall, and some spawning in other seasons. The oocyte sizes show a bimodal distribution during the spring, summer, and fall. Small primary oocytes are found in spent or partially spawned ovaries from January to July, and development begins after all the ova are released. The number of spent individuals in the population gradually increases until reaching a maximum in late fall, indicating that each individual spawns only once per year. H. ruber from Tipara Reef, have a single period of gametogenesis and a single spawning season.[10]

In general, gametogenesis begins during the period of decreasing water temperatures and spawning occurs at the minimum or the initial rise of the water temperature.[47]

Haliotis roei

There are three categories of oocyte size frequency found in the ovary of H. roei. A bimodal frequency with large oocytes (150 to 250 μm) and small primary oocytes (25 to 50 μm) in the ovary, a unimodal frequency with only small to medium oocytes, and a unimodal frequency with only mature oocytes. The bimodal frequency indicates that the individual was collected during spawning. The gonads never have a spent stage since primary oocytes develop as soon as the mature oocytes are spawned. H. roei is considered to spawn throughout the year since individuals may spawn sporadically over a long period and are not synchronized with other individuals in the population.[10]

Haliotis laevigata

Timing of the spawning period in H. laevigata appears to be controlled by the food supply. Spawning ends by March in regions where food is scarce by late summer, but continues until late May if the food supply is plentiful.[16] Gametogenesis is also closely correlated with changes in the supply of food rather than water temperature alone.[10]

Haliotis cyclobates

In H. cyclobates, small oocytes (25 to 75 μm) attached to the trabeculae, begin to develop in January and grow about 10 μm/month during the fall. In winter, the oocytes rapidly develop and grow 25 μm/month. The oocytes reach their maximum size by December. The mature oocytes are released at a smaller size (150 to 225 μm in diameter) than in other species of abalone.[10]

Europe
Haliotis tuberculata

Spawning in H. tuberculata populations is not synchronous but occurs over a relatively short period (August to September). Spawning time varies from island to island in the Channel Islands. After spawning, the gonads become very soft and filled with a gelatinous substance, and some gonads are hollow inside. The ovary contains very few developing primary oocytes and resorption of ova occurs from December to February. Oocytes develop slowly from February to April, and then rapidly increase in quantity and size, until maximum density is reached by August. Males usually become ripe earlier than females.[48]

Table 7
MINIMUM SIZE OR AGE OF SEXUAL
MATURITY

Species	Size (mm)	Age (year)	Ref.
Haliotis australis	55		9
H. coccinea canariensis	29 (female)		68
H. corrugata		3	13
H. cyclobates	40—48		10
H. fulgens		5	13
H. iris	60	4	9
H. kamtschatkana	50	3	46
H. laevigata	75—77	3	10
	86—98	3	10
	98—120	3	10
H. roei	45—55		10
H. ruber	60—69		47
	72—92	3	10
	80—108	3	10
	92—112	3	10
H. rufescens	39.5 (female)		25
	84.5 (male)		25
	50 (male)		2
	55 (female)		2
H. scalaris	45—58		10
	57—62		10
	60—70		10
H. sorenseni		4	13
H. tuberculata	49 (female)		48
	40 (male)		48

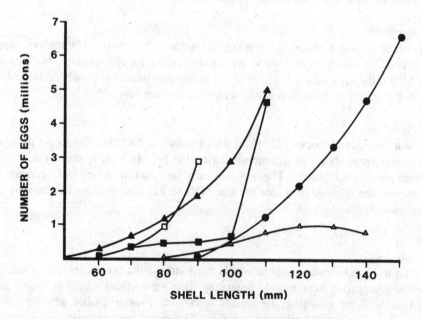

FIGURE 2 .   Relationship between shell length and fecundity. (●) *H. iris*;[9] (△) *H. iris*;[49] (□) *H. australis*;[9] (■) *H. tuberculata*;[34] (▲) *H. tuberculata*.[48]

Table 8
FECUNDITY OF MAJOR CULTURED ABALONE
SPECIES

Species	Fecundity	Ref.
H. australis	3.0×10^6 (91 mm)	9
	4.6×10^4 (62 mm)	9
	$(0.0005 W^a + 1.124) \times 10^6$	9
H. coccinea canariensis	7×10^4 (49 mm)	68
H. corrugata	$0.44 - 2.32 \times 10^6$	13
H. fulgens	$2.05 - 3.45 \times 10^6$	13
H. gigantea	$(1.3 - 6.3) W \times 10^3$	69
H. iris	1.3×10^3 (68 mm)	9
	$(0.0170 W - 1.528) \times 10^6$	9
	11.0×10^6 (155 mm)	9
	3×10^6 (125 mm)	70
H. kamtschatkana	2.3×10^6 (135 mm)	71
H. midae	$(0.0198 W - 2.196) \times 10^6$	9
H. rufescens	1.0×10^5 (111 mm)	2
	1.1×10^7 (182 mm)	2
	$0.6 - 12.6 \times 10^6$	25
	$6.5 - 19.2 \times 10^4$/g of ovary	25
H. sorenseni	$3.69 - 6.53 \times 10^6$	13
H. tuberculata	6.2×10^5 (80 mm)	22
	5.0×10^6 (110 mm)	34
	5.0×10^5 (70 mm)	34
	3.8×10^4 (49 mm)	48
	5.1×10^6 (115 mm)	48
	3.8×10^5 (61 mm)	72
	3.4×10^6 (101 mm)	72

^a W = body weight in grams.

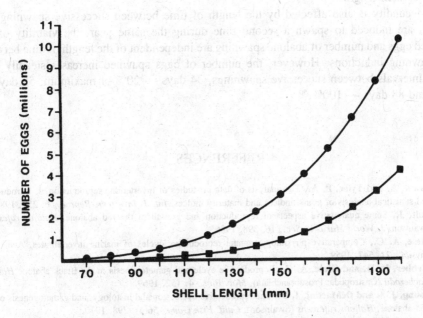

FIGURE 3 . Relationship between shell length and fecundity of *H. rufescens*; (●) individuals conditioned in the laboratory; (■) individuals collected from the wild. (From Ault, J., *J. World Maricult. Soc.*, 16, 398, 1985. With permission.)

MINIMUM SIZE OF SEXUAL MATURITY AND FECUNDITY

The best method to estimate the size of sexual maturity and fecundity is to induce spawning in fully conditioned animals of each shell-length size. The amount of error is reduced when fecundity is estimated from induced spawning, since only viable, spawnable eggs are counted for the estimation.[2] A sexually mature animal is defined as an individual having either spermatozoa or primary oocytes.[25] Table 7 shows the minimum size of maturity in several abalone species. Even though an individual is mature, it may have an insignificant input to the total number of eggs released in the population. Individuals close to the minimum size of sexual maturity probably have low fecundities (10^2 to 10^3), and do not contribute much to the total reproductive effort of the population.[9,49] Figure 2 and Table 8 show the relationship between shell length and fecundity in the important abalone species in the world.

Gonad histology can also give an estimation of sexual maturity and fecundity. A histological section can be used from any position of the ovary. These estimates will be higher than the actual fecundity since the total number of eggs in the gonad will not be released during spawning.[9] There are many sources of error when using live or preserved ovary tissue to estimate fecundity. Errors can be made in weighing tissue, counting eggs, teasing out oocytes, and viability of oocytes.[2]

The fecundity of individuals with the same shell length is not a constant and can vary greatly between individuals. Ault[2] found a nonlinear relationship of fecundity to shell length, with fecundity increasing rapidly with shell length. There is considerable variability in the fecundity of larger animals.[49] In addition, laboratory-conditioned animals had higher fecundities than field-conditioned individuals of equivalent shell length (Figure 3). The development of oocytes by the germinal epithelium is a surface phenomenon, and the amount of development is proportional to the available surface area. The number and convolutions of trabeculae were considerably more extensive in laboratory-conditioned individuals and increased folding may account for the significant increases in fecundity.[2] There is not always a fecundity/shell length relationship. The fecundity of *H. australis* is independent of body weight or shell length.[9]

The fecundity is also affected by the length of time between successive spawnings. If animals are induced to spawn a second time during the same year, the viability of the spawned eggs and number of abalone spawning are independent of the length of time between the spawning inductions. However, the number of eggs spawned increases steadily with longer intervals between successive spawnings, 34 days — 20% of maximum, 55 days — 61%, and 88 days — 100%.[50]

REFERENCES

1. **Grant, A. and Tyler, P. A.,** The analysis of data in studies of invertebrate reproduction. I. Introduction and statistical analysis of gonad indices and maturity indices, *Int. J. Invertebr. Reprod.,* 6, 259, 1983.
2. **Ault, J.,** Some quantitative aspects of reproduction and growth of the red abalone, *Haliotis rufescens* Swainson, *J. World Maricult. Soc.,* 16, 398, 1985.
3. **Giese, A. C.,** Comparative physiology: annual reproductive cycles of marine invertebrates, *Ann. Rev. Physiol.,* 21, 547, 1959.
4. **Webber, H. H. and Giese, A. C.,** Reproductive cycle and gametogenesis in the Black abalone *Haliotis cracherodii* (Gastropoda: Prosobranchiata), *Mar. Biol.,* 4, 152, 1969.
5. **Young, J. S. and DeMartini, J. D.,** The reproductive cycle, gonadal histology, and gametogenesis of the Red abalone, *Haliotis rufescens* (Swainson), *Calif. Fish Game,* 56(4), 298, 1970.
6. **Boolootian, R. A., Farmanfarmaian, A., and Giese. A. C.,** On the reproductive cycle and breeding habits of two western species of *Haliotis, Biol. Bull.,* 122, 183, 1962.
7. **Ino, T. and Harada, K.,** On the spawning of abalone in the vicinity of Ibargi prefecture, *Bull. Tokai. Reg. Fish. Res. Lab.,* 31, 275, 1961.

8. **Newman, G. G.**, Reproduction of the South African abalone *Haliotis midae, Div. Sea Fish. Invest. Rep.*, 64, 1, 1967.

9. **Poore, G. C. B.**, Ecology of New Zealand abalones, *Haliotis* species (Mollusca: Gastropoda). IV. Reproduction, *N. Z. J. Mar. Freshwater Res.*, 7(1 and 2), 67, 1973.

10. **Shepherd, S. A. and Laws, H. M.**, Studies on southern Australian abalone (genus *Haliotis*). II. Reproduction of five species, *Aust. J. Mar. Freshwater Res.*, 25, 49, 1974.

11. **Price, P. S.**, Aspects of the reproductive cycle of the red abalone, *Haliotis rufescens*, Master's thesis, San Diego State University, California, 1974.

12. **Hahn, K.**, The reproductive cycle and gonadal histology of the pinto abalone, *Haliotis kamtschatkana* Jonas, and the flat abalone, *Haliotis walallensis* Stearns, *Adv. Invertebr. Reprod.*, 2, 387, 1981.

13. **Tutschulte, T. and Connell, J. H.**, Reproductive biology of three species of abalones *(Haliotis)* in southern California, *The Veliger*, 23(3), 195, 1981.

14. **Webber, H. H.**, Changes in metabolite composition during the reproductive cycle of the abalone *Haliotis cracherodii* (Gastropoda: Prosobranchia), *Physiol. Zool.*, 43, 213, 1970.

15. **Olive, P. J. W. and Garwood, P. R.**, The importance of long term endogenous rhythms in the maintenance of reproductive cycles of marine invertebrates: a reappraisal, *Int. J. Invertebr. Reprod.*, 6, 339, 1983.

16. **Shepherd, S. A.**, Breeding, larval development and culture of abalone, *Aust. Fish.*, 35(4), 7, 1976.

17. **Carlisle, J. G.**, The technique of inducing spawning in *Haliotis rufescens* Swainson, *Science*, 102, 566, 1945.

18. **Carlisle, J. G.**, Spawning and early life history of *Haliotis rufescens* Swainson, *Nautilus*, 76, 44, 1962.

19. **Ino, T.**, Biological studies on the propagation of Japanese abalone (genus *Haliotis*), *Bull. Tokai. Reg. Fish. Res. Lab.*, 5, 1, 1952.

20. **Ino, T.**, *Fisheries in Japan, Abalone and Oyster*, Japan Marine Products Photo Materials Assoc., Tokyo, 1980, 165.

21. **Kurita, M., Sato, R., Ishikawa, Y., Shiihara, H., and Ono, S.**, Artificial seed production of abalone, *Haliotis discus* (Reeve,) (transl.), *Bull. Oita Pref. Fish. Exp. St.*, 10, 78, 1978.

22. **Koike, Y.**, Biological and ecological studies on the propagation of the ormer, *Haliotis tuberculata* Linnaeus. I. Larval development and growth of juveniles, *La Mer*, 16(3), 124, 1978.

23. **Owen, B., McLean, J. H., and Meyer, R. J.**, Hybridization in the eastern Pacific abalones *(Haliotis)*, *Bull. L. A. Co. Mus. Nat. Hist. Sci.*, 9, 1, 1971.

24. **Fretter, V. and Graham, A.**, *Reproduction, in Physiology of Mollusca*, Vol. 1, Wilbur, K. M. and Yonge, C. M., Eds., Academic Press, New York, 1964, 1.

25. **Giorgi, A. E. and DeMartini, J. D.**, A study of the reproductive biology of the Red abalone, *Haliotis rufescens* Swainson, near Mendocino, California, *Calif. Fish Game*, 63(2), 80, 1977.

26. **Olley, J. and Thrower, S. J.**, Abalone — an esoteric food, in *Advances in Food Research*, Vol. 23, Chichester, C. O., Mrak, I. M., and Stewart, G. F., Eds., Academic Press, New York, 1977, 144,.

27. **Morse, D. E., Duncan, H., Hooker, N., Morse, A.**, Hydrogen peroxide induces spawning in mollusks, with activation of prostaglandin endoperoxide synthetase, *Science*, 196, 298, 1977.

28. **Crofts, D. R.**, Haliotis, *Liverpool Mar. Biol. Commi. Mem. Typ. Br. Mar. Plants Anim.*, 29, 1, 1929.

29. **Bolognari, A.**, Ricerche sulla sessualita di *Haliotis lamellosa* Lam., *Arch. Zool. Ital.*, Italy, 38, 361, 1954.

30. **Bolognari, A.**, Alcuni aspetti dell ovogensi di *Haliotis lamellosa* Lam., *Boll. Accad. Gioenia Di Sci. Nat.*, Catania Ser., 4(2), 420, 1954.

31. **Tomita, K.**, The testis maturation of the abalone, *Haliotis discus hannai* Ino in Rebun Island, Hokkaido, Japan, *Sci. Rep. Hokkaido Fish. Exp. St.*, 3, 56, 1968.

32. **Tomita, K.**, The maturation of the ovaries of the abalone, *Haliotis discus hannai* Ino, in Rebun Island, Hokkaido, Japan, *Sci. Rep. Hokkaido Fish. Exp. St.*, 7, 1, 1967.

33. **Yahata, T. and Takano, K**, On the maturation of the gonad of the abalone, *Haliotis discus hannai*. I. A comparison of the gonadal maturation of the abalone from Matsumae and Rebun in Hokkaido, *Bull. Fac. Fish. Hokkaido Univ.*, 21(3), 193, 1970.

34. **Girard, A.**, La reproduction de l'ormeau *Haliotis tuberculata* L., *Rev. Trav. Inst. Peches Marit.*, 36(2), 163, 1972.

35. **Lee, T. Y.**, Histological study on gametogenesis and reproductive cycle of the Korean coasts, *Bull. Pusan Fish. Coll.*, 14(1), 59, 1974.

36. **Lee, T. Y.**, Gametogenesis and reproductive cycle of abalones, *Publ. Mar. Lab. Busan Fish. Coll.*, 7, 21, 1974.

37. **Boolootian, R. A.**, Reproductive physiology, in *Physiology of Echinodermata*, Boolootian, R. A., Ed., Interscience, New York, 1966, 561.

38. **Grant, A. and Tyler, P. A.**, The analysis of data in studies of invertebrate reproduction. II. The analysis of oocytes size/frequency data, and comparison of different types of data, *Inter. J. Invertebr. Reprod.*, 6, 271, 1983.

39. **Uki, N. and Kikuchi, S.**, Regulation of maturation and spawning of an abalone, *Haliotis* (Gastropoda) by external environmental factors, *Aquaculture*, 39, 247, 1984.

40. **Kafuku, T. and Ikenoue, H.,** Abalone *(Haliotis (Nordotis) discus)* culture, in *Modern Methods of Aquaculture in Japan, Developments in Aquaculture and Fisheries Science,* Vol. 11, Kafuku, T. and Ikenoue, H., Eds., Kodansha, Tokyo, 1983, 172.

41. **Leighton, D.L.,** The influence of temperature on larval and juvenile growth in three species of southern California abalones, *Fish. Bull.,* 72 (4), 1137, 1974.

42. **Cota, I. F.,** Fecundacion artificial y dessarrollo embrionario de *Haliotis fulgens* Philippi, 1845 y *Haliotis rufescens* Swainson, 1822 en condiciones aquario, Thesis, Univ. Auton. Baja Calif., Ocean., 1970.

43. **Sevilla, M. L., Hernandez, H., Mondragon, E., Farfan, O. N., Giovanini, A., and Hernandez, A.,** Estudio histologico comparativo de algunos moluscos de importancia economica en Mexico, *Inst. Nac. Invest. Biol. Pesqs., Dir. Gener. Pesca. S. I. C. Serv. Trav. D. v III,* 22, 1, 1965.

44. **Leighton, D. L. and Lewis, C. A.,** Experimental hybridization in abalones, *Int. J. Invertebr. Reprod.,* 5, 273, 1982.

45. **Breen, P. A. and Adkins, B. E.,** Spawning in a British Columbia population of Northern abalone, *Haliotis kamtschatkana, The Veliger,* 23(2), 177, 1980.

46. **Quayle, D. B.,** Growth, morphometry and breeding in the British Columbia abalone *(Haliotis kamtschatkana* Jonas), *Fish. Res. Board Can. Tech. Rep.,* 279, 1, 1971.

47. **Harrison, A. J. and Grant, J. F.,** Progress in abalone research, *Tasmanian Fish. Res.,* 5(1), 1, 1971.

48. **Hayashi, I.,** The reproductive biology of the ormer, *Haliotis tuberculata, J. Mar. Biol. Assoc. U.K.,* 60, 415, 1980.

49. **Sainsbury, K. J.,** Population dynamics and fishery management of the paua, *Haliotis iris.* I. Population structure, growth, reproduction, and mortality, *N. Z. J. Mar. Freshwater Res.,* 16, 147, 1982.

50. **Nishikawa, N., Obara, A., and Ito, Y.,** A method to induce artificial spawning of the abalone, *Haliotis discus hannai* Ino, throughout the year, *J. Hokkaido Fish. Sci. Inst.,* 31(5), 21, 1974.

51. **Leighton, D. and Boolootian, R. A.,** Diet and growth in the black abalone, *Haliotis cracherodii, Ecology,* 44(2), 227, 1963.

52. **Quintanella, M. O.,** Informe preliminar de tas investigaciones sobre la biologia y pesca del abulon commercial de las islos de Cedros, Benitos y Guadalupe, Baja California, *Mex. Inst. Nac. Invest. Biol. Pesq.,* 17, 766, 1966.

53. **Kishinoue, K.,** Study on abalone. I. *Reports Fisheries Investigations,* 3 (1 and 2), 1, 1894. (Cited in **Ino, T.,** *Fisheries in Japan, Abalone and Oyster,* Japan Marine Products Photo Materials Assoc., Tokyo, 1980, 165.)

54. **Tago, K.,** Distribution of the genus Haliotis in Japan, *Fish. Prot.,* 43, 352, 1931. (Cited in **Boolootian, R.A., Farmanfarmaian, A., and Giese, A. C.,** *Biol. Bull.,* 122, 183, 1962.)

55. **Sakai, S.,** On the shell formation of the annual ring on the shell of the abalone, *Haliotis discus* var. *hannai* Ino, *Tohoku J. Agric. Res.,* 11, 239, 1960.

56. **Kanno, H. and Kikuchi, S.,** On the rearing of *Anadara broughtonii* (Schrenk) and *Haliotis discus hannai* Ino, *Bull. Mar. Biol. Stn. Asamushi,* Tohoku Univ., 11 (2), 71, 1962.

57. **Sakai, S.,** Ecological studies on the abalone, *Haliotis discus hannai* Ino. III. Study on the mechanism of production of the abalone in the region of Onagawa Bay, *Bull. Jpn. Soc. Sci. Fish.,* 28(9), 891, 1962.

58. **Oba, T.,** Studies on the propagation of an abalone, *Haliotis diversicolor supertexta* Lischke. I. On the spawning habits, *Bull. Jpn. Soc. Sci. Fish.,* 30 (9), 742, 1964.

59. **Bonnot, P.,** Abalones in California, *Calif. Fish Game,* 16 (1), 15, 1930.

60. **Bonnot, P.,** California abalones, *Calif. Fish Game,* 26 (3), 200, 1940.

61. **Bonnot, P.,** The abalones of California, *Calif. Fish Game,* 34 (4), 141, 1948.

62. **Croker R.,** Abalones, *Calif. Fish Game Fish. Bull.,* 30, 58, 1931.

63. **Cox, K. W.,** Review of the abalone in California, *Calif. Fish Game,* 46 (4), 381, 1960.

64. **Crofts, D. R.,** Haliotis, *Liverpool Mar. Biol. Comm. Mem. Typ. Br. Mar. Plants Anim.,* 29, 1, 1929.

65. **Forster, G. R.,** Observation on the ormer population of Guernsey, *J. Mar. Biol. Assoc. U.K.,* 42, 493, 1962.

66. **Shepherd, S.A.,** Studies on southern Australian abalone (genus *Haliotis*) I. Ecology of five sympatric species, *Aust. J. Mar. Freshwater Res.,* 24(3), 217, 1973.

67. **Shepherd, S. A., Clarkson, P. S., and Turner, J. A.,** Studies on southern Australian abalone (genus *Haliotis*). V. Spawning, settlement and early growth of *H. scalaris, J. Exp. Mar. Biol. Ecol.,* 109, 61, 1985.

68. **Pena, J.B.,** Preliminary study on the induction of. artificial spawning in *Haliotis coccinea canariensis* Nordsieck (1975), *Aquaculture,* 52, 35, 1986.

69. **Kikuchi, S. and Uki, N.,** Technical study on artificial spawning of abalone, genus *Haliotis* VI. On sexual maturation of *Haliotis gigantea* Gmelin under artificial conditions, *Bull. Tohoku Reg. Fish Res. Lab.,* 35, 85, 1975.

70. **Tong, L. J.,** The potential for aquaculture of paua in New Zealand, in *Proceedings Paua Fisheries Workshop,* Fish. Res. Div. Pub. No. 41, Akroyd, J. M., Murray, T. E., and Tayler, J. L., Eds., N. Z. Min. Agric. Fish., Wellington, 1982, 36.

71. **Caldwell, M. E.,** Spawning, Early Development and Hybridization of *Haliotis kamtschatkana* Jonas, Master's thesis, University of Washington, Seattle 1981.
72. **Hayashi, I.,** Small scale laboratory culture of the ormer, *Haliotis tuberculata, J. Mar. Biol. Assoc. U. K.,* 62, 835, 1982.

ARTIFICIAL INDUCTION OF CONDITIONING (GONAD MATURATION)

Kirk O. Hahn

INTRODUCTION

The term "conditioning" applies to the laboratory process used to induce gonad ripeness and spawnability (usually out of the normal spawning season) in adult animals. Conditioning is a broad term which encompasses several biological processes; gametogenesis, synthesis of hormones controlling gametogenesis, and spawning, and in the female, vitellogenesis and synthesis of hormones controlling vitellogenesis. In addition to the gonad being ripe and spawnable, the animal must be in the proper physiological state (e.g., good nutrition, water quality, photoperiod, and water temperature) to promote spawning.

There is a two-fold need for methods to condition abalone. Conditioning is important to aquaculturists for year-round, efficient culture of juvenile abalone and to researchers for studying the reproduction of abalone. One of the biggest difficulties in investigating the biological, physiological or technical aspects of abalone reproduction is the short period of natural spawning. In addition, induction of spawning is not always easy in the laboratory.[1]

The relationship between external environmental factors and gonad maturation in *Haliotis* spp. has been studied to establish conditioning methods for use in seed production. Several methods have been proposed for conditioning abalone out of the natural spawning season. The most common conditioning method consists of giving adults good living conditions (tailored to the ecology of the particular species) and *ad libitum* feeding. In a few species, methods have been designed to take advantage of special biological characteristics of the species being cultured.

It is important the animals are always fed *ad libitum* during experiments to study exogenous factors controlling gonad maturation. Food levels play an important role in gonad development and maturation, and the lack of sufficient food could obscure experimental results (Figure 1).[2] The water temperature is believed to be the main exogenous factor regulating reproductive cycles in marine invertebrates, and should probably be investigated first in any study on conditioning of a new species.[3] In the laboratory, simply controlling water temperature and feeding level is not always successful in inducing gonad maturation in all species of abalone. In some species, additional environmental factors or hormone treatments will have to be studied to achieve rapid gonad maturation in a shorter time.[4]

Current methods for conditioning several economically important abalone species will be described in this chapter. These methods should be used as a starting point for developing conditioning techniques for other abalone species.

SPECIES

Haliotis discus hannai

Ezo awabi, *H. discus hannai*, is the only abalone species for which the exogenous factor controlling gonad maturation is known and the level of gonad maturation can be quantified. Kikuchi and Uki[1] showed that the water temperature-controlled gonad maturation but the effect of the water temperature was not simply additive, that is, an animal did not become mature twice as fast at 20°C than at 10°C. *H. discus hannai* appeared to require a critical minimum water temperature before gonad maturation began.

Kikuchi and Uki[1] first studied the relationship between water temperature and the gonad index (a subjective measurement of the gonad development). Initially, an index was developed to judge the level of gonad maturation without killing the animal. The gonad was

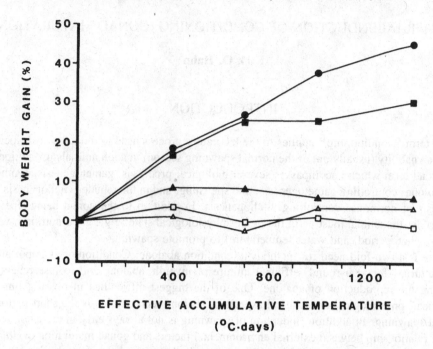

FIGURE 1. Change in body weight of *H. discus hannai* exposed to five different food levels. Feeding levels — (●) satiation; (■) 65%; (▲) 40%; (△) 26%; and (□) 17%. Adult individuals were fed *Undaria pinnatifida* and *Laminaria religiosa* under constant water temperature — 17°C and photoperiod — 12 light/12 dark. (From Uki, N. and Kikuchi, S., *Bull. Tohoku Reg. Fish. Res. Lab.*, 45, 45, 1982. With permission.)

observed macroscopically and subjectively assigned a value from 0 to 3. Each abalone was held horizontally upside down and the gonad was viewed by looking straight across the edge of the shell. Gonad stage was rated "0" when the sex of the individual could not be distinguished (immature), "1" when the gonad was concave into the shell, "2" when the gonad was even with the shell edge, and "3" when the gonad was bulging above the shell edge.[5] The gonad index was developed as a semiquantitative measurement of gonad growth. However, the gonad index does not directly measure the gonad maturation or spawnability.[4]

Groups of animals were raised at different water temperatures and the average gonad index value was measured for each group at regular intervals. Each group of animals showed an initial linear increase in the average gonad index before reaching a plateau at the maximum value, 3. The rate of increase in gonad index (calculated from the initial portion of the graph which was assumed to be linear) increased in correlation with higher water temperatures and showed a clear linear relationship with the water temperature. The formula for the linear relationship between the rate of increase of gonad index (Y) and water temperature (T) is: $Y = 0.00597\ T - 0.04527\ (T \leq 20°C)$.[1] When this equation is solved for $Y = 0$ (i.e., no gonad maturation after time = infinity), then $T = 7.6°C$. This temperature, called the biological zero point, is the theoretical minimum temperature at which gonad growth and development begins.[4] Although the animal can live at this temperature, gonad maturation is initiated only when the water temperature rises above this value.[1]

This relationship allowed the gonad maturation of *H. discus hannai* to be quantified by calculating the daily exposure to water temperatures above 7.6°C. However, it was not the absolute water temperature but only the difference between the water temperature and 7.6°C, which contributed to gonad maturation. For example, if the water temperature was 20°C, only 12.4°C actually contributed to the gonad maturation. The quantity above the biologial zero point had an additive effect toward gonad maturation. The difference between the

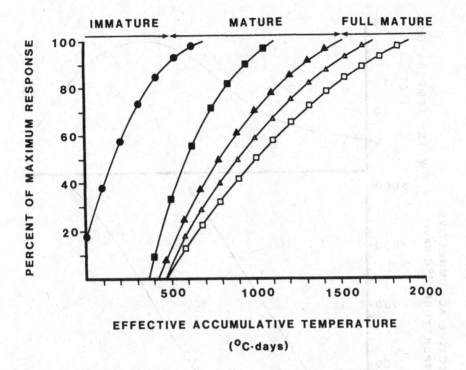

FIGURE 2 . Gonad maturation of *H. discus hannai* related to the EAT. Gonad maturation is classified into three stages. (●) Gonad index (100% = 3); (■) spawning rate (male); (▲) spawning rate (female); (△) duration of ejeculation (100% = 400 min.); (□) number of eggs spawned per 100 g of abalone weight (100% = 2 × 10⁶). (From Kikuchi, S. and Uki, N., *Bull. Tohoku Reg. Fish. Res. Lab.*, 33, 69, 1974. With permission.)

biological zero point and the water temperature has been called the effective temperature and summation of this value during gonad maturation is called the effective accumulative temperature (EAT). Therefore, if the effective temperature is measured daily and the EAT is calculated from all days since the water temperature rose above the biological zero point, the level of gonad maturation can be determined.

$$Y_n = \sum_{i=1}^{n} t_i - \theta$$

where Y_n [°C-days] = EAT, n [days] = number of days after the water temperature rose above θ, t_i [°C] = daily water temperature to which the animal is exposed (values of $t_i <$ θ are not added) and θ [°C] = biological zero point for gonad development. The mean gonad index of the population is directly proportional to EAT until reaching 2.5, after which it gradually approaches 3.0 independently of temperature (Figure 2).[4]

Spawning capacity for both sexes of *H. discus hannai* gradually appears after 500°C-days. The spawning rate reaches 100% in males at 1000°C-days and females at 1500°C-days. The fertilization rate is over 90% for all gametes released irrespective of the EAT. Thus, spawned gametes are not released until fully mature and viable.[4] The number of spawned eggs increases proportionally with EAT from 500 to 1500°C-days and then levels off.[4,6] The duration of ejaculation (indicator of number of sperm released) reaches a maximum at 1200°C-days.[4]

Gonad maturation can be divided into three stages from observed relationships with the EAT.[4]

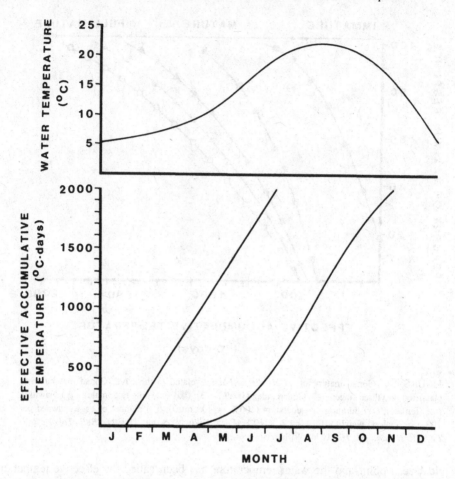

FIGURE 3.   Change in ambient water temperature along the northeastern coast of Honshu, Japan, and the correponding effective accumulative temperature for *H. discus hannai* in the wild and laboratory-conditioning tanks. The upper diagram shows the change of average temperature at Enoshima Island in Miyagi prefecture. The sigmoid curve in the lower figure shows the increase in EAT calculated from ambient water temperatures starting from January when abalone gonads are in the recovery stage. The straight line indicates an example of the increase in EAT when the abalone is reared under a constant temperature of 20°C from February. (From Uki, N. and Kikuchi, S., *Aquaculture*, 39, 247, 1984. With permission.)

1.   Immature stage (0 to 500°C-days) — the gonad index ranges from 0 to 3 as the gonad volume increases but spawning cannot be induced.
2.   Mature stage (500 to 1500°C-days) — the gonad index reaches 3. The spawning rate and quantity of gametes released rises with increasing EAT. Animals with an EAT between 1000 to 1500°C-days are sufficiently ripe for hatchery production.
3.   Fully mature stage (> 1500°C-days) — gonad development reaches its maximum level and there is a reliable high rate of spawning.

When EAT is calculated from the ambient water temperature, it can be seen that the natural spawning season (late August to mid October, with the peak period from early September to early October) corresponds to an EAT of 1200 to 1800°C-days and the peak period occurs at 1300 to 1700°C-days. This finding confirms the laboratory results and indicates the water temperature is the principle factor which regulates the reproductive cycle (Figure 3).[4]

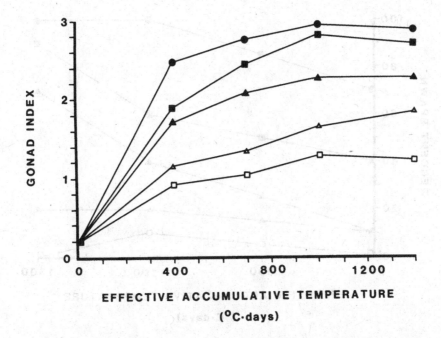

FIGURE 4 . Change in gonad index of *H. discus hannai* exposed to five different food levels. Feeding levels — (●) satiation; (■) 65% (▲) 40% △ 26%; and (□) 17%. Adult individuals were fed *Undaria pinnatifida* and *Laminaria religiosa* under constant water temperature — 17°C and photoperiod — 12 light/12 dark. (From Uki, N. and Kikuchi, S., *Bull. Tohoku Reg. Fish. Res. Lab.*, 45, 45, 1982. With permission.)

Besides the relationship between water temperature and gonad maturation, there is naturally a relationship between food consumption and gonad maturation.[2] Separating the two factors is difficult in poikilotherms, since higher water temperatures cause an increase in feeding. The observed gonad index and rate of spawning is proportional to the feeding level, and the feeding level is more important in females than males (Figures 4 and 5). The daily feeding rate must exceed 5% of body weight before EAT correlates to the gonad maturation level.[2,4] Daily feeding rate increases at an almost constant rate during gonad development, but decreases after the gonad is fully developed. The average daily feeding rate (percent of live body weight) at different temperatures was 6.3% — 22.4°C, 8.3% — 20.1°C, 7.1% — 17.1°C, 6.5% — 14.3°C, 4.7% — 11.5°C, and 3.2% — 8.2°C.[4] Provided the food is adequate, EAT gives a reproducible relationship to gonad maturation in *H. discus hannai*. However, the specific water temperature (7.6 to 20°C) to which the abalone is exposed shows no other direct effect.[4]

Ezo awabi should usually be induced to spawn before they reach 2000°C-days. Once an animal reaches 2000°C-days, even slight fluctuations in water temperature in the conditioning tanks can induce spawning. The permissible conditioning level for spawning is considered from 1000 to 2000°C-days which is an interval of about 80 days at 20°C or even longer at lower temperatures.[4]

Haliotis discus

H. discus also shows a relationship between EAT and gonad maturation, but is less responsive to artificial conditioning than *H. discus hannai* and requires 2 to 3 times the EAT to attain the same level of gonad maturation as *H. discus hannai* (Figure 6).[7] There is a general tendency for populations which inhabit warmer water to require a higher EAT than populations of the same species or subspecies which inhabit colder water.[4] The equation

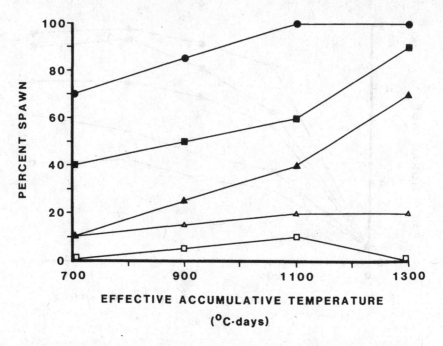

FIGURE 5 .   Spawning rate with UV-irradiated sea water induction on *H. discus hannai* exposed to five food levels. Five individuals of each sex were tested at each feeding level. Feeding levels — (●) satiation; (■) 65%; (▲) 40%; (△) 26%; and (□) 17%. Adult individuals were fed *Undaria pinnatifida* and *Laminaria religiosa* under constant water temperature — 17°C and photoperiod — 12 light/12 dark. (From Uki, N., and Kikuchi, S., *Bull. Tohoku Reg. Fish. Res. Lab.*, 45, 45, 1982. With permission.)

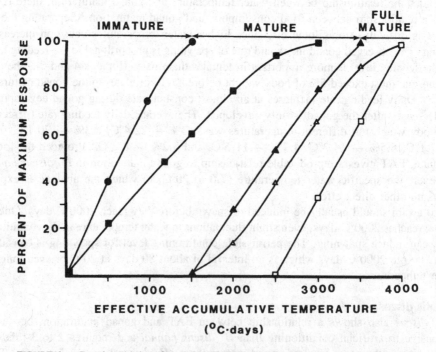

FIGURE 6 .   Gonad maturation of *H. discus* related to the EAT. Gonad maturation is classified into three stages. (●) Distinction rate of sex by eye; (■) gonad index (100% = 3); (▲) spawning rate (male and female); (△) duration of ejaculation (100% = 600 min.); (□) number of eggs spawned per 100 g of abalone weight (100% = 4×10^4). (From Kikuchi, S. and Uki, N., *Bull. Tohoku Reg. Fish. Res. Lab.*, 34, 77, 1974. With permission.)

representing the relationship between gonad index and water temperature is $Y = 0.00102$ $T - 0.00548$ and when solved for $Y = 0$, $T = 5.3°C$ (biological zero point).[8]

1. Immature stage (0 to 1800°C-days) — during this period the gonad index ranges from 0 to 2. Distinguishing the sex becomes easier as the EAT increases but induction of spawning is impossible when the EAT is below 1800°C-days.
2. Mature stage (1800 to 3500°C-days) — the gonad index increases from 2 to 3 and induction of spawning is possible during the second half of this period.
3. Fully mature stage (>3500°C-days) — spawning rate and number of gametes spawned are maximal. Reliable spawning with large amounts of gametes is possible with an EAT above 3500°C-days.[8]

Haliotis gigantea

No relationship was found between the gonad index or gonad maturation, and water temperature. Gonad maturation, however, was related to the length of time cultured. The sexes could gradually be distinguished after 120 days of culture, regardless of the water temperature at which the animal was raised. Spawning could be induced after 220 days at all water temperatures. The daily increase in gonad index for this species is smaller than in *H. discus hannai* and *H. discus*.[9]

Haliotis rufescens

The conditioning of red abalone *(H. rufescens)*, appears to be controlled by the food level. There is no direct correlation between gonad maturation and an exogenous physical factor (e.g., water temperature, photoperiod). Red abalone can easily be conditioned by raising them under ideal conditions with *ad libitum* feeding. Approximately 3 to 4 months after spawning, the animals are fully mature and can be induced to spawn again.[10]

If the brood stock is collected from the wild, for best results, they should be acclimatized for several weeks with abundant food and good rearing conditions (proper water circulation and temperature) before being used for spawning.[11] The spawnability of *H. rufescens* collected from nature depends on the size of the individual. Ault[10] found spawning could only be induced in females larger than 110 mm and males larger than 65 mm, when the animals were collected from the wild.

Laboratory conditioning of red abalone with *ad libitum* feeding can reduce the minimum size of spawnability and increase the frequency of successful induction of spawning. The minimum size of spawning went from 100 mm in wild-caught animals to 55 mm in laboratory-conditioned animals. Also, the ratio of spawners to nonspawners, in animals less than 100 mm, went from 8/93 for wild-caught animals to 22/12 for laboratory-conditioned animals. The fecundity of laboratory-conditioned animals was greater, with a 100-mm laboratory-conditioned female having a fecundity equivalent to a 140-mm wild-caught female.[10] The *ad libitum* feeding and higher water temperatures in the laboratory are probably the main factors for the difference in spawning and fecundity.[10]

After complete spawning, females reared with ideal conditions have extensive ovarian growth and proliferating oocytes within 15 to 40 days postspawning and mature oocytes comprise 85% of the ovary by 75 days postspawning.[10] The time interval an individual requires to become ripe again can be reduced from 120 days (wild) to between 75 and 90 days (laboratory conditioned). Successful induction of spawning is possible after this time period. It is possible, in both females and males, to spawn and recondition individuals three times withing a 10-month period.[10]

Ebert and Houk[12] developed a gonad index for Californian species, similar to the index used in Japan, to follow conditioning and determine the gonad ripeness by gross examination of the gonad size and color. In their method, stage ''0'' is assigned when the individual is immature, sex cannot be determined, and the digestive gland can easily be seen through the gonad tissue and appears grayish brown. Stage "1" is assigned when gametogenesis has begun. The newly developed gametes appear as patchy areas on the surface of the digestive gland. Recognizing males is easy at this stage because the testis is creamy white in color; however, recognizing females is more difficult because ovarian tissue is not green at this stage. Stage "2" is assigned when gonad tissue with gametes completely surrounds the conical appendage and the sex can be easily determined. Gonad size has not increased and is still not very large. Stage "3" is identical to stage "2" except the gonad has increased in size and the bulk extends to the tip. Brood stock used for spawning should ideally be stage "3" but individuals between "2" and "3" can be used for spawning induction.[12]

Haliotis fulgens, Haliotis corrugata, and Haliotis cracherodii

H. fulgens can be easily conditioned during winter, when normally spent, by raising them in sea water at 20 to 24°C.[13] Accelerated gametogenesis, gonad maturation, and spawning in the warm water species *(H. fulgens, H. corrugata,* and *H. cracherodii)* of California requires *ad libitum* feeding at elevated temperatures (1 to 2 months at 18 to 25°C, depending upon species, feeding rate, water quality, and water exchange rate).[11,14]

CONDITIONING TANKS

Rearing tanks used for conditioning must not cause any stress to the adult animals during the process. A good indicator of stress is the growth rate. If growth rate decreases at any time during conditioning, then tank conditions should be considered unsatisfactory.[15] Animal growth in nature is affected by many factors both external — temperature and food, and internal — growth stage and spawning period. When animals are raised under artificial conditions, additional factors can affect the growth rate. These factors include amount of dissolved oxygen, concentration of excreta, rapid changes in water temperaure, water flow, and light. Therefore, rearing tanks must be designed to eliminate any factors which would be detrimental to growth. A shell growth 50 μm/day is considered a satisfactory growth rate and an indication of suitable rearing conditions.[15]

The most important factor affecting the growth rate is the amount of excreta in the tanks, and rapid removal of excreta is critical for maintaining maximum growth rate. Growth is inhibited by high levels of metabolic products and reduced water quality (e.g., bacteria growth). Levels of metabolic wastes can be reduced by having sufficient water flow to the conditioning tank. Kikuchi and Uki[15] found the growth rate is not reduced from a build up of metabolic products, if the water flow in the tank is sufficient to maintain 80% oxygen saturation in the tank . This flow rate can be calculated with the following equation on the basis of oxygen consumption of *H discus hannai*:

$$Q = \frac{R_w}{S_t * (1 - 0.8)}$$

where Q (ℓ/kg/ hr) = water flow requirement, R_w (mℓ/kg/hr) = volume of oxygen consumed per kilogram of *H. discus hannai* having an average weight of "w" grams and S_t (mℓ/ℓ) = volume of dissolved oxygen at its saturation point in sea water at a temperature of t° C.[15] The oxygen consumption, R (mℓ O_2/h) of *H. discus hannai* is calculated with the formula R = M*W^b*A^t, where the temperature, "t", is between 8 to 20°C and the weight, "w", is from 5 to 150 g, and M = 0.0210, b = 0.8025, and A = 1.0963 (Figure 7).[15,16]

The decrease in oxygen concentration in the tank is assumed to be entirely from the respiration of the animals, and the equation disregards oxygen input from the water surface or aeration. It should be noted that this calculation is used to establish an appropriate water

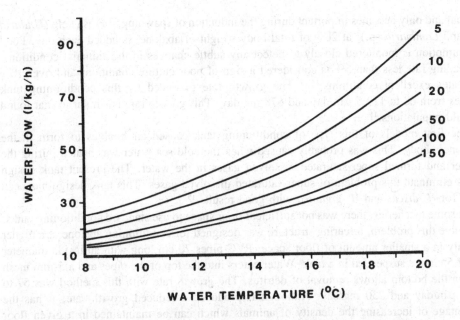

FIGURE 7 . Water flow necessary to maintain 80% oxygen saturation of sea water at a given temperature in abalone-rearing tank. Water supply is assumed to be 100% oxygen saturated and oxygen decrease the tank is caused by respiration of abalone only, regardless of oxygen supply from water surface and aeration under the conditions of chlorinity = 19 ‰, 760 mb atmospheric pressure. Number on right side of each curve indicates the average weight (g) of the individuals in the rearing tank. (From Uki, N. and Kikuchi, S., *Bull. Tohoku Reg. Fish. Res. Lab.*, 43, 47, 1981. With permission.)

flow to remove excreta produced by the animals and is not intended necessarily to maintain a particular level of dissolved oxygen. At the calculated flow rate, aeration of the water has no influence on the growth rate of adult abalone.[15]

It is critical that ideal culture conditions are maintained in the conditioning tank, since a disease outbreak or mass spawning would cause the loss of the entire tank.[17] Several designs of conditioning tanks have been tested in Japan. The best design (good water flow and water circulation by aeration) for *H. discus hannai* (and abalone in general) is rectangular (150 cm L × 90 cm W × 80 cm H), constructed of black PVC-impregnated canvas, and supported by a steel frame. The water level is approximately 70 cm, which is a volume of approximately 1000 ℓ.[15] Sea water filtered to between 1 and 5 μm for prevention of disease and fouling, flows continuously into the bottom of the tank on one side, and along the other side, aeration is supplied at 20 ℓ/min through a 140-cm pipe (20 mm diameter) with 1-mm holes at 10-cm intervals. The combination of aeration and water flow creates an air curtain and causes a circular water motion. This water action automatically removes excreta and dissolved gases, and helps maintain the kelp in good shape.[15] Three V-shaped shelters are arranged parallel to the rotating water flow on the bottom of the tank. The shelters are made of 0.5-cm-thick PVC sheets, 60 cm long and 45 cm wide. The tank is covered with a black-colored canvas which excludes all direct and reflected light.[15]

The adults (50 individuals per tank) are conditioned for 3 to 6 months with the water temperature maintained at a constant 20°C with no fluctuation in temperature. The animals are acclimatized to an artificial photoperiod of 12 hr light (1 a.m.to 1 p.m)/ 12 hr dark (1 p.m.to 1 a.m.) by a 40-W light bulb suspended from the roof of the tank. This wattage of light bulb produces a light intensity of 150 lux at the bottom of the tank. This light intensity is equivalent to the natural light intensity at a water depth of 7 to 8 m, the preferred water depth of *H. discus hannai*. The shifted photoperiod has no effect on the conditioning of the animal and only becomes important during the induction of spawning.[17] Fresh kelp *(Undaria* sp. and *Laminaria* sp.), at 20% of total body weight of abalone, is added each day. Food consumption is monitored closely to detect any subtle changes in the animals' conditions. A feeding rate less than 5% is considered a sign of poor culture conditions and over 10% indicates excellent conditions.[15,17] The growth rate produced by this conditioning tank ranges from 68 to 119.5 μm/day and 678 mg/day. This growth rate is similar to that found in wild populations.[15]

The first models of this type of conditioning tank caused air bubbles to form in the abalone's body. This was probably caused when the cold sea water was heated during the winter and formed super-saturated dissolved gases in the water. The present tank design helps eliminate this problem of super-saturated dissolved gases. This tank design has been used for *H. discus* and *H. gigantea* with good results.[15]

At some hatcheries, there was not sufficient floor space to use standard conditioning tanks. To solve this problem, a rearing structure was designed for conditioning abalone at a higher density in a smaller amount of floor space. PVC pipes 70 cm long with an inside diameter of 12.5 cm are suspended in a tank. Water enters into the top of the pipes and a 5-mm mesh net on the bottom allows removal of detritus. The growth rate with this method was 57 to 83.9 μm/day and 250 mg/day. Although it produced a reduced growth rate, it has the advantage of increasing the density of animals which can be maintained in a given floor space.[15]

Brood stock does not necessarily have to be maintained in large tanks. At the Granite Canyon Laboratory (California Fish and Game) the brood stock is maintained in 15-ℓ polyethylene plastic containers (0.34 m² surface area, including the lid). The containers are cleaned every week and excess *Macrocystis* sp. is supplied at the same time. Broodstock is maintained separately by sex. The containers receive 3 to 4 ℓ/min of ambient (9 to 15°C), sand-filtered (20 μm), UV-irradiated sea water.[12]

REFERENCES

1. **Kikuchi, S. and Uki, N.,** Technical study on artificial spawning of abalone, genus *Haliotis.* I. Relation between water temperature and advancing sexual maturity of *Haliotis discus hannai* Ino, *Bull. Tohoku Reg. Fish. Res. Lab.,* 33, 69, 1974.
2. **Uki, N. and Kikuchi, S.,** Influence of food levels on maturation and spawning of the abalone, *Haliotis discus hannai* related to effective accumulative temperature, *Bull. Tohoku Reg. Fish. Res. Lab.,* 45, 45, 1982.
3. **Giese, A. C.,** Comparative physiology: annual reproductive cycles of marine invertebrates, *Annu. Rev. Physiol.,* 21, 547, 1959.
4. **Uki, N. and Kikuchi, S.,** Regulation of maturation and spawning of an abalone, *Haliotis* (Gastropoda) by external environmental factors, *Aquaculture,* 39, 247, 1984.
5. **Kumabe, L.,** Methods used at the Oyster Research Institute, unpublished report, 1981.
6. **Nishikawa, N., Obara, A., and Ito, Y.,** A method to induce artificial spawning of the abalone, *Haliotis discus hannai* Ino, throughout the year, *J. Hokkaido Fish. Sci. Inst.,* 31(5), 21, 1974.
7. **Kafuku, T., and Ikenoue, H.,** Abalone *(Haliotis (Nordotis) discus)* culture, in *Modern Methods of Aquaculture in Japan, Developments in Aquaculture and Fisheries Science,* Vol. 11, Kafuku, T. and Ikenoue, H., Eds., Kodansha, Tokyo, 1983, 172.
8. **Kikuchi, S. and Uki, N.,** Technical study on artificial spawning of abalone, genus *Haliotis.* V. Relation between water temperature and advancing sexual maturity of *Haliotis discus* Reeve, *Bull. Tohoku Reg. Fish. Res. Lab.,* 34, 77, 1974.
9. **Kikuchi, S. and Uki, N.,** Technical study on artificial spawning of abalone, genus *Haliotis.* VI. On sexual maturation of *Haliotis gigantea* Gmelin under artificial conditions, *Bull. Tohoku Reg. Fish. Res. Lab.,* 35, 85, 1975.

10. **Ault, J.,** Some quantitative aspects of reproduction and growth of the red abalone, *Haliotis rufescens* Swainson, *J. World Maricult. Soc.,* 16, 398, 1985.

11. **Hooker, N. and Morse, D. E.,** Abalone: the emerging development of commercial cultivation in the United States, in *Crustacean and Mollusk Aquaculture in the United States,* Huner, J. V. and Brown, E. E., Eds., AVI Publishing Co., 1985, 365.

12. **Ebert, E. E. and Houk, J. L.,** Elements and innovations in the cultivation of Red abalone *Haliotis rufescens, Aquaculture,* 39, 375, 1984.

13. **Leighton, D. L., Byhower, M. J., Kelly, J. C., Hooker, G. N., and Morse, D. E.,** Acceleration of development and growth in young Green abalone *(Haliotis fulgens)* using warmed effluent seawater, *J. World Maricult. Soc.,* 12 (1), 170, 1981.

14. **Morse, D. E.,** Biochemical and genetic engineering for improved production of abalones and other valuable molluscs, *Aquaculture,* 39, 263, 1984.

15. **Uki, N. and Kikuchi, S.,** Technical study on artificial spawning of abalone, genus *Haliotis.* VII. Comparative examinations of rearing apparatus for conditioning adult abalone, *Bull. Tohoku Reg. Fish. Res. Lab.,* 43, 47, 1981.

16. **Uki, N. and Kikuchi, S.,** Oxygen consumption of the abalone, *Haliotis discus hannai* in relation to body size and temperature, *Bull. Tohoku Reg. Fish. Res. Lab.,* 35, 73, 1975.

17. **Cuthbertson, A.,** *The Abalone Culture Handbook,* Tasmanian Fisheries Development Authority, Hobart, Australia, 1985, 1.

ARTIFICIAL INDUCTION OF SPAWNING AND FERTILIZATION

Kirk O. Hahn

INTRODUCTION

Artificial induction of spawning is usually the first step in attempting to culture a new species. The control of spawning, and therefore control over the production of larvae, is essential for the efficient use of hatchery facilities and determines the success of the aquaculture operation. Artificial induction of spawning allows the production of larvae over a longer period than would be possible relying solely on natural spontaneous spawning.

Induction of spontaneous spawning in nature is not completely understood, and several endogenous and exogenous factors have been proposed for its control. Spawning might be caused by a sudden change in water temperature (increase or decrease), exposure to air during low tide, photoperiod, lunar cycle, release of gametes from other individuals in the population, or some combination of these or others.[1-11]

As a general rule, males begin spawning slightly earlier and require less stimulus to induce spawning than females. It is frequently possible to predict slightly in advance when a female will spawn. About 5 to 10 min before spawning, the female will elevate the shell until the gonad is visible and will extend a tentacle out the third respiratory pore.[12] The number of gametes released depends on the size of the individual and the level of gonad development. A large, fully gravid red abalone (*Haliotis rufescens*) can release as many as 10 million eggs or 10^{12} sperm.[6] The release of gametes during spawning can cause 5 to 10% reduction in body weight.[1,13]

Mature oocytes in the ovary do not undergo germinal vesicle breakdown until spawning is induced. The nuclear membrane and nucleoli become indistinct as the oocyte separates from the trabeculae and enters the ovarian lumen. Spawned gametes are squeezed out of the gonad by rapid contractions of the columnar muscle, which compresses the conical appendage between the foot and shell.[14] The oocytes are in the middle of the first meiotic reduction division when spawned from the gonad.[15] Abalone have no genital duct. Gametes first enter the right kidney through a longitudinal slit in the roof, exit through the renal duct into the gill chamber, and are carried by water currents out the second and third respiratory pores.[16,17] Mature eggs are extruded separately and free from each other, while immature eggs come out in clumps.[18] The eggs are heavier than sea water and quickly sink after release.[14] A mature abalone egg is approximately 250 μm in diameter and usually dark green. A mature sperm has three parts — head, middle segment, and filament. The head is an elongated cone with an acrosome at the top. The middle segment is 1 μm in width, 8 μm in length, cylindrical, and slightly longer than the head. The filament propels the sperm and is about 50 μm long.[18]

Individuals can easily be sexed by the color of the gonad although there are no secondary sexual characteristics. The gonad can be examined by pulling the foot away from the right side of the shell. The color of the testis has been described as white, cream, beige, yellow, greenish white, greenish cream, or salmon, and the color of the ovary as dark green, brown, gray, or violet.[19-23] Immature or spent individuals are very difficult to sex and the gonad may appear dark brown, cream or yellow. In some species (e.g., *H. rufescens*) the gonad is covered by a darkly pigmented tissue making it very difficult to sex the animal. In these species it is necessary to look at the region at the bottom of the conical appendage where the gonad begins and the pigmentation is reduced.[24]

Several methods of artificially inducing spawning have been tried over the years, some successful and others of limited use. The most commonly used methods will be presented and their usefulness and efficiency evaluated for an aquaculture operation.

INDUCTION OF SPAWNING

Gamete Stripping

Gamete stripping, used by Kishinoue in 1893, was the first attempt to artificially obtain viable gametes from abalone. After fertilization, some of the zygotes developed to the 2- to 4-cell stage before dying.[25,26]

In the first place, the flesh of every male abalone was separated from the shell, the position of the gonad was pressed with a finger, and the sperm flowing out of the hole of the kidney was received in a vessel containing a small amount of sea water. On the other hand, eggs were obtained from every female abalone in the same manner as mentioned above. Both of these samples obtained were filtered through pieces of victoria lawn, respectively, and mixed with each other. However, in many cases the outer membranes of gonad were broken on a victoria lawn filter, sea water was poured on them to wash down gametes into a glass receiver, and they were made to fertilize there, because mature gonads were small in number.[25,26]

Gamete stripping is usually not successful. Even if the eggs are mature, they are unfertilizable when taken directly from the gonad before undergoing the final maturation step that occurs immediately before spawning.[1,15]

The only reliable way to obtain viable, fertilizable eggs is by induction of spawning. Artificial induction of spawning was unsuccessful until Murayama[7] in 1935 added sperm to a bucket containing mature females and induced them to spawn. Several researchers have reported the induction of spawning by adding newly spawned gametes from the opposite sex into the water.[2-4,6-8] This still leaves unanswered the question, "What induced the first individual to spawn?"

Desiccation

If adults are very ripe and about to spawn in nature, sometimes just removing individuals from the water for about 1 hr is sufficient and spawning begins as soon as they are returned to the water. This method is unreliable and has no biological significance because these animals are subtidal and would never experience this stimulus in nature. The desiccation method is also inefficient because it causes the release of large quantities of immature gametes.[2,3,6,27]

Thermal Shock

Ino[18] was the first person to use thermal shock to induce spawning. This method consists of placing ripe males and females together in the same tank and raising the water temperature 3 to 6°C above ambient. The amount of temperature increase depends on the ambient temperature, since care must be taken not to raise the water temperature to a lethal level. In addition, it is important to avoid too great a difference in water temperature between ambient and spawning water during the colder months. This is especially true when inducing spawning in *H. gigantea* and *H. discus*.[25] Water temperature is rapidly increased during the first 30 min and then gradually decreased to ambient over 2.5 hr. Ino[18] also induced spawning by increasing the pH of the water.

A variation of this method is to place animals directly into warmer sea water without exposing them to a gradual increase in temperature. The animals are first removed from the water for 60 to 90 min and then returned to sea water which is 2.4 to 4.0° C above ambient. Males spawn 1 to 2 hr after stimulation and females spawn 2 to 4 hr after stimulation using this method.[28]

Koike[29] used the desiccation and thermal shock experienced by wild caught *H. tuberculata* brood stock during transport to the laboratory as his method of inducing spawning. Brood stock were transported to the lab without water and experienced 40 min of desiccation. During the trip the abalone slowly warmed to approximately ambient air temperature (~29°C).

Upon arrival at the lab, the animals were immediately placed in spawning tanks with water maintained at 27°C, so there would not be a large temperature difference from the air temperature. The water temperature was gradually decreased by adding 21°C sea water. After 2.25 hr, the water temperature in the tanks decreased to about 21°C and spawning began.[29]

Thermal shock, like desiccation, is an unsatisfactory method because it causes the release of immature gametes. It is also difficult to assure the simultaneous release of eggs and sperm.[2,3,6,18,27]

Ultraviolet Irradiated Sea Water

The irradiation of sea water with ultraviolet (UV) light is a fast and reliable method for induction of spawning. The use of UV light to induce spawning was a serendipitous discovery by Shōgo Kikuchi and Nagahisa Uki at the Tohoku Regional Fisheries Research Laboratory. In 1972, Kikuchi and Uki, and Tetsuo Seki at the Oyster Research Institute (ORI) facility located next to the Tohoku Electric Power Station were trying, at their separate facilities in Shiogama, to develop a reliable method to induce spawning in abalone. Kikuchi and Uki found several methods that induced spawning; however, Seki was unable to repeat the results at his facility (even though the two labs were only about 3 km apart).

After close examination, the only differences between the two laboratories were the filtration and water sterilization systems. The Tohoku Regional Fisheries Research Laboratory had a very good sand filter system and sterilized the water with 4 UV lights in series. Conversely, the ORI facility had only a two stage sand filter capable of filtering to 5μm and 1 UV light for water sterilization. The water of Matsushima Bay is fairly dirty, so Kikuchi wanted the water at his laboratory to be very clean. The filter system consisted of eight separate sand filters in series and could clean 50,000 ℓ of sea water per hour. The water would pass through one filter, be air lifted to the next filter, and continue down the series. After this filtration, the water passed through four UV lights in series to be sure "all" bacteria were killed. Since the water was very clean and there was a large amount of UV irradiation, the abalone were induced to spawn. Inducement of spawning with UV light depends on two factors — the water quality at the facility and the amount of UV irradiation. The Tohoku Regional Fisheries Research Laboratory uses 4 UV lights, the ORI facility at the Tohoku Electric Power Station uses 11 lights, and the ORI facility at Mohne Bay uses 2 lights.[30]

The quantity (intensity × flow rate of water) of UV light (2537 Å) needed to induce spawning in abalone depends on the quality and condition of the sea water. UV light will be dispersed or deflected by particles in the water column and the actual amount of UV light acting on the water molecules will be decreased. A supplementary 1-μm filter should be used to remove particles from the sea water before it enters the UV system. The optimum amount of UV irradiation needed for spawning at the ORI is 2.42 W/ℓ.[12]

Although UV systems manufactured in Japan and U.S. are very similar in strength and structure, comparison between the systems is difficult because criteria used for their evaluation are very different. In the U.S., the primary function of UV systems is water purification and sterilization. U.S. companies describe their systems by the amount of lamp strength per unit of exposed area or by the various microorganism species which are killed by the UV irradiation. Even though UV irradiation is important for water sterilization in Japan, the primary purpose of UV systems is to induce spawning. The critical measure of the system is the actual strength of the UV light reaching the water. The UV system at the ORI has a strength of 50,000 mW/cm².[12]

The exact mechanism has never been discovered how UV-irradiated sea water induces spawning in abalone. Water temperature, salinity, and pH during spawning inducement have been insignificant; however, the salinity and pH may be significant at extreme values.[12] UV

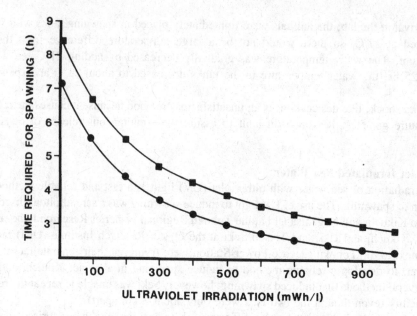

FIGURE 1.   The hyperbolic line describing the relation between the time required from the start of stimulating to ejacultion of spawning (y) and the dosage of UV rays (x). (■) Female, (●) male. (From Kikuchi, S. and Uki, N., *Bull. Tohoku Reg. Fish. Res. Lab.*, 33, 79, 1974. With permission.)

irradiation is believed to cause the energetic decomposition of the water molecule to produce a hydroperoxy free radical, $H00^{\cdot}$, or peroxy diradical, $^{\cdot}00^{\cdot}$.[6,31]

UV irradiation of sea water is the single major factor causing predictable and reliable spawning. Before the UV irradiation method was developed, typical spawning success rate was approximately 40 to 60%, although the rate was higher during the natural spawning season. The UV-irradiated sea water treatment, however, does not always produce 100% spawning. Spawning failure is usually caused by some error of technique or methodology. Also, incomplete filtration of the sea water may leave particles in the water column which will diffract UV light and decrease the theorized photochemical reaction with sea water. The net amount of UV irradiation may not be sufficient to induce spawning.[12]

The time between initiation of UV stimulation and spawning is proportional to the amount of UV irradiation. Shorter intervals between initiation and spawning are found with increasing amounts of UV irradiation. There is, however, an absolute finite minimum time interval from initiation of artificial induction of spawning to release of the oocytes, due to the final oocyte maturation before spawning. The relationship between the time interval from initiation to spawning, Y (hours), and the UV irradiation strength, x (mWhr/ℓ), can be represented by the formula,

$$Y = \frac{c}{x+a} = b$$

where a, b, and c are constants.[32] The constant a is the quantity of effective environmental stimuli, other than UV irradiation, expressed as UV units of irradiation, b = Y, when x = infinity (the minimum time required for spawning at the strongest stimulus), and c is the sensitivity of the animal to the stimulus. The constants for *H. discus hannai* are, male: a = 231.3, b = 1.064, and c = 1422.4, and female: a = 278.5, b = 1.350, and c = 1978.7. The values of b calculated empirically and measured experimentally are almost the same in each sex (males — 1 hr 4 min, females — 1 hr 21 min) (Figure 1). UV irradiation of 800 mWhr/ℓ was determined to be the ideal strength. At this level, spawning is induced in males by 3 hr 18 min and females by 3 hr 42 min, and there is no harm to the adult abalone or the gametes.[30]

During the initial attempts of using UV-irradiated sea water, it was noted the adults would spawn approximately 3 hr after initiation of stimulation and spawn a second time at the beginning of night or the photoperiod dark phase. The first spawning was induced by the UV irradiation but the second spawning appeared to be induced by an endogenous rhythm synchronized by the photoperiod. If UV irradiation was begun 7 to 9 hr before the start of the dark phase of a 12:12 photoperiod, 61% of the males and 38% females spawned within 4 to 5 hr, and the duration of spawning was usually about 30 min, with some lasting 1hr. Many individuals spawned repeatedly: males, 52% — one time, 35% — two times, and 12% — three times; and females, 75% — one time, 23% — two times, and 2% — three times. Thirty-two percent of the males that spawned for the first time later than 4 hr after initiation and 20% of second spawnings for all males were grouped around 1 hr after the beginning of the dark phase.[32]

If adults are collected from the wild and spawning induction is initiated 1 hr 30 min before the beginning of night, the average time to spawning is 1 hr 27 min in males and 1 hr 36 min in females. However, if adults collected from the wild are first conditioned for 13 days with the dark period shifted to begin at 1 p.m. (12:12 photoperiod) and initiation of spawning begins 1 hr before the artificial dark period, the average time to spawning is males — 1 hr 20 min and females — 1 hr 45 min. The spawning is well synchronized and all individuals release their gametes within a range of ± 10 to 20 min.[32]

These results support the hypothesis of photoperiod control of spawning synchronization.[31] It can be concluded from these data that the natural timing of spawning is approximately 1 hr after sunset. It is possible to take advantage of this photoperiod synchronization to achieve better control over fertilization and increase the quantity of spawned eggs by synchronizing the spawnings induced by UV irradiation and the photoperiod.[12] The best time to begin stimulation with UV irradiation is 1 to 2 hr before the beginning of the artificial dark phase.[32] The absolute synchronization of spawning in both sexes is not necessary since gametes can be maintained for 3 hr at 17°C.[33] Successful induction and synchronization of the sexes during spawning can be achieved with UV irradiation and entrainment with an artificial out-of-phase photoperiod.[31]

Spawning begins even without the lights actually going out at the beginning of the entrained dark phase. This indicates that the endogenous photoperiodic rhythm and not the actual ending of the light phase induces spawning. The beginning of the light period and not the dark period probably starts the cascade to induction of spawning in nature. Since the chain of events leading to spawning must begin 2 to 3 hr before the onset of the dark phase, the animal must start "counting" at the beginning of the light phase to synchronize spawning.

In some cases there is an indication that spawning can be artificially induced solely by changing the photoperiod. Spawning of *H. discus hannai* in Korea could be induced prematurely in June by artificially reducing the day length until it matched the photoperiod found in October, the natural spawning period (Figure 2).[34]

Hydrogen Peroxide

Besides UV irradiation of sea water, the hydroperoxy free radical, HOO·, or peroxy diradical, ·OO· can be produced chemically by the addition of peroxide to water. Hydrogen peroxide is the cheapest and most commonly available peroxide; however, reagent grade (30%) must be used, since 3% pharmaceutical solutions are unstable and give unreliable results.[6]

Adults should be fully gravid and fed *ad libitum* for 2 weeks prior to spawning. Gravid individuals are placed singly in containers with enough water (12 to 18° C for red abalone)

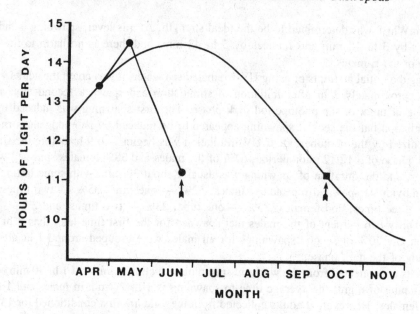

FIGURE 2.   (●) Artificial photoperiod before inducing spawning (arrow) in *H. discus hannai*. (■) Natural photoperiod and spawning period (arrow) in the nearby sea. (From Kim, Y. and Cho, C., *Bull. Korean Fish. Soc.*, 9(1), 61, 1976. With permission.)

to cover them and the water is gently aerated. The pH of the water is first increased to 9.1 by adding 6.6 mℓ of 2 *M* tris-(hydroxymethylamino)methane (Tris-base, mol wt = 121.1) for each liter of water in the container. (To make a 2-*M* solution, dissolve 24.2 g of Tris in about 75 mℓ of water. After it has dissolved, add water to a final volume of 100 mℓ.)

Reagent grade hydrogen peroxide (30%) is diluted to make a 6% working solution. (To make a 6% solution, add 20 mℓ of 30% hydrogen peroxide to 80 mℓ of distilled water. This gives a final volume of 100 mℓ of 6% hydrogen peroxide.) Care should be taken when working with reagent grade hydrogen peroxide because it is caustic, unstable, and should be stored at O to 4° C. Fifteen minutes after adding the Tris, 3 mℓ of freshly prepared 6% hydrogen peroxide solution is added for each liter of water in the container.[27] Water in the container is then thoroughly stirred to mix the solution.

The animal is exposed to the hydrogen peroxide/Tris solution for 2.5 hr. After this period, the water is decanted and the container is thoroughly rinsed with clean isothermal sea water. It is very important to remove the hydrogen peroxide and Tris, and rinse out the container before spawning begins because the presence of these chemicals will cause gametes to be nonviable. After rinsing, the container is filled with clean isothermal sea water so as not to disturb the animal. Spawning will usually begin within 2.5 to 3.5 hr following the addition of hydrogen peroxide (2.5 hr at 18° C, 3.5 hr at 12° C).[27]

Tris is not mandatory for successful artificial induction of spawning. Tris only increases the pH and is not required for any specific chemical reaction. Tris increases the percentage of individuals that spawn at a specific hydrogen peroxide concentration. Sodium hydroxide (1 × 10⁻³ *M*) has the same effect as Tris when used to increase the pH to 9.1. A high pH enhances the decomposition of the added hydrogen peroxide into highly reactive and short-lived, free-radical oxidants which cause the induction of spawning.[6,27]

Spawning of warm-water species (*H. fulgens*, *H. corrugata*, and *H. cracherodii*) in California is usually difficult when the animals are collected directly from the wild. Once these animals have been conditioned in the laboratory, spawning of gravid animals can be reliably induced with hydrogen peroxide at a higher water temperature (20 to 25°C, depending on species) than used for red abalone, 15°C (Table 1).[14,35]

Table 1
SPAWNING SUCCESS IN CALIFORNIA ABALONES USING
HYDROGEN PEROXIDE

Species	Sex	Occasions spawned		Individuals spawned		Mean response time (hr)
		Occasions tested	%	Individuals tested	%	
H. sorenseni	Male	4/4	100.0	6/6	100.0	1.7
	Female	2/6	33.3	2/12	16.7	2.3
H. rufescens	Male	57/58	98.3	61/74	82.4	2.5
	Female	39/40	97.5	49/59	83.0	2.9
H. corrugata	Male	10/14	71.4	14/24	58.3	2.6
	Female	5/13	38.5	6/39	15.4	2.8
H. fulgens	Male	7/11	63.6	9/23	39.1	3.0
	Female	3/9	33.3	5/18	27.8	3.8

(From Leighton, D.L. and Lewis, C.A., *Int. J. Invertebr. Reprod.*, 5, 273, 1982. With permission.)

Although the hormonal control of reproduction in abalone and other prosobranch gastropods is not known, it is suspected that a prostaglandin-like chemical may be involved in the induction of spawning. Morse[6] was able to induce spawning by adding the vertebrate hormones prostaglandin E or F (3×10^{-12} M) to the water. These hormones, however, only induced spawning in 38 to 47% of gravid red abalone. Hydrogen peroxide and its organic peroxide derivatives probably act as donors of electronically activated oxygen suitable as substrate for the cyclooxygenase-catalyzed addition of oxygen to form prostaglandins. Aspirin (a prostaglandin inhibitor) inhibits spawning when it is added before the hydrogen peroxide. Aspirin inhibits the cyclooxygenase activity, suggesting that this enzyme is necessary to induce spawning in the presence of hydrogen peroxide.[6]

Yahata[36] found the pedal-pleural and visceral ganglia were involved in inducing spawning in *H. discus hannai*. These ganglia may produce and release factor(s) that induce spawning in abalone. Histological examination of the pedal-plural ganglion showed several cell types, producing abundant neurosecretory material that appears to vary in correlation to the reproductive cycle.[24]

EVALUATION OF ARTIFICIAL SPAWNING INDUCTION METHODS

Desiccation and thermal shock have been demonstrated to have limited success in inducing spawning. These methods should not be used alone due to their unreliability. The two other methods, UV irradiation and hydrogen peroxide, are very reliable and induce spawning in ~ 100% of ripe individuals. Both methods probably induce spawning by the same mechanism, even though the active molecule is produced differently. These methods should be considered comparable in reliability and success in inducing spawning. However, each method has advantages and disadvantages that should be considered before choosing the method to be used.

The major disadvantage to using UV irradiation is not the method but the initial cost of the equipment. These costs might be prohibitive to a small culture operation or if brood stock are spawned infrequently. If a UV system is purchased for water sterilization, a slightly larger investment can enhance the system to include the capacity to induce spawning. There are many advantages to using UV irradiation for induction of spawning. The most important is that the method is completely safe to the spawned gametes, and water does not have to be changed during the spawning induction. Also, UV systems can be electronically controlled by timers to activate before the normal working day starts. The UV method is very simple and employees can run the entire process error-free with minimal training. The system can easily be altered to add the capacity to change water temperature during the process, thus increasing the reliability of spawning.

Ebert and Houk[37] consider UV-irradiation induction of spawning one of the most significant contributions to abalone aquaculture. They prefer this method because it requires no chemical additions, the UV unit can be preprogramed with a timer clock, the abalone are ready to spawn when the normal workday begins, and the entire day can be used to process the freshly fertilized eggs and begin larval cultures. Also, larvae will hatch out early on the following day, which is very important for the timing of future culture steps.[37]

The hydrogen peroxide method has the advantage of being very cheap. The chemicals necessary for this procedure can easily be obtained and stored for a relatively long period (months to a year for hydrogen peroxide and years for Tris) without deteriorating. The major disadvantages of the hydrogen peroxide method are due to the nature of the chemical. The correct concentration of chemicals must be used because the chemicals are harmful to the spawned gametes. Making the stock chemical solutions is fairly easy but requires a little knowledge of chemistry and careful training of employees. It is very easy to make a mistake while calculating the amounts needed to obtain a 6% hydrogen peroxide solution or a 2 M Tris solution.

Both hydrogen peroxide and Tris are very toxic to the spawned gametes and must be removed before spawning actually begins. This is not difficult but requires that a person be present in case the abalone begins spawning before the water is routinely changed (i.e., 2.5 hr). If the animal spawns early, a portion or all the gametes from that individual will be lost.

In summary, if the abalone culture operation is small, the advantages of hydrogen peroxide (i.e., low initial cost) may exceed the advantages of UV irradiation. However, if the culture operation is medium to large, with spawning occurring on a regular basis, the advantages of UV irradiation far surpass those of hydrogen peroxide.

FERTILIZATION

Fertilization of the gametes is a critical step in the hatchery process. The techniques used should guarantee rapid fertilization of short duration, high fertilization rate ($\sim$ 100%) with excellent survival, and low abnormality or polyspermy. If all the eggs are fertilized during a narrow time period (e.g., 2 min), then larval development will be almost synchronous.[38] It is important the eggs are fertilized within 1 hr after being released.[12,33] This is usually possible since the males tend to spawn first.

Synchronous Fertilization Protocol

A sample of sperm suspension is taken from the top layer of water in the spawning tank of each male, being careful not to disturb the feces and mucus which has settled to the bottom. The samples from each male are mixed to guard against the possibility of using nonviable sperm for fertilization, and the sperm mixture is set aside for use during the fertilization process. The sperm concentration can be determined either with a hemacytometer

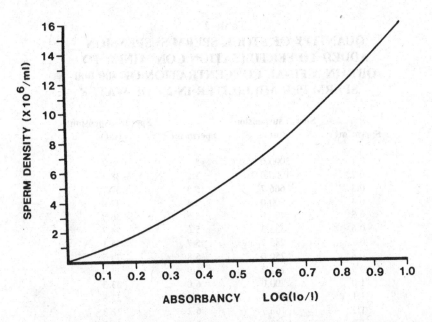

FIGURE 3. Relation between sperm density and the absorbency through 10.5-cm column of sea water. (From Kikuchi, S. and Uki, N., *Bull. Tohoku Reg. Fish. Res. Lab.*, 34, 67, 1974. With permission.)

or spectrophotometer (optical density at 340 nm) (Figure 3).[39] A hemacytometer will give a direct count of the density in the sperm suspension and does not require any large equipment. A 5-mℓ subsample of the suspension is placed in a test tube and 0.5 mℓ of Lugol/eosin solution is added with a graduated Pasteur pipette. [To prepare a stock solution of Lugol/eosin: first, dissolve 1 g of potassium iodide in 100 mℓ of distilled water and then add 1 g of iodine crystals (the chemicals will go into solution easily if added in this order). To this solution, add 0.1 g eosin Y (C.I. 45380).] Lugol/eosin is extremely harmful to gametes and larvae, and should be kept away from the actual hatchery room.[24]

A drop of sperm suspension with Lugol/eosin is allowed to flow by suction under the cover glass of the hemacytometer. Sperm are counted in ten diagonal squares (the smallest squares on the hemacytometer) and the average number of sperm per square is calculated. [Note: all sperm within each square and sperm laying on two sides and one corner (selected before counting) of the square are counted.] Average number of sperm per square is multiplied by 4.4×10^6 to obtain the sperm concentration (sperm per milliliter) in the stock sperm suspension. Sperm concentration is multiplied by 1.25×10^{-9} to calculate the number of milliliters of stock sperm suspension needed for fertilization (final concentration of 400,000 sperm per milliliter in 2 ℓ of water). Table 2 gives the number of milliliters of stock sperm suspension needed for fertilization with average densities of 0.1 to 10 sperm per square on the hemacytometer. This amount of sperm suspension is needed to obtain the correct concentration for complete fertilization. The final sperm concentration in the fertilization container is above the optimum concentration for fertilization with fresh sperm (Figure 4).[40] In practice, the sperm suspension used for fertilization will contain sperm of varying age, so the higher sperm concentration ensures an adequate concentration of viable sperm.[24]

Poor fertilization rates can be a serious problem for the culturist. The most common cause of poor fertilization is improper sperm concentration. The sperm concentration is critical since there must be enough sperm to ensure 100% fertilization but not too many to cause the dissolving of the egg membrane. The suitable range of final sperm concentration is from 100,000 to 1,730,000 sperm per milliliter with 200,000 sperm per milliliter being optimum with freshly spawned sperm.[40]

Table 2
QUANTITY OF STOCK SPERM SUSPENSION ADDED TO FERTILIZATION CONTAINER TO OBTAIN A FINAL CONCENTRATION OF 400,000 SPERM PER MILLILITER IN 2 ℓ OF WATER

Sperm/mℓ	Sperm suspension (mℓ)	Sperm/mℓ	Sperm suspension (mℓ)
0.1	2000.0	5.1	39.2
0.2	1000.0	5.2	38.5
0.3	666.7	5.3	37.7
0.4	500.0	5.4	37.0
0.5	400.0	5.5	36.4
0.6	333.3	5.6	34.7
0.7	285.7	5.7	35.1
0.8	250.0	5.8	34.5
0.9	222.2	5.9	33.9
1.0	200.0	6.0	33.3
1.1	181.8	6.1	32.8
1.2	166.7	6.2	32.3
1.3	153.8	6.3	31.7
1.4	142.9	6.4	31.3
1.5	133.3	6.5	30.8
1.6	125.0	6.6	30.3
1.7	117.6	6.7	29.9
1.8	111.1	6.8	29.4
1.9	105.3	6.9	29.0
2.0	100.0	7.0	28.6
2.1	95.2	7.1	28.2
2.2	90.9	7.2	27.8
2.3	87.0	7.3	27.4
2.4	83.3	7.4	27.0
2.5	80.0	7.5	26.7
2.6	76.9	7.6	26.3
2.7	74.1	7.7	26.0
2.8	71.4	7.8	25.6
2.9	69.0	7.9	25.3
3.0	66.7	8.0	25.0
3.1	64.5	8.1	24.7
3.2	62.5	8.2	24.4
3.3	60.6	8.3	24.1
3.4	58.8	8.4	23.8
3.5	57.1	8.5	23.5
3.6	55.6	8.6	23.3
3.7	54.1	8.7	23.0
3.8	52.6	8.8	22.7
3.9	51.3	8.9	22.5
4.0	50.0	9.0	22.2
4.1	48.8	9.1	22.0
4.2	47.6	9.2	21.7
4.3	46.5	9.3	21.5
4.4	45.5	9.4	21.3
4.5	44.4	9.5	21.1
4.6	43.5	9.6	20.8
4.7	42.6	9.7	20.6
4.8	41.7	9.8	20.4
4.9	40.8	9.9	20.2
5.0	40.0	10.0	20.0

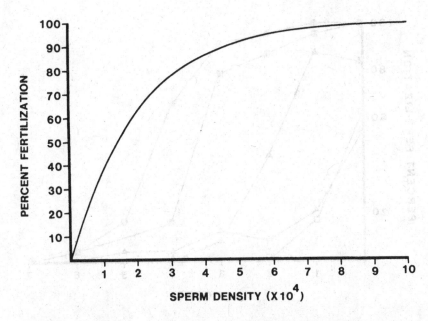

FIGURE 4.   Relation between number of sperm per milliliter and fertilization rate; r = 1 − (1 − 0.00005)^d. d = sperm density (× 10^4/ml). (From Kikuchi, S. and Uki, N., *Bull. Tohoku Reg. Fish. Res. Lab.*, 34, 67, 1974. With permission.)

By the time the sperm concentration in the stock sperm suspension has been calculated, the females should have finished spawning. The eggs are siphoned out of the tank and filtered through a 276-μm nitex screen to collect feces, debris, and mucus. Spawned eggs are filtered into a common container and the total number of eggs is calculated. The water in the container is increased to a convenient level (e.g., 15 ℓ) for counting the eggs (approximately 10 eggs per milliliter). The water is thoroughly mixed with both forward/backward and side to side motions. After the water is mixed well, water is washed in and out of a pipette (10- or 25-mℓ pipette works best), and a 5-mℓ sample is quickly taken. The sample is placed in a watch glass to count the eggs. Lightly tapping the side of the watch glass is helpful before counting the eggs. This procedure will cause the eggs to form rows in the bottom of the watch glass which facilitates counting. The total number of eggs in the container is calculated by multiplying the number of eggs in the watch glass by 3000 (15,000 mℓ/5 mℓ).

A monolayer of eggs (approximately 1400 eggs per square centimeter of bottom surface area) should be put into each fertilization container to ensure complete fertilization. After placing the correct number of eggs into each container, the calculated volume of sperm suspension necessary for final concentration of 400,000 sperm per milliliter (in 2 ℓ) is added to the container and the water volume is *quickly* increased to 2 ℓ. The egg-sperm suspension is allowed to sit undisturbed for 2 min. At 2 min, the water in the container is violently agitated by twisting the container back and forth with the handle, and the water level is quickly raised to the top with clean sea water. This will help stop fertilization. It is important for fertilization to rapidly occur within a narrow time period, which will synchronize the larval development and simplify future procedures.

The eggs are allowed to settle for 15 min, and then the water is decanted off with a smooth continous motion. Shining a light sideways across the spout makes the eggs visible in the milky water. Decanting is stopped when the eggs reach the spout edge, and the eggs are retained in the container. The container is refilled with clean isothermal sea water. A balance must be found between removing as much water as possible at each decant and retaining the fertilized eggs in the container. The container is decanted/refilled every 10

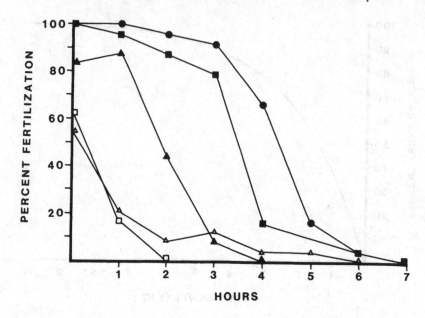

FIGURE 5. Fertilization rate of *H. rufescens* (mean %) related to water temperature and time post-spawning. (△) 9°C, (■) 12°C, (●) 15°C, (▲) 18°C, (□) 21°C. (From Ebert, E. E. and Hamilton, R. M., *Calif. Fish Game*, 69(2), 115, 1983. With permission.)

min. for a total of ten times or until all the excess sperm is removed. After the final decant, the water temperature is measured and the container is placed in a dark temperature-controlled room.

Some laboratories spawn male and female abalone in the same container which allows fertilization to occur at the same time. The disadvantage of this method is a loss of control over the fertilization process because polyspermy and fertilization will occur over a wide time interval. After the adults have finished spawning, they are removed and the sea water is allowed to stand for 1 to 2 hr after the eggs have settled. The sea water is then changed several times with clean isothermal sea water to remove excess sperm, sedimented feces, and mucus excreted by the adults. The developing eggs are kept in clean, stagnant (no water flow) sea water until hatching.[14]

Gamete Viability

As discussed above, either UV irradiation or hydrogen peroxide are reliable methods to induce spawning. The viability of spawned gametes is independent of the length of time between successive spawning inductions and has no effect on the rate of fertilization.[41] However, spawning of the brood stock may not be synchronous, which can delay fertilization and result in a loss in gamete viability.[42]

Duration of gamete viability is a function of the ambient water temperature and can be highly variable at the extreme permissible temperatures.[33,42] The fertilization rate of freshly spawned red abalone *(H. rufescens)* gametes (0 hr old) at 9°C ranges from 13 to 97% (average — 54%) and at 21°C ranges from 20 to 85% (average — 65%). Gamete viability is most affected at high temperatures, with no fertilization possible after 1 hr at 21°C. Cooler water temperatures are less severe, although fertilization rate declines after 1 hr, fertilization is still possible after 5 hr at 9°C (Figure 5). Gamete viability at intermediate temperatures (12 to 18°C) is more uniform and allows higher fertilization rates for longer times, although there is still variability. The fertilization rate ranges from 39 to 92% after 4 hr at 15°C.[42]

The gametes from *H. discus hannai* show the same relationship between water temperature and duration of fertility as *H. rufescens* (Figure 6). The duration of high fertility increases

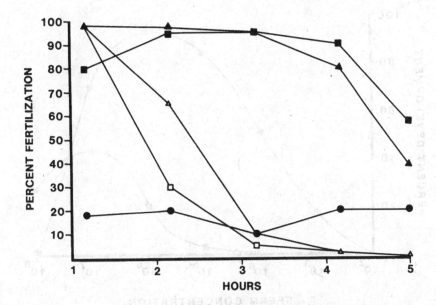

FIGURE 6. Fertilization rate of *H. discus hannai* (mean %) related to water temperature and time post-spawning. Duration of the fertility related to the temperature. (●) 9°C, (▲) 13.6°C, (■) 17.3°C, (△) 22.7°C, (□) 26.2°C. (From Kikuchi, S. and Uki, N., *Bull. Tohoku Reg. Fish. Res. Lab.*, 34, 73, 1974. With permission.)

from 1 hr at 26°C to 4 hr at 14°C. This trend reverses at low temperatures (9°C) which never show high fertility. Under normal conditions, high fertility will last for 1 to 4 hr after spawning.[33]

The viability of sperm decreases more rapidly than ova. When 6-hr old ova are fertilized with 6-hr old sperm, the fertilization is 3%. However, when 6-hr old ova are fertilized with freshly spawned sperm, the fertilization rate is 72%. Sperm viability might be extended by storing them at reduced densities, lower temperatures, with aeration or some combination of methods. The optimal temperature for a high fertilization rate with delayed mixing of the gametes is approximately 15°C for red abalone. This is also the optimal temperature for rearing the larvae and juveniles.[42]

HYBRIDS

Hybrid abalone have potential for stock improvement in aquaculture and fishery enhancement. Morphological characteristics of hybrid abalone are intermediate between the two parent species.[8] The characteristics selected for in hybrids are faster growth, adaptation to environmental conditions, and better quality of meat than found in either parent. These potential benefits have stimulated research on hybrids but the production of first-generation hybrids has been hindered by large variability in fertilization rates, larval development, and juvenile survival.[43]

Rapid fertilization of eggs after spawning is critical to produce hybrids. Immediately adding sperm to freshly spawned eggs is essential for producing the largest quantity of hybrid embryos. The fertilization rate of the eggs rapidly decreases after spawning. If the eggs are fertilized within 5 min after spawning, fertilization rate is 30 to 60%, but decreases to 20% after 60 min. Homospecific (parents same species) fertilization rate is usually greater than 90%, even if fertilization is delayed for 2 hr (Figure 7).[43]

A block to fertilization by foreign sperm quickly develops in the egg, reducing the fertilization rate of heterospecific (parents different species) sperm. The addition of heter-

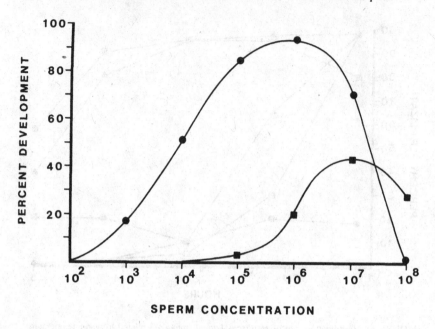

FIGURE 7. Decline in fertilizability of ova after spawning in homospecific combination of gametes. In all obsevations, developing zygotes were held at 17 ± 2°C and sperms for delayed fertilizations were retained at 15 ± 1°C. (▲) = homospecific, *H. rufescens* female × *H. rufescens* male. (■) = heterospecific, *H. sorenseni* female × *H. rufescens* male. (△) = heterospecific, *H. corrugata* female × *H. rufescens* male. (□) = heterospecific, *H. rufescens* female × *H. fulgens* male. (From Leighton, D. L. and Lewis, C. A., *Int. J. Invertebr. Reprod.*, 5, 273, 1982. With permission.)

ospecific sperm after this critical time causes a general block of fertilization that precludes subsequent fertilization by either hetero- or homospecific sperm. The block to fertilization by homospecific sperm only develops if the egg is first exposed to heterospecific sperm (Table 3).[43]

The sperm concentration giving the maximum fertilization rate with the lowest abnormality was about 10^6 sperm per milliliter for homologous fertilizations and 10^7 sperm per milliliter for heterologous fertilizations (Figure 8). Higher sperm concentrations usually caused lysis of the vitelline layer and abnormal development of zygotes.[43]

Normally, an abalone sperm binds at its anterior tip to the egg vitelline layer, the acrosome granule opens (acrosome reaction), soluble contents are released and dissolve a hole in the vitelline layer, and the sperm passes through the hole to fuse with the plasma membrane.[44,45] Heterospecific sperm do not acrosome-react or enter the egg vitelline layer if the egg has aged. This could be due to the rapid development of blocks to penetration of the vitelline layer by the egg to inhibit fertilization by heterospecific sperms. Lysins isolated from the sperm of red *(H. rufescens)* or green *(H. fulgens)* abalone may be partially specific in degrading the egg vitelline layer of the respective species.[43]

The vitelline layer and surrounding jelly coat of the egg appear to be essential for normal fertilization. Removal of these envelopes prior to fertilization render the eggs unfertilizable. Substances present in these external egg layers are required to induce the acrosome reaction that is necessary for the fusion of the sperm to the egg.[43]

Several species of abalone have been experimentally crossed to produce hybrid larvae (Table 4). The most closely related California abalone species (as determined by antibody reaction of the hemocyanin) are the species that form hybrids most successfully.[46] The most successful crosses are between red abalone eggs and pink *(H. corrugata)*, green *(H. fulgens)*, or white *(H. sorenseni)* abalone sperm. There is evidence of heterosis (a greater vigor or

Table 3
TWO-STEP FERTILIZATION OF RED ABALONE EGGS WITH HOMOSPECIFIC AND HETEROSPECIFIC SPERMS

	Female	Male	Fertilization (min after egg release) First	Second	Development (%)
I[a]	Red	Green	8	—	23.1
		Green × red	8	40	26.2
		Red	10	—	85.3
II[b]	Red	Green	45	—	4.0
		Green × red	45	75	7.3
		Red	50	—	64.6

[a] *H. rufescens* (red) eggs were promptly fertilized by *H. fulgens* (green) sperms, or for the homospecific control, by red sperms. After removing excess sperms, one hybrid group (ca. 500 eggs) was combined with red sperms 30 min later.
[b] Eggs were aged for 45 min before their first exposure to sperms.

(From Leighton, D. L. and Lewis, C. A., *Int. J. Invertebr. Reprod.*, 5, 273, 1982. With permission.)

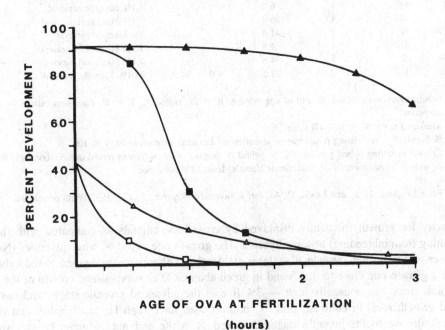

FIGURE 8. The effect of sperm concentration on fertilization and subsequent normal development in homospecific and heterospecific combination of gametes. At sperm concentrations greater than 10^8/mℓ lysis of the vitelline layer occurred. (●) = Homospecific fertilization, *H. rufescens* female × *H. rufescens* male. (■) = heterospecific cross, *H. corrugata* female × *H. rufescens* male. (From Leighton, D. L. and Lewis, C. A., *Int. J. Invertebr. Reprod.*, 5, 273, 1982. With permission.)

Table 4
GAMETE COMBINATIONS AND FERTILIZATION SUCCESS[a]

Female × male		No. of spawnings	Development (%)[b]		Comments
			Mean	Range	
Homospecific					
R	R	17	92.8	82.4 — 100.0	
G	G	2	94.0	93.0 — 95.0	
P	P	3	89.2	86.5 — 94.2	
S	S	1	98.0		
Heterospecific					
R	G	8	32.1	21.7 — 53.3	1% Juvenile survival[c]
R	P	4	16.4	4.3 — 24.0	5% Juvenile survival
R	S	2	29.7	19.4 — 40.0	5% Juvenile survival
G	R	4	29.8	9.5 — 61.1	1% Juvenile survival
G	P	3	36.1	9.0 — 72.3	1% Juvenile survival
G	S	1	19.7		1% Juvenile survival
P	R	10	23.2	7.5 — 45.7	5% Juvenile survival
P	G	7	35.0	10.5 — 84.3	0.1% Juvenile survival
S	R	2	96.3	94.0 — 98.7	Highest hybrid fertilization rate 5% Juvenile survival
S	G	1	10.0		0.1% Postlarval survival
Hybrid crosses[d]					
R	GR	1	6.0		0.1% Juvenile survival
S	GR	1	26.4		0.1% Postlarval survival
RW	R	1	41.4		5% Juvenile survival
RW	GR	1	4.8		0.1% Postlarval survival
GR	R	1	34.9		0.1% Postlarval survival
GR	GR	2	22.5	6.0 — 38.9	0.01% Juvenile survival

[a] Fertilizations were within 30 min of egg release. R = *H. rufescens*, P = *H. corrugata*, and S = *H. sorenseni*.

[b] Measured at the 4– to 16– cell stage.

[c] % Survival = (no. living postlarvae or juveniles/no. larvae at time of settling) × 100.

[d] Crosses involving hybrid parents: GR = hybrid *H. fulgens* × *H. rufescens* reared in the laboratory. RW = hybrid *H. rufescens* × *H. walallensis* (female) from field collection.

From Leighton, D. L. and Lewis, C. A., *Int. J. Invertebr. Reprod.*, 5, 273, 1982. With permission.

capacity for growth frequently displayed by cross-bred animals as compared with those resulting from inbreeding) in some hybrids. The growth rate of red × white juveniles (Note: crosses will be written, female × male) is greater than either parent and green × red hybrids have a growth rate close to that found in green abalone. The survival and growth of the red × pink cross was unusually high — 3% live to the advanced juvenile stages and have a good growth rate. In contrast, pink × pink crosses have high larval mortality and only 0.01% survive to the juvenile stage. Both red × white and red × green crosses show heterosis with the hybrids having a greater growth rate than one or both parent species. Offspring from crossing first generation (red × green) abalone hybrids had a high survival (85 to 95%) but juvenile mortality was very high (over 99%).[43]

The species selected as the source for eggs or sperm is an important consideration when producing hybrids. Red × white crosses have excellent postlarval survival rates as compared to the reciprocal white × red cross postlarvae, despite the fact that the latter cross has the highest hybrid fertilization rate.[43]

REFERENCES

1. **Boolootian, R. A., Farmanfarmaian, A., and Giese. A. C.**, On the reproductive cycle and breeding habits of two western species of *Haliotis*, *Biol. Bull.*, 122, 183, 1962.
2. **Carlisle, J. G.**, The technique of inducing spawning in *Haliotis rufescens* Swainson, *Science*, 102, 566, 1945.
3. **Carlisle, J. G.**, Spawning and early life history of *Haliotis rufescens* Swainson, *Nautilus*, 76, 44, 1962.
4. **Giorgi, A. E. and DeMartini, J. D.**, A study of the reproductive biology of the Red abalone, *Haliotis rufescens* Swainson, near Mendocino, California, *Calif. Fish Game*, 63(2), 80, 1977.
5. **Hahn, K.**, The reproductive cycle and gonadal histology of the Pinto abalone, *Haliotis kamtschatkana* Jonas, and the Flat abalone, *Haliotis walallensis* Stearns, *Adv. Invertebr. Reprod.*, 11, 387, 1981.
6. **Morse, D. E., Duncan, H., Hooker, N., Morse, A.**, Hydrogen peroxide induces spawning in mollusks, with activation of prostaglandin endoperoxide synthetase, *Science*, 196, 298, 1977.
7. **Murayama, S.**, On the development of the Japanese abalone, *Haliotis gigantea*, *J. Coll. Agric. Tokyo Imp. Univ.*, 13, 227, 1935.
8. **Owen, B., McLean, J. H., and Meyer, R. J.**, Hybridization in the eastern Pacific abalones *(Haliotis)*, *Bull. L. A. Co. Mus. Nat. Hist. Sci.*, 9, 1, 1971.
9. **Tutschulte, T.**, The Comparative Ecology of Three Sympatric Abalones, Ph.D. thesis, University of California, San Diego, 1976.
10. **Webber, H. H. and Giese, A. C.**, Reproductive cycle and gametogenesis in the Black abalone *Haliotis cracherodii* (Gastropoda: Prosobranchiata), *Mar. Biol.*, 4, 152, 1969.
11. **Young, J. S. and DeMartini, J. D.**, The reproductive cycle, gonadal histology, and gametogenesis of the Red abalone, *Haliotis rufescens* (Swainson), *Calif. Fish Game*, 56(4), 298, 1970.
12. **Kumabe, L.**, Methods used at the Oyster Research Institute, unpublished report, 1981.
13. **Newman, G. G.**, Growth of the South African abalone *H. midae*, *Div. Sea Fish. Union So. Afr. Invest. Rep.*, 67, 1, 1968.
14. **Hooker, N. and Morse, D. E.**, Abalone: the emerging development of commercial cultivation in the United States, in *Crustacean and Mollusk Aquaculture in the United States*, Huner, J. V. and Brown, E.E., Eds., AVI Publ. Co., 1985, 365.
15. **Takashima, F., Okuno, M., Nishimura, K., and Nomura, M.**, Gametogenesis and reproductive cycle in *Haliotis diversicolor diversicolor* Reeve, *J. Tokyo Univ. Fish.*, 65(1), 1, 1978.
16. **Newman, G.G.**, Reproduction of the South African abalone *Haliotis midae*, *Div. Sea Fish. Invest. Rep.*, 64, 1, 1967.
17. **Crofts, D. R.**, Haliotis, *Liverpool Mar. Biol. Comm. Mem. Typ. Br. Mar. Plants Anim.*, 29, 1, 1929.
18. **Ino, T.**, Biological studies on the propagation of Japanese abalone (genus *Haliotis*), *Bull. Tokai Reg. Fish. Res. Lab.*, 5, 1, 1952.
19. **Cox, K. W.**, Review of the abalone in California, *Calif. Fish Game*, 46(4), 381, 1960.
20. **Pena, J. B.**, Preliminary study on the induction of artificial spawning in *Haliotis coccinea canariensis* Nordsieck (1975), *Aquaculture*, 52, 35, 1986.
21. **Poore, G. C. B.**, Ecology of New Zealand abalones, *Haliotis* species (Mollusca: Gastropoda). IV. Reproduction, *N. Z. J. Mar. Freshwater Res.*, 7(1,2), 67, 1973.
22. **Shepherd, S. A. and Laws, H. M.**, Studies on southern Australian abalone (genus *Haliotis*). II. Reproduction of five species, *Aust. J. Mar. Freshwater Res.*, 25, 49, 1974.
23. **Stephensen, T. A.**, Notes on *Haliotis tuberculata*, *J. Mar. Biol. Assoc. U. K.*, 13(2), 480, 1924.
24. **Hahn, K.**, personal observation, 1986.
25. **Ino, T.**, *Fisheries in Japan, Abalone and Oyster*, Japan Marine Products Photo Materials Assoc., Tokyo, 1980, 165.
26. **Kishinoue, K.**, Study on abalone. II, *Rep. Fish. Invest.*, 4 (2), 1, 1895. (Cited in Ino, T., *Fisheries in Japan, Abalone and Oyster*, Japan Marine Products Photo Materials Assoc., Tokyo, 1980, 165.)
27. **Morse, D. E., Hooker, N., and Morse, A.**, Chemical control of reproduction in bivalve and gastropod molluscs. III. An inexpensive technique for mariculture of many species, *Proc. World Maricult. Soc.*, 9, 543, 1978.
28. **Kurita, M., Sato, R., Ishikawa, Y., Shiihara, H., and Ono, S.**, Artificial seed production of abalone, *Haliotis discus* (Reeve), *Bull. Oita Pref. Fish. Exp. St.*, 10, 78, 1978. (Translated in Mottet, M.G., Summaries of Japanese papers on hatchery technology and intermediate rearing facilities for clams, scallops, and abalones, *Prog. Rep., Dept. Fish. Washington*, 203, 10, 1984.)
29. **Koike, Y.**, Biological and ecological studies on the propagation of the ormer, *Haliotis tuberculata* Linnaeus. I. Larval development and growth of juveniles, *La Mer*, 16(3), 124, 1978.
30. **Kikuchi, S. and Uki, N.**, Technical study on artificial spawning of abalone, genus *Haliotis*. II. Effect of irradiated sea water with ultraviolet rays on inducing to spawn, *Bull. Tohoku Reg. Fish. Res. Lab.*, 33, 79, 1974.

31. **Uki, N. and Kikuchi, S.,** Regulation of maturation and spawning of an abalone, *Haliotis* (Gastropoda) by external environmental factors, *Aquaculture, 39,* 247, 1984.

32. **Uki, N. and Kikuchi, S.,** Technical study on artificial spawning of abalone, genus *Haliotis.* VIII. Characteristics of spawning behavior of *H. discus hannai* induced by ultraviolet irradiation stimulus, *Bull. Tohoku Reg. Fish. Res. Lab., 44,* 83, 1982.

33. **Kikuchi, S. and Uki, N.,** Technical study on artifical spawning of abalone, genus *Haliotis.* IV. Duration of fertility related to temperature, *Bull. Tohoku Reg. Fish. Res. Lab., 34,* 73, 1974.

34. **Kim, Y. and Cho, C.,** Technical study on the artificial precocious breeding of abalone, *Haliotis discus hannai* Ino, *Bull. Korean Fish. Soc.,* 9(1), 61, 1976.

35. **Morse, D. E.,** Biochemical and genetic engineering for improved production of abalones and other valuable molluscs, *Aquaculture, 39,* 263, 1984.

36. **Yahata, T.,** Induced spawning of abalone (*Nordotis discus* Reeve) injected with ganglional suspensions, *Bull. Jpn. Soc. Sci. Fish.,* 39(11), 1117, 1973.

37. **Ebert, E. E. and Houk, J. L.,** Elements and innovations in the cultivation of Red abalone *Haliotis rufescens, Aquaculture, 39,* 375, 1984.

38. **Seki, T.,** personal communication, 1981.

39. **Vacquier, V. D. and Payne, J.E.,** Methods for quantitating sea urchin sperm-binding, *Exp. Cell Res., 82,* 227, 1973.

40. **Kikuchi, S. and Uki, N.,** Technical study on artificial spawning of abalone, genus *Haliotis.* III. Reasonable sperm density for fertilization, *Bull. Tohoku Reg. Fish. Res. Lab., 34,* 67, 1974.

41. **Nishikawa, N., Obara, A., and Ito, Y.,** A method to induce artificial spawning of the abalone, *Haliotis discus hannai* Ino, throughout the year, *J. Hokkaido Fish. Sci. Instit.,* 31(5), 21, 1974.

42. **Ebert, E. E. and Hamilton, R. M.,** Ova fertility relative to temperature and to the time of gamete mixing in the red abalone, *Haliotis rufescens, Calif. Fish Game,* 69(2), 115, 1983.

43. **Leighton, D. L., and Lewis, C. A.,** Experimental hybridization in abalones, *Int. J. Invertebr. Reprod., 5,* 273, 1982.

44. **Lewis, C. A., Leighton, D. L., and Vacquier, V. D.,** Morphology of abalone spermatozoa before and after the acrosome reaction, *J. Ultrastruct. Res., 72,* 39, 1980.

45. **Lewis, C. A., Talbot, C. F., and Vacquier, V. D.,** A protein from abalone sperm dissolves the egg vitelline layer by a nonenzymatic mechanism, *Dev. Biol., 92,* 227, 1982.

46. **Meyer, R. J.,** Hemocynanins and the systematics of California *Haliotis,* Ph.D. thesis, Stanford University, California, 1967.

47. **Gove, P. B.,** *Webster's Third New International Dictionary of the English Language Unabridged,* G. & C. Merriam Co., Springfield, Mass., 1968, 1063.

LARVAL DEVELOPMENT OF ABALONE

Kirk O. Hahn

INTRODUCTION

Larval development is a gradual process that does not occur in discrete stepwise stages. However, various stages can be recognized during larval and post-larval development (Table 1). Larval development rate is measured by the time required for larvae to exhibit features distinctive to each stage. Settlement, metamorphosis, and deposition of peristomal shell marks the transition from larval to post-larval development. The post-larval period continues until formation of the first respiratory pore (notch stage) at about 1 to 3 months age. From notch stage until sexual maturity, the abalone is called a juvenile.[1]

BIOLOGICAL ZERO POINT

As expected for a poikilotherm, larval development in *Haliotis* spp. occurs at a faster rate with higher water temperatures. The development rate is simply a function of water temperature and time, since abalone larvae do not feed before settlement. Larval development, however, is not twice as fast at double the water temperature. Instead, there appears to be a critical water temperature necessary before initiation of larval development. This critical minimum water temperature (biological zero point) is defined as the water temperature at which larval development is possible but would require infinity to complete development. The biological zero-point temperature is unique to each species and may be an important factor in the natural geographical distribution of abalone.[2] Knowledge of the biological zero point is important to the culturist because it allows him to determine when each larval stage will occur during development.

Seki and Kan-no[2] were the first researchers to calculate the biological zero point of an abalone species. They calculated this value empirically from the larval development rate. Immediately after fertilization, eggs were placed in containers with thermostatically controlled water temperatures. Seki and Kan-no reared *H. discus hannai* larvae at 13, 16, 20, and 22°C, and noted the time of occurrence of hatch out, larval retractor muscle, torsion, and formation of epipodal tentacle (Figure 1). [Usually four or five water temperatures are selected over the normal range of temperatures experienced by the animal (e.g., 7 to 22°C) and the time of occurrence of three or four major larval stages (e.g., hatch out, torsion, settlement) is noted at each water temperature.] The *H. discus hannai* larvae were observed every 3 hr until they began crawling behavior, which precedes settlement and metamorphosis. At 20°C, the water temperature when natural spawning occurs, hatch out was at 12.6 hr, 90° twisting of the cephalopedal mass (torsion) was at 32.1 hr, formation of first epipodal tentacle was at 64.9 hr, and completion of larval development was at 99 hr.[2] Seki and Kan-no[2] plotted the water temperature vs. the time (1/ time) for each stage. Theoretically, there should be a linear relationship between water temperature and 1/time for each stage, and importantly, all the lines should cross the X–axis at the same point, the biological zero point.

The equations of the relationships between appearance of the four major larval stages in *H. discus hannai* and water temperature are shown in Table 2. The average biological zero point (1/t = 0) calculated from these formulas was 7.6°C.[2] It is interesting to note that the biological zero point for larval development is the same value as the biological zero point for gonad maturation, although each value is calculated independently.[3]

Once the biological zero point is known, larval development can be quantified by cal-

Table 1
TIMING OF LARVAL DEVELOPMENT STAGES

Species	H. discus hannai[2]	H. discus[6]	H. sieboldii[5]	H. gigantea[6]	H. diversicolor supertexta[16]	H. corrugata[1]
Water temperature (°C)	20	16—17	16—17	16—18	26.2	16
Larval stages						
First polar body (min)	—a	7—8	—	—	—	—
Second polar body (min)	—	7—8	—	—	—	—
First cleavage (min)	—	100	40	—	—	—
Second cleavage (hr)	3.2	2	2	—	—	—

Species	H. rufescens[1]	H. rufescens[27]	H. rufescens[9]	H. rufescens[4]	H. rufescens[8]	H. fulgens[1]
Water temperature (°C)	17	18	15	17	15	17
Larval stages						
First polar body	—	—	—	—	—	—
Second polar body	—	—	—	—	1	—
First cleavage (hr)	—	—	2	—	2	—
Second cleavage (hr)	—	—	—	—	4	—

Species	H. sorensensi[11]	H. kamtschatkana[28]	H. iris[29]	H. ruber[14]	H. tuberculata[13]	H. coccinea canariensis[15]
Water temperature (°C)	15	10—13	18	15.5	20	15
Larval stages						
First polar body	—	—	—	—	Few minutes	—
Second polar body	—	—	—	—	Few minutes	—
First cleavage (hr)	—	2	1	3.2	1	1.5
Second cleavage (hr)	—	2.5—3.0	—	4.3	2	2.5

Species	H. discus hannai[2]	H. discus[6]	H. sieboldi[5]	H. gigantea[6]	H. diversicolor supertexta[16]	H. fulgens[1]	H. corrugata[1]
Third cleavage (hr)	—	—	3	—	—	—	—
Fourth cleavage (hr)	—	5	5	—	—	—	—
Morula (hr)	—	6-7	8	—	—	—	—
Blastula (hr)	—	12	10	—	—	—	—
Gastrula (hr)	8.1	13	15	14	4.7	—	—
Prototrochal cilia (hr)	11.7	—	18	—	—	—	—
Stomodeum (hr)	—	17	—	—	—	—	—
Prototrochal girdle (hr)	—	—	—	—	—	14	—
Hatch out (hr)	13.3	20	18	21-22	6	—	17

Species	H. rufescens[1]	H. rufescens[27]	H. rufescens[9]	H. rufescens[4]	H. rusfescens[8]
Third cleavage (hr)	—	—	—	—	—
Fourth cleavage (hr)	—	—	—	—	—
Morula (hr)	6	5	—	—	8
Blastula (hr)	14	10-16	—	15	18
Gastrula (hr)	—	—	—	18	—
Prototrochal cilia (hr)	—	—	—	—	—
Stomodeum (hr)	—	—	15	21	—
Prototrochal girdle (hr)	—	—	—	—	—
Hatch out (hr)	24	16-24	20	25	20

Species	H. sorenseni[11]	H. kamtschatkana[28]	H. iris[29]	H. ruber[14]	H. tuberculata[13]	H. coccinea canariensis[15]
Third cleavage (hr)	—	—	—	—	2.3	4
Fourth cleavage (hr)	—	—	—	—	4	—
Morula (hr)	—	—	—	21.3	5	11
Blastula (hr)	—	—	—	—	—	—
Gastrula (hr)	—	—	—	—	—	—
Prototrochal cilia (hr)	—	19	—	—	10	—

Table 1 (continued)
TIMING OF LARVAL DEVELOPMENT STAGES

	H. sorenseni[11]	H. kamtschatkana[28]	H. iris[29]	H. ruber[14]	H. tuberculata[13]	H. coccinea canariensis[15]	H. corrugata[1]
Stomodeum (hr)	—	—	—	—	—	—	—
Prototrochal girdle (hr)	—	—	—	28.3	—	15	—
Hatch out (hr)	24	24	13	—	13	21	—

Species	H. discus hannai[2]	H. discus[5]	H. sieboldii[5]	H. gigantea[6]	H. diversicolor supertexta[16]		
Beginning of larval shell (hr)	16.1	27	24	—	—		
Completion of velum (hr)	18.5	—	29	—	—		
Larval retractor muscle (hr)	23.4	—	29	—	—		
Integumental attachment (hr)	—	—	—	—	—		
Protrusion of foot mass (hr)	26.2	—	—	—	—		
Completion of larval shell (hr)	31.5	—	48	—	—		
Torsion — 90° twisting (hr)	—	45-46	35	40-43	13		

	H. rufescens[1]	H. rufescens[27]	H. rufescens[9]	H. rufescens[4]	H. rusfescens[8]	H. fulgens[1]	H. corrugata[1]
Beginning of larval shell (hr)	—	—	—	29	—	—	—
Completion of velum (hr)	—	—	—	31	24	—	—
Larval retractor muscle (hr)	—	—	—	36	—	—	—

	H. sorenseni[11]	H. kamtschatkana[28]	H. iris[29]	H. ruber[14]	H. tuberculata[13]	H. coccinea canariensis[15]
Integumental attachment (hr)	—	—	—	—	48	—
Protrusion of foot mass (hr)	—	—	—	—	—	—
Completion of larval shell (hr)	—	—	30	44	—	48
Torsion — 90° twisting (hr)	44	—	—	48	—	—
Beginning of larval shell (hr)	—	—	18	—	18	—
Completion of velum (hr)	48	—	—	—	—	25
Larval retractor muscle (hr)	—	—	—	—	20.5	—
Integumental attachment (hr)	—	—	—	—	—	—
Protrusion of foot mass (hr)	—	—	—	—	—	—
Completion of larval shell (hr)	—	30	36	—	—	—
Torsion — 90° twisting (hr)	—	72	—	—	—	52

Species	H. discus hannai[2]	H. discus[6]	H. sieboldii[5]	H. gigantea[6]	H. diversicolor supertexta[16]
180° Rotation of foot mass (hr)	—	—	—	—	—
Long spines on end of metapodium (hr)	—	—	48	—	—
Operculum (hr)	33.1	—	42	—	—
Fine cilia on foot (hr)	37.9	—	48	—	—
Vertical groove in velum (hr)	42.7	60	48	—	—

Table 1 (continued)
TIMING OF LARVAL DEVELOPMENT STAGES

Species	H. discus hannai[2]	H. discus[6]	H. sieboldii[5]	H. gigantea[6]	H. diversicolor supertexta[16]	H. corrugata[1]
Eye spot (hr)	45.2	60	60	240	17-19	—
Propodium (hr)	49.2	—	—	—	—	—
Cephalic tentacle (hr)	51.6	84	60	240	17-19	—

	H. rufescens[1]	H. rufescens[27]	H. rufescens[9]	H. rufescens[4]	H. rusfescens[8]	H. fulgens[1]
180° Rotation of foot mass (hr)	—	—	—	—	65	—
Long spines on end of metapodium (hr)	—	—	—	—	—	—
Operculum (hr)	—	—	—	—	—	—
Fine cilia on foot (hr)	—	—	—	—	70	—
Vertical groove in velum (hr)	—	—	—	—	—	—
Eye spot (hr)	72	—	—	—	—	96
Propodium (hr)	—	—	—	—	—	—
Cephalic tentacle (hr)	—	—	—	—	90	—

	H. sorenseni[11]	H. kamtschatkana[28]	H. iris[29]	H. ruber[14]	H. tuberculata[13]	H. coccinea canariensis[15]
180° Rotation of foot mass (hr)	—	—	—	—	—	—
Long spines on end of metapodium (hr)	—	—	—	—	—	—
Operculum (hr)	—	—	—	—	35-38	—
Fine cilia on foot (hr)	—	—	—	—	72	—
Vertical groove in velum (hr)	—	—	—	—	—	—
Eye spot (hr)	—	—	—	—	24-36	—
Propodium (hr)	—	—	—	—	—	—
Cephalic tentacle (hr)	72	—	—	—	24-36	—

Species	H. discus hannai[2]	H. discus[6]	H. sieboldii[5]	H. gigantea[6]	H. diversicolor supertexta[16]	H. rusfescens[8]	H. fulgens[1]	H. corrugata[1]
Cilia on propodium (hr)	54.4	—	—	—	—	—	—	—
Cilia in mantle cavity (hr)	60.1	—	96	—	—	—	—	—
Apophysis on propodium (hr)	62.5	—	—	—	—	—	—	—
First epipodial tentacle (hr)	64.1	120	96	—	38	—	—	—
Otolith (hr)	66.1	154	96	—	—	—	—	—
Short spine on cephalic tentacle (hr)	70.2	154	120	—	—	—	—	—
Snout protrusion (hr)	74.2	216	120	—	—	—	—	—

Species	H. rufescens[1]	H. rufescens[27]	H. rufescens[9]	H. rufescens[4]	H. rusfescens[8]	H. tuberculata[13]	H. ruber[14]	H. iris[29]
Cilia on propodium (hr)	—	—	—	82	—	—	—	—
Cilia in mantle cavity (hr)	—	—	—	—	—	—	—	—
Apophysis on propodium (hr)	—	—	—	—	—	—	—	—
First epipodial tentacle (hr)	—	—	—	—	—	—	—	—
Otolith (hr)	120	—	—	—	—	—	—	—
Short spine on cephalic tentacle (hr)	—	—	—	—	—	—	—	—
Snout protrusion (hr)	—	—	—	—	—	—	—	—

Species	H. sorensoni[11]	H. kamtschatkana[28]	H. coccinea canariensis[15]
Cilia on propodium (hr)	—	—	—
Cilia in mantle cavity (hr)	—	—	—

Table 1 (continued)
TIMING OF LARVAL DEVELOPMENT STAGES

Species	H. sorenseni[11]	H. kamtschatkana[28]	H. iris[29]	H. ruber[14]	H. tuberculata[13]	H. coccinea canariensis[15]
Apophysis on propodium (hr)	—	—	—	—	—	—
First epipodial tentacle (hr)	—	—	—	—	84	—
Otolith (hr)	—	—	—	—	—	—
Short spine on cephalic tentacle (hr)	—	—	—	—	—	—
Snout protrusion (hr)	—	—	—	—	—	—

Species	H. discus hannai[2]	H. discus[6]	H. sieboldii[8]	H. gigantea[6]	H. diversicolor supertexta[16]	H. rufescens[4]	H. rufsescens[8]
Two tubules on cephalic tentacle (hr)	—	—	120	—	—	—	—
Ciliary process in mantle cavity (hr)	76.2	—	—	—	—	—	—
Third tubule on cephalic tentacle (hr)	82.3	—	—	—	—	—	—
Retractor muscle drawn in mantle cavity (hr)	92.7	—	—	—	—	—	—
Fourth tubule on cephalic tentacle (hr)	97.6	—	144-168	—	—	—	—

Species	H. rufescens[1]	H. rufescens[27]	H. rufescens[9]	H. rufescens[4]	H. rufsescens[8]	H. fulgens[1]	H. corrugata[1]
Two tubules on cephalic tentacle (hr)	—	—	—	90	142	—	144
Ciliary process in mantle cavity (hr)	—	—	—	—	—	—	—

Third tubule on cephalic tentacle (hr)
Retractor muscle drawn in mantle cavity (hr)
Fourth tubule on cephalic tentacle (hr)

192

360

175

144

168-192

H. coccinea canariensis[15]

H. tuberculata[13]

108

120

H. ruber[14]

H. iris[29]

108

H. kamtschatkana[28]

H. sorenseni[11]

Two tubules on cephalic tentacle (hr)
Ciliary process in mantle cavity (hr)
Third tubule on cephalic tentacle (hr)
Retractor muscle drawn in mantle cavity (hr)
Fourth tubule on cephalic tentacle (hr)

240

— = Time of larval stage not measured.

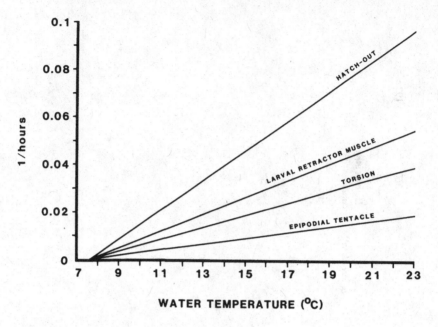

FIGURE 1.   Relationship between water temperature and beginning (1/time) of the four important larval stages in *H. discus hannai*. The biological zero point is the x-intercept of the four lines. (From Seki, T. and Kan-no, H., *Bull. Tohoku Reg. Fish. Res. Lab.*, 38, 143, 1977. With permission.)

culating the hourly exposure to water temperatures above that point. However, it is not the absolute water temperature but only the difference between the water temperature and biological zero point. For example in *H. discus hannai*, if the water temperature was 20°C, only 12.4°C (20°C − 7.6°C) actually contributes to larval development, and the quantity above the biological zero point has an additive effect toward larval development. The difference between the water temperature and biological zero point is called the effective temperature and summation of this value during larval development is called the effective accumulative temperature (EAT).

$$Y_t = \sum_{i=1}^{n} T_i - \Theta$$

where Y_t (°C · hr) = EAT, t (hours) = number of hours larvae are exposed to temperature T_i, T_i (°C) = water temperature in larval-rearing container (values of $T_i < \Theta$ are not added) and Θ (°C) = biological zero point for larval development.

In simple terms, if a constant water temperature is used during larval development, the EAT is the effective temperature multiplied by the number of hours of larval development. A larva raised in 20°C water for 24 hr will have an EAT of (24 − 7.6) × 24 hr = 297.6°C-hr.[2]

The biological zero point of *H. discus* is 8.5°C and *H. gigantea* is 9.0°C. The equations of the relationships between water temperature and timing of the major larval stages are shown in Table 2.[2]

Seki and Kan-no[2] calculated the biological zero point for *H. rufescens* (8.5°C), *H. corrugata* (5.7°C), and *H. fulgens* (9.9°C) from data published by Leighton.[1] Holsinger[4] conducted experiments to specifically calculate the biological zero point for *H. rufescens* larvae from northern California, and found the biological zero point was 0.9°C (Figure 2). The different biological zero points of *H. rufescens* calculated by Seki and Kan-no[2] and Holsinger[4]

Table 2
EQUATIONS OF RELATIONSHIP BETWEEN WATER TEMPERATURE AND BEGINNING (1/TIME) OF THE IMPORTANT LARVAL STAGES

H. discus hannai

Hatch out	$1/t^a = 0.00640\ T^b - 0.0502$
Larval retractor muscle	$1/t = 0.00361\ T - 0.0278$
Torsion	$1/t = 0.00252\ T - 0.0187$
First epipodial tentacle	$1/t = 0.00124\ T - 0.0093$

H. discus

Hatch out	$1/t = 0.006440\ T - 0.0555$
Torsion	$1/t = 0.003083\ T - 0.0268$
Eye spot	$1/t = 0.001826\ T - 0.0152$

H. gigantea

Hatch out	$1/t = 0.007147\ T - 0.0648$
Torsion	$1/t = 0.002985\ T - 0.0261$
Eye spot	$1/t = 0.001942\ T - 0.0178$

[a] t = hours.
[b] T = °C.

(From Seki, T. and Kan-no, H., *Bull. Tohoku Reg. Fish. Res. Lab.*, 38, 143, 1977. With permission).

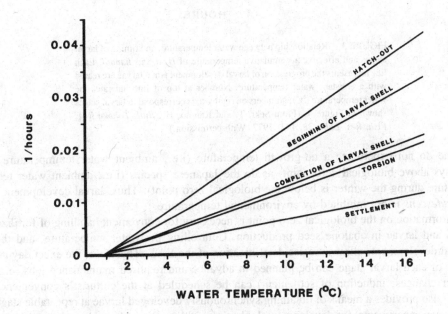

FIGURE 2. Relationship between water temperature and beginning (1/time) of important larval stages in *H. rufescens*. The biological zero point is the x-intercept of the lines. (From Holsinger, L. M., unpublished data, 1984.)

could be due to the fact that the abalone were collected from different regions of California. The abalone in Leighton's[1] study were collected from southern California (water temperature ranges from 14 to 16°C) and in Holsinger's study were collected from northern California (water temperature ranges from 9 to 14°C). These biological zero points indicate *H. rufescens*

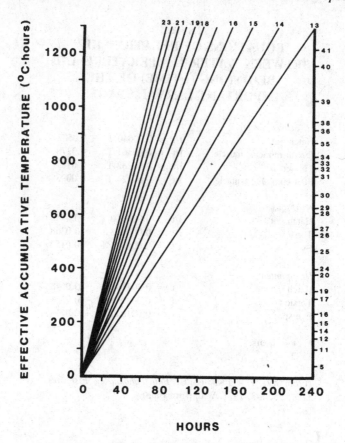

FIGURE 3.   Relationship between water temperature, and timing of larval stages and effective accumulative temperature of *H. discus hannai*. Each line represents the progression of larval development when larvae are reared with a constant water temperature. Number at top of line indicates the water temperature (°C). Numbers on right y-axis correspond to larval stages shown in Figure 4. (From Seki, T. and Kan-no, H., *Bull. Tohoku Reg. Fish. Res. Lab.*, 38, 143, 1977. With permission.)

larvae do not experience a no-growth temperature (i.e., ambient water temmperature is always above biological zero point) as do the Japanese species (i.e., ambient water temperature during the winter is below the biological zero point). Thus, larval development of *H. rufescens* is not inhibited by environmental temperature.[4]

Information on the biological zero point is necessary for convenient handling of fertilized eggs and larvae in abalone seed production. Controlling the water temperature, and thus larval development, allows precise determination of the EAT. Therefore, the exact day and hour of each larval stage can be planned in advance and required maintenance jobs (e.g., water changes, induction of settlement) can be scheduled at the culturist's convenience. EAT also provides a means of obtaining synchronously developed larvae at repeatable stages (between experiments) for developmental, physiological, ecological, and biochemical studies on larval development in Haliotidae.[2]

LARVAL DEVELOPMENT

Haliotis discus hannai

There are 41 distinct larval stages, with recognizable external features, from fertilization until initiation of metamorphosis in *H. discus hannai*.[2] Figure 3 shows the time required for each larval stage at water temperatures from 13 to 23°C. Shortly after fertilization (stage 1), the first polar body is discharged (stage 2) followed quickly by discharge of the second polar body (stage 3) (Figure 4A to C). Cleavage begins after discharge of the polar bodies and development progresses to the gastrula (stages 4 to 10) (Figure 4C to E). Cilia grow along the top of the embryo forming the prototrochal girdle and apical tuft, and begin beating (stage 11) (Figure 4F). The cilia cause the embryo to rotate intermittently inside the egg membrane. The stomodeum forms (stage 12) and cilia (24 cilia cells) along the prototrochal girdle are completely formed (stage 13) (Figure 4G). At this stage the embryo is classified as a trochophore larva. The larva begins moving more frequently inside the egg membrane, the egg membrane becomes thinner, and finally bursts (stage 14). The apical cilia aid the larva in bursting the egg membrane during hatch out. The hatched larva immediately swims to the water surface.

Soon after hatch out, the larval shell begins to be secreted at the back of the larva (stage 15). The trochophore larva continues to develop until it becomes a veliger larva. The larva is classified as veliger when the apical region becomes flat and the velum is completely developed with long cilia present (stage 16) (Figure 4H). The larval shell grows from dorsal to ventral, until it covers the body to just below the velum.[2]

The larval retractor muscle forms (stage 17), followed by formation of an integumental attachment to the larval shell (stage 18) (Figure 4I to J). The foot mass protrudes to the top of the shell (stage 19) at completion of the larval shell (stage 20) (Figure 4K). During torsion, the cephalo-pedal mass rotates 90°, followed by the top of the mantle membrane tearing off from the top of the larval shell (stage 21). The velum and cephalo-pedal mass rotate between the region of the body covered by the larval shell and the "waist". The region destined to become the mouth and the foot continues to rotate, until the cephalo-pedal mass is rotated 180° from its original position (stage 22) (Figure 4L).[2]

There are three pairs of long spines at the posterior end of the metapodium after torsion (stage 23) (Figure 4M). These long spines were first described in abalone larvae by Seki and Kan-no.[2] The operculum forms (stage 24) and at this time, the cephalo-pedal mass can be retracted into the shell. In succession, fine cilia develop on the foot sole and begin beating (stage 25), a vertical groove forms in the velum (stage 26), the eye spot appears (stage 27), the propodium forms (stage 28), a cephalic tentacle forms on the velum (stage 29), and cilia begin growing on the propodium (stage 30) (Figure 4N to P). Cilia form in the mantle cavity up to the anterior edge of the velum and the cilia begin beating (stage 31). The propodium twists to the side and an apophysis appears on the propodium (stage 32). A pair of epipodial tentacles forms on both sides of the foot under the operculum (stage 33) and the larva can crawl on a surface with its foot at this stage. From this point on, the larva actively crawls with its foot, but does not stop swimming behavior unless suitable settlement substratum is present.[2]

The otolith forms and is clearly visible (stage 34), short spines appear on the cephalic tentacles (stage 35), the snout begins to protrude from underneath the velum (stage 36), two tubules appear on the cephalic tentacle (stage 37), a ciliary process forms on the roof of the mantle cavity (stage 38), and a third tubule forms on the cephalic tentacle (stage 39) (Figure 4Q to R). The larval retractor muscle attached to the larval shell draws the enlarged mantle cavity toward the back of the shell (stage 40). Larval development is completed with the formation of the fourth tubule on the cephalic tentacle (stage 41). Although the veliger shows crawling behavior after the formation of the first epipodal tentacle, it is not until the fourth tubule is formed on each of the cephalic tentacles that the veliger shows the crawling, exploratory movements characteristic of settling larvae.[2] The length of the *H. discus hannai* larval period is approximately 4 to 5 days at 20°C and 3 days at 23 to 24°C.[5] Once the larva has reached an EAT of 1030°C-hr it can settle and undergo metamorphosis. A larva can remain arrested at this stage for up to 14 days if no suitable settlement substratum is present.[2]

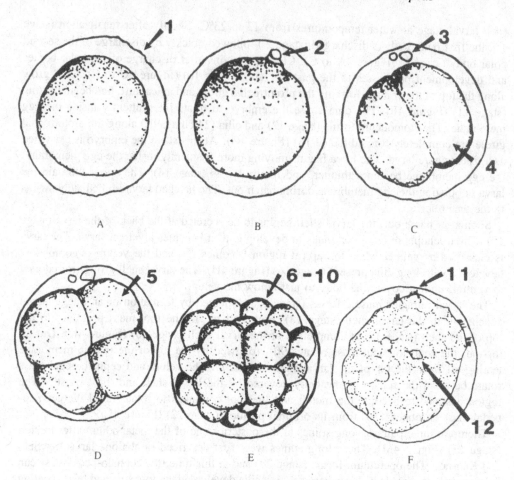

FIGURE 4.   Drawing of the larval stages in *H. discus hannai*. (From Seki, T. and Kan-no, H., *Bull. Tohoku Reg. Fish. Res. Lab.*, 38, 143, 1977. With permission.) (Drawings by M. Brody.) (A) Stage 1 — fertilization (0°C-hr). (B) Stage 2 — discharge of first polar body. (C) Stage 3 — discharge of second polar body. Stage 4 — first cleavage. (D) Stage 5 — second cleavage (40°C-hr). (E) Stages 6 to 10 — third cleavage to gastrula. (F) Stage 11 — appearance of prototrochal cilia (100°C-hr). Stage 12 — formation of stomodeum (140°C-hr). (G) Stage 13 — formation of prototrochal girdle. Stage 14 — hatch out (170°C-hr). Stage 15 — beginning of larval shell secretion (200°C-hr). (H) Stage 16 — completion of velum (230°C-hr). (I) Stage 17 — appearance of larval retractor muscle (290°C-hr). (J) Stage 18 — formation of integumental attachment with the larval shell. (K) Stage 19 — protrusion of foot mass (320°C-hr). Stage 20 — completion of larval shell (385°C-hr). Stage 21 — torsion, 90° twisting of cephalo-pedal mass. (L) Stage 22 — 180° rotation of foot mass. (M) Stage 23 — appearance of long spines on the end of metapodium. Stage 24 — formation of operculum (405°C-hr). (N) Stage 25 — appearance of fine cilia on the foot sole (475°C-hr). Stage 26 — vertical groove formation in the velum (535°C-hr). Stage 27 — appearance of eye spot (555°C-hr). (O) Stage 28 — formation of propodium (610°C-hr). Stage 29 — appearance of cephalic tentacle (635°C-hr). (P) Stage 30 — growing of cilia on the propodium (675°C-hr). Stage 31 — appearance of cilia in the mantle cavity (745°C-hr). Stage 32 — formation of apophysis on the propodium (775°C-hr). Stage 33 — formation of first epipodal tentacle (795°C-hr). (Q) Stage 34 — appearance of otolith (815°C-hr). Stage 35 — appearance of short spine on the cephalic tentacle (865°C-hr). Stage 36 — snout protrusion (915°C-hr). Stage 37 — appearance of two tubules on the cephalic tentacle. (R) Stage 38 — formation of ciliary process in the roof of mantle cavity (945°C-hr). Stage 39 — appearance of third tubule on the cephalic tentacle (1030°C-hr). Stage 40 — larval retractor muscle drawn into enlarged mantle cavity (1160°C-hr). Stage 41 — appearance of fourth tubule on the cephalic tentacle (1220°C-hr).

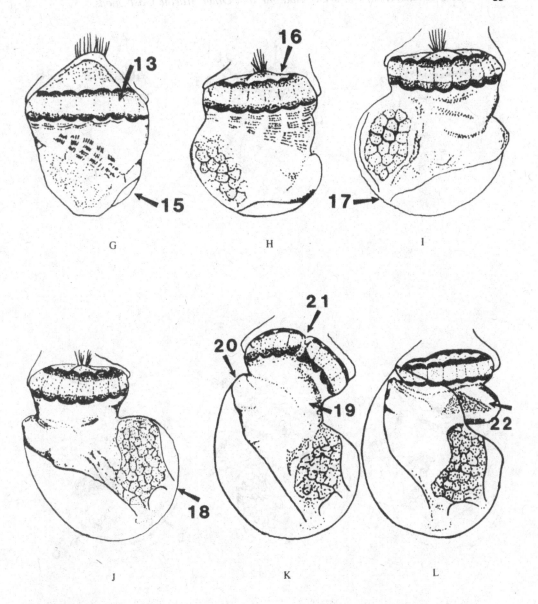

Haliotis discus

Larval development of *H. discus* is very similar to *H. discus hannai*.[6] The prototrochal girdle consists of 24 cells, which gives the characteristic shape to the trochophore larva. Forty to fifty cilia grow in clusters at the base of each cell along the prototrochal girdle. Cilia are also found on the top of the trochophore larva. The trochophore larva measures 200 μm (length) by 178 μm (breadth). After the larval shell is completely formed, the length of the veliger is 290 μm. The shell surface is densely sculptured with ''x'' and ''y'' markings and dots.[6]

The velum begins to regress after day 9 of development at 16 to 17°C, and the snout forms near the mouth from tentacular projections on the velum. By day 10, the foot has developed and some larvae are able to crawl up the vertical wall of the culture tank.[6] The length of the *H. discus* larval period is 5 to 10 days at 16 to 20°C.[5]

The peristomal shell begins to form very soon after metamorphosis (approximately day 11) and grows at the right side of the larval shell aperture. The cilia are lost from the velum, and the eye stalks and eyes develop. The green pigmentation in the velum begins to fade and the extreme edges of the velum join to form the snout, with the green pigmentation

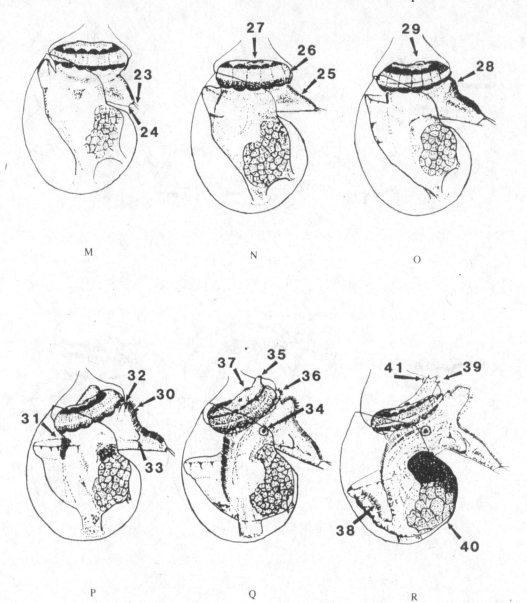

concentrating in the surrounding area. The radula begins rasping and the heart begins beating at this stage.[6]

Peristomal shell covers the entire right margin of the larval shell by day 13. The next day, the cilia on the pedal sole disappear, except in the anterior region. Ino[6] states the operculum is still present at this stage but Murayama[7] reports it is lost after metamorphosis. The peristomal shell has grown 170 μm by day 16, giving the larva a total shell length of 310 μm. Shell length is 420 μm by day 21 and 540 μm after day 35. The shell becomes flatter as it grows since the peristomal shell is formed only on the right side of the median line of the shell. Green pigmentation remains only in the stomach, snout, mantle margin and foot margin by day 21. The body mass appears white, except for the yellow digestive gland, to the naked eye.[6]

The left ctenidium appears in the gill cavity by day 32. Also a disk-shaped projection with cilia appears on the right side, behind the base of the right eye stalk. Approximately eight days later, a second projection, similar to the first projection, forms behind the original projection (shell length is approximately 720 μm). Ino[6] called these projections the "ciliary lobes". Both the first and second ciliary lobes begin to disappear after the initial formation of the first respiratory pore and are absent by the time the second respiratory pore is formed. The ciliary lobes appear to assist in respiration. They are located on the body at the points where the future respiratory pores will be formed and once the pores are formed, they are absorbed and disappear. Additionally, the ciliary lobes circulate water by ciliary action from left to right, and help expel water before the respiratory pores form.[6]

Larval development of *H. sieboldii* is almost the same as *H. discus* during the earlier stages. The phototaxis of the larvae is weaker than *H. discus* and the larvae swim in the middle layer of water. The body is green except for a light brown color within the shell. The green pigmentation is darker on the upper third and lower margin of each prototrochal girdle cell. The larval shell has a diameter of 290 μm, and the shape and size of the completed shell is similar to *H. discus* but the shell is not densely sculptured with markings. After completion of the larval shell, small cilia appear at the base of the long cilia on the velum girdle.[6]

The anterior cells of the velum are reduced in size by day 5 and become transparent projections. Sword-shaped projections also develop at the anterior edge of the foot, and cephalic tentacular processes appear. The larvae begin testing the substratum and crawling after day 6. The next day, the larvae have all settled and the cephalic projections have small setae. The velum forms the snout and pigmentation fades in the anterior region of the velum after day 10. The peristomal shell begins growing and the operculum has already disappeared in some individuals. Ciliary lobes form behind the right eye stalk and the cilia disappears on the pedal sole, except at the anterior edge. The radula is functional after day 16. The post-larva is approximately 420 μm in shell length after day 20.[6]

Holsinger[4] conducted a careful study of *H. rufescens* larval development and the effect of water temperature on the timing of the larval stages. She used a synchronized fertilization method to ensure uniformly developing larvae and raised the larvae in the dark. Batches of larvae were raised at seven different water temperatures (10, 12, 13, 15, 16, 17, 20, and 23°C). The time from fertilization to settlement ranged from 4 to 7 days, depending on the water temperature (Figure 5).[4]

The larvae were examined every 3 hr to determine the larval stages. Twenty-one stages were observed from gastrula to settlement (Table 3). Five of the stages previously had not been described in *H. rufescens;* (1) long spines on end of metapodium, (2) cilia on foot sole, (3) vertical groove in velum overlaying the foot, (4) cilia on propodium, and (5) cilia in mantle cavity.[4]

Normal development to hatch out occurred from 10 to 20°C, but survival to settlement was only possible from 10 to 17°C. Temperatures below 10°C were not tested and normal larval development is probably possible at some temperatures below 10°C. At 20°C, larval growth was normal only for the first 15 hr. At this point, some embryos began irregular growth which formed globular larvae and tiny ciliated bodies. Most larvae at 20°C continued normal development until 30.5 hr (beginning of shell formation), but died immediately after reaching this stage. At 23°C, normal development occurred for the first 12 hr, but all larvae died by 15 hr.[4]

These results were in contrast to findings of Leighton[1] where normal development to hatch out occurred between 10 to 23°C and survival to settlement occurred between 13 to 20°C

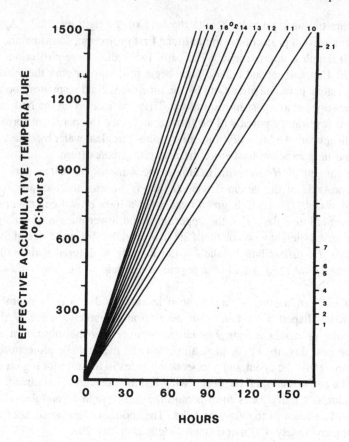

FIGURE 5.   Relationship between water temperature, and timing of larval stages and effective accumulative temperature of *H. rufescens*. Each line represents the progression of larval development when larvae are reared with a constant water temperature. Number at top of line indicates the water temperature (°C). Numbers on right y-axis correspond to larval stages listed in Table 3. (From Holsinger, L. M., unpublished data, 1984. With permission.)

(Figures 6 and 7). These differences could be due to the fact that the abalone were collected from different regions of California. Leighton[1] used abalone from southern California where the water temperature ranges from 14 to 16°C, while Holsinger[4] used abalone from northern California where the water temperature ranges from 9 to 14°C. Therefore, the larvae of red abalone in northern California may be adapted to colder temperatures.[4]

Scott[8] in a follow-up study to Holsinger described 32 stages during larval development. Four of the features previously had not been described for *H. rufescens*, although they had been described in *H. discus hannai*, and *H. tuberculata*.[8] The newly described stages were (1) formation of apical cilia, (2) subsequent degeneration of the apical cilia (present about 30 hr in red abalone), (3) cilia on metapodium, and (4) spines on operculum.

Scott[8] did not find several stages in *H. rufescens* which had been described by Seki and Kan-no[2] in *H. discus hannai* (first epipodal tentacle, otolith, protrusion of snout, and cilia on roof of mantle cavity). However, these stages could have been present but not recognized by Scott,[8] since he was not using live specimens.[8]

H. rufescens trochophore larvae measure 160 μm × 195 μm while in the egg case. After development into a veliger larvae and completion of the larval shell, the larvae measure 210 μm × 270 μm.[9] It takes approximately 6 days of rearing at 15°C for the larvae to develop to the stage necessary for settlement and metamorphosis. Survival rate to settlement is usually

Table 3
LARVAL STAGES IN *H. RUFESCENS*

Stage	Feature
1	Gastrula (240°C-hr)
2	Appearance of prototrochal cilia (290°C-hr)
3	Formation of prototrochal girdle (350°C-hr)
4	Hatch out (400°C-hr)
5	Beginning of larval shell (460°C-hr)
6	Completion of velum (500°C-hr)
7	Appearance of retractor muscle (580°C-hr)
8	Appearance of integumental attachment to larval shell
9	Protrusion of foot mass
10	Completion of larval shell (710°C-hr)
11	Torsion (770°C-hr)
12	Formation of operculum
13	Appearance of long spines on end of metapodium
14	Appearance of fine cilia on foot sole
15	Vertical groove formation in velum
16	Appearance of eye spot
17	Formation of propodium
18	Cephalic tentacle
19	Cilia growing on propodium (1300°C-hr)
20	Cilia in mantle cavity
21	Competent for settlement (1450°C-hr)

(From Holsinger, L. M., unpublished data, 1984. With permission.)

close to 90%. The larval stage necessary for settlement is the same as reported by Seki and Kan-no[2] for *H. discus hannai*. The veligers have four branches on the cephalic tentacles, and the fully developed foot is able to pull the larva upright and move by ciliary action. Ebert and Houk called this stage the "gliding stage". Veligers at this stage were competent for settlement.[9]

Haliotis fulgens

Leighton[1] found larvae could be cultured from 10 to 25°C, with the best development at the higher temperatures (Figure 8). Larvae became abnormal and died above 25°C.[1] High larval survival is possible at water temperatures from 20 to 28°C, but the optimum temperature is from 20 to 24°C.[10] The optimum temperature for larval development of *H. fulgens* shows very little variation (±1.5°C) from different spawnings or parents.[1] The time from fertilization to settlement at different temperatures within the physiological range were from 3.5 days at 24°C to 12 days at 14°C. Larvae raised at 12°C did not settle within 2 weeks which was the length of the experiment. *H. fulgens* settle the earliest at 25.5°C (less than 3 days) but survival of the post larvae is poor. Larvae raised at 22 to 23°C settle after 4 days and have good growth (Figure 9).[10]

Haliotis corrugata

The optimum temperature for larval development of *H. corrugata* showed very little variation (±1°C) from different spawnings or parents. The most rapid growth and best survival was at 21 to 22°C (Figure 10). *H. corrugata* settled within 3.5 days at 22 to 23°C but the post larvae did not survive.[1]

Haliotis sorenseni

The trochophore larvae are subcylindrical in outline, have a distinct prototroch, and are

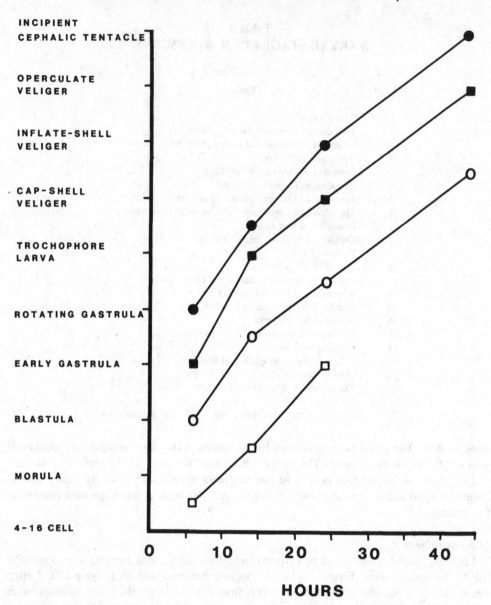

FIGURE 6.   Relationship between water temperature and development of *H. rufescens* eggs. (□) 10°C; (○) 14°C; (■) 17°C; (●) 20°C. (From Leighton, D. L., *Fish. Bull.*, 72 (4), 1137, 1984. With permission.)

yellow-tan in color. Tissue pigmentation is predominantly beige, velar fringes are yellow, and digestive gland is maroon. The maroon-colored digestive gland in *H. sorenseni* larvae appears to be distinctive of this species, since other species have either green- or brown-colored digestive glands. Larval development was most rapid at higher temperatures but survival was reduced at above 20°C. During settlement, velar dystrophy with loss of ciliated fringes takes place gradually and the larvae retain the ability for several days after initial settlement, to swim if dislodged from the substratum.[11]

Some individuals settled as early as day 7 at 20°C, day 8 at 18°C, and day 9 to 10 at 15°C (Figure 9). Almost all larvae raised at 18 and 20°C, but none of the larvae raised at 10°C, settled by day 15. Most larvae at 10°C were dead after day 25 and never developed past the late veliger stages. The larvae raised at 12°C died between day 25 and 30. Settlement

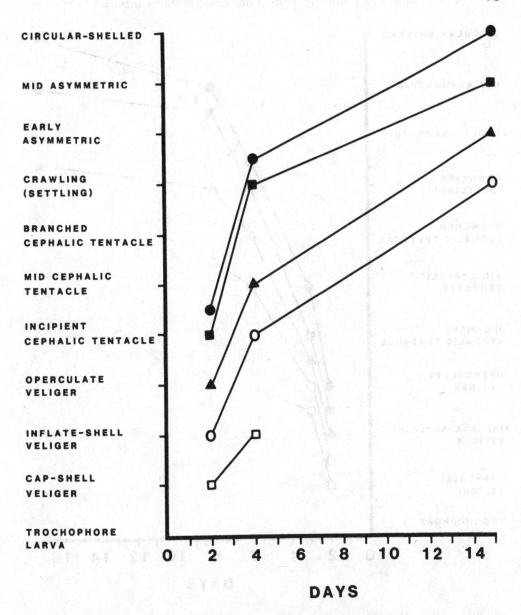

FIGURE 7. Relationship between water temperature and development of *H. rufescens* larvae. Fertilized eggs were reared at ambient water temperature until hatch out. (▲) 9°C, (■) 12°C; (●) 14°C; (△) 15°C; (□) 16°C; (●) 17°C. (From Leighton, D. L., *Fish. Bull.*, 72 (4), 1137, 1974. With permission.)

rates varied; about 5% did not settle at 18°C. Post-larval survival at day 36 was best at 16 and 18°C, somewhat less at 20°C, and least at 14°C. Therefore, the optimum water temperature is 16 to 18°C.[11]

Haliotis kamtschatkana

Results of a study by Beaudry[12] on the effects of water temperature on larval development in *H. kamtschatkana* were hard to interpret. Larvae reared at 14°C, the lowest temperature tested, had the highest number of normal larvae and survival from fertilization to trochophore larva. Larvae reared at 21°C, showed signs of thermal stress and all larvae died. Although 14°C produced the best development to early larval stages, larvae reared at 16°C had the most rapid growth after settlement and larvae reared at 18.5°C had the highest overall survival

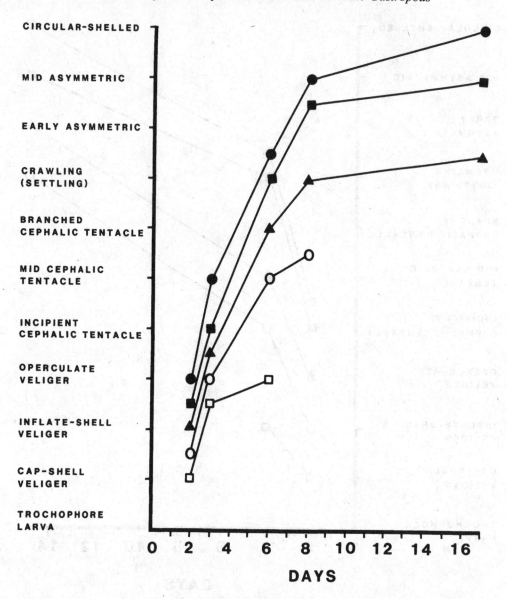

FIGURE 8.   Relationship between water temperature and development of *H. fulgens* larvae. Fertilized eggs were reared at ambient water temperature until hatch out. (□) 12°C; (○) 15°C; (▲) 17°C; (■) 20°C; (●) 23°C. (From Leighton, D. L., *Fish. Bull.*, 72 (4), 1137, 1974. With permission.)

from fertilization to 2 months old. The increased overall survival at higher temperatures could be due to shorter settlement times, reducing the larval period when most mortalities occur.[12]

Haliotis tuberculata

The first polar body appears a few minutes after fertilization in *H. tuberculata*, followed by the second polar body, but many times it is not observed. Cell cleavage is total, unequal, and spiral. The first cleavage is along the vertical axis of the egg with the second cleavage along the same axis. The third cleavage is in the horizontal plane just above the axis. At the third division, there is differentiation between the micromeres and macromeres, and the micromeres move around the macromeres in a clockwise direction. At the fourth division, the direction of division is counterclockwise.[13]

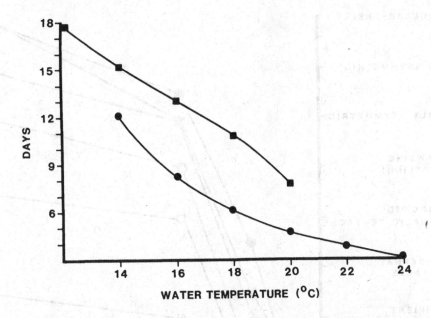

FIGURE 9. Relationship between water temperature and time to settlement. (●) *H. fulgens*,[10] (■) *H. sorenseni*.[11]

An apical tuft of cilia is present just before hatching until just before the velum divides in the veliger. After day 4 to 5, the statocyst can be observed between the head and foot, with five or six grains grouped together in the center forming the statolith. After day 6, cilia on the velum disappear and the peristomal shell begins secretion along the right side of the larval shell aperture. The snout is well formed, larvae can creep around actively, and the heart is beating. Juveniles are 370 μm with 180 μm of new shell growth after day 9. Movement of the radula is first observed at this stage. Settlement occurs within 3.5 to 5 days after fertilization.[13]

COMPARISON OF LARVAL CHARACTERISTICS

Egg Diameter

The size of the unfertilized egg in *H. discus*, *H. sieboldii* and *H. gigantea* is approximately 220 μm in diameter, including the egg membrane.[6] Unfertilized eggs are spherical and enclosed within a 400 to 500-μm-thick gelatinous coating.[6,14] After fertilization, the perivitelline space between the outer layer and egg membrane increases in size. The increase in egg diameter, due to the perivitelline space, is different in each species. Diameter of the fertilized egg is 230 μm in *H. discus*,[6] 280 μm in *H. sieboldii*,[6] 270 μm in *H. gigantea*,[6] 230 μm in *H. iris*,[14] 200 μm in *H. ruber*,[14] 200 μm in *H. sorenseni*,[11] and 210 μm (yolk — 170 μm) in *H. tuberculata*.[13]

Pigmentation

Pigmentation of the egg is dark at the animal pole and light at the vegetable pole. Pigments derived from maternal yolk appear to be retained by *Haliotis* spp. trochophore and veliger larvae. Among California species, ovarian tissue is dark green in *H. rufescens*, *H. cracherodii*, *H. walallensis*, *H. assimilis*, and *H. kamtschatkana*, and the larvae of these species are green. The other three species, *H. fulgens*, *H. sorenseni*, and *H. corrugata*, produce brown, beige, and olive eggs, respectively, and their larvae reflect these pigments. *H. fulgens* larvae are generally brown with green velar margins, *H. sorenseni* larvae are beige with

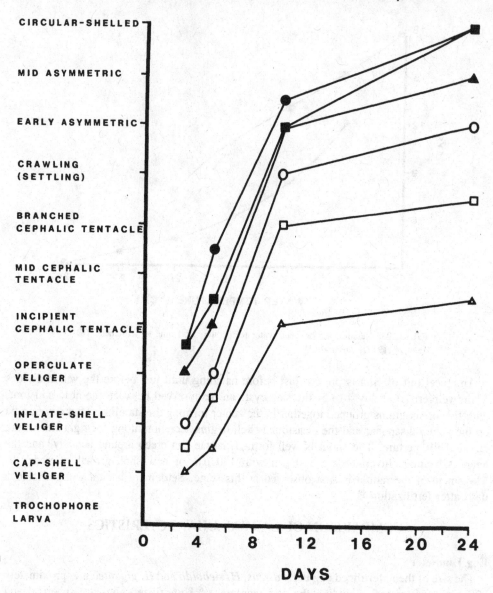

FIGURE 10.   Relationship between water temperature and development of *H. corrugata* larvae. Fertilized eggs were reared at ambient water temperature until hatch out. (□) 11°C; (○) 14°C; (▲) 16°C; (■) 19°C; (●) 22°C. (From Leighton, D.L., *Fish. Bull.*, 72 (4), 1137, 1974. With permission.)

yellowish velar margins, and *H. corrugata* larvae are light yellow-green with velar fringes of a darker shade of green.[11] *H. coccinea canariensis*, an abalone found in the Canary Islands, has violet-colored eggs, dark at the animal pole and yellowish at the vegetative pole. The larva has a violet-colored foot, velum, and cephalic tentacles, and an orange-yellowish-colored visceral mass.[15]

Phototaxis

The amount of phototaxis exhibited by larvae is species specific. *H. discus*[6] shows strong phototaxis, *H. sieboldii*[6] moderately strong phototaxis and *H. diversicolor supertexta*[16] little phototaxis.[13] The phototaxis of larvae can be conveniently used to separate larvae during culture.[5]

The larval stage when retraction occurs is also species specific. *H. discus*[6] and *H.*

diversicolor supertexta[16] can only retract into the larval shell at a late larval stage (division of the velum, and appearance of the eyes and tentacle rudiments), while *H. tuberculata*[13] is able to retract into the larval shell at an early stage.[13]

Larval Stages

There are no differences in the development from fertilization to trochophore larva between *H. discus hannai*, when compared with *H. gigantea*,[7] *H. tuberculata*,[17,18] *H. sieboldii*,[6] *H. discus*,[6] and *H. diversicolor supertexta*.[2,16]

H. discus hannai, *H. sieboldii*, *H. discus*, and *H. diversicolor supertexta* form the larval retractor muscle prior to formation of the integumental attachment to the larval shell. Also, protrusion of the foot mass and development of the operculum occurs after these stages. However, Crofts[17,18] reported that *H. tuberculata* protrudes the foot mass and develops the operculum before forming the larval retractor muscle and integumental attachment. Also, cilia are only observed when the larval retractor muscle is formed and then quickly disappear. Conversely, *H. discus hannai*, *H. sieboldii*, *H. discus*, and *H. diversicolor supertexta*, form cilia during early larval development and the cilia are retained.[2]

There are many differences in larval development from torsion to metamorphosis. Seki and Kan-no[2] found the larval shell of *H. discus hannai* was completed before torsion occurred. In *H. sieboldii*, Ino[6] found torsion occurred first and then the larval shell was completed about 13 hr later. Three long spines on the end of the metapodium are seen only in *H. discus hannai*.[2] *H. sieboldii* develops only one spine.[2] Also, Seki and Kan-no[2] found the spines on the metapodium were formed before the operculum, while Ino[6] found the reverse order in *H. sieboldii*. The cilia on the propodium of *H. discus hannai* are more active and move longer than the cilia on the foot sole. This has also been reported in *H. sieboldii* and *H. discus*.[2]

H. discus hannai develops all the larval stages from stage 32 to 41 before metamorphosis. However, the first epipodial tentacle in *H. gigantea* is observed only after metamorphosis starts (approximately 4 weeks after fertilization). The otolith in *H. tuberculata* is observed before the formation of the vertical groove in the velum and cephalic tentacles, and the first epipodial tentacle appears on the right of the foot after metamorphosis. The ciliary process on the roof of the mantle cavity and the enlargement of the mantle cavity are observed after metamorphosis.[2]

H. discus forms snout protrusions and tubules on the cephalic tentacles, but there is no enlargement of mantle cavity to the back of the body. *H. sieboldii* also did not have an enlargement of the mantle cavity but other development was very similar to *H. discus hannai*.

H. diversicolor supertexta has stage 32 (apophysis on the propodium), stage 33 (formation of first epipodial tentacle), stage 35 (appearance of short spine on the cephalic tentacle), stage 36 (snout protrusion), and stage 39 (appearance of third tubule on the cephalic tentacle before metamorphosis). However, stage 31 (appearance of cilia in the mantle cavity), stage 34 (appearance of otolith), and stage 38 (formation of ciliary process on the roof on mantle cavity) are not seen before or after metamorphosis.[2]

Larval Shell Morphology

Murayama reported the larval shell is 360 μm in *H. gigantea* but is 290 μm in *H. discus* and *H. sieboldii*. The shell surface is different in *H. discus* and *H. sieboldii*.[6] Larval shell size differs among the five species in Japan. They are, in order from large to small, *H. diversicolor aquatilis*, *H. discus hannai*, *H. gigantea*, *H. discus*, and *H. madaka*. The differences in shell size are statistically significant except between *H. discus* and *H. gigantea*. The general shape of the shells do not differ except that *H. diversicolor aquatilis* has a slightly longer shell on the adapical side. Shell sculpture patterns are very similar in form, with spiral ridges and a higher density of surface sculptures on the adapical side than on the abapical side. However, differences in sculptures become more apparent when viewed with scanning electron microscopy. From the size and pattern of shell sculptures it may be possible to identify veliger larvae in nature.

HATCHERY TECHNIQUES

Haliotis discus hannai

Immaculate cleanliness must be maintained at all times and all possible sources of contamination must be eliminated during the larval development period. The two most probable times of introducing contaminants into the larval rearing container are during removal of the eggs from the spawning container and transfer of the trochophore larvae after hatch out to a clean larval-rearing container. The Japanese culturists are very careful not to introduce any contaminants (e.g., mucus, feces, dirt, etc.) into the container used for fertilization and larval development during the collection of the spawned eggs.

A successful culture and high yield is assured by changing the rearing conditions at three important stages during larval development.

Hatch Out

At most culture facilities in Japan, embryos are reared in 15-ℓ plastic containers during the period from fertilization to hatch out. There is no water flow or aeration in the containers during this period and the larval culture tanks are kept in a temperature-controlled enclosed room. The time of hatch out, as well as all developmental stages, is determined by the water temperature. The room temperature is maintained at approximately 20°C, which controls the development rate and ensures that the necessary larval culture procedures occur during normal working hours. If the larvae were raised at ambient air temperature, the warm temperatures during late summer would increase the rate of larval development and the larvae would hatch out during the night. Also, some of the required water changes in the larval tanks would occur during the night.[19]

If fertilization is rapid and synchronous, hatch out occurs over a short time interval. After the trochophore larvae hatch out, the rearing container contains normally developed trochophore larvae, and also egg membranes, unfertilized eggs, and abnormally developed larvae, which can cause contamination of the larval-rearing container. Therefore, it is necessary to know precisely when hatch out occurs, so the trochophore larvae are decanted as soon as possible from the empty egg cases and unfertilized eggs. Separation of the trochophore larvae from the discarded egg membranes reduces bacterial contamination from decomposition of the egg membranes.[20] Enzymes secreted by the larva to facilitate hatching may also contribute to lowering water quality.

The trochophore larvae swim toward the surface immediately after hatch out. Introduction of contaminants is avoided by only transferring larvae swimming in the top two thirds of water. If precautions are taken to avoid introducing contamination, the need for antibiotics is eliminated during the larval-rearing period.[21] Decanting of trochophore larvae is done in a dark room with the aid of a flashlight. One person shines the light across the mouth of the container while a second person decants the container in one smooth continuous motion. The water from the fertilization container is decanted directly into another container and a micron filter is not used, since the filter would harm or kill the trochophore larvae. This fraction should be absolutely free of all egg cases. Egg cases in the decanted fraction will defeat the purpose of changing the water. The water remaining in the original container is allowed to settle and is decanted a second time. The larvae in this fraction will probably be contaminated with some egg cases but care is taken to minimize contamination. Some laboratories assume the trochophore larvae in the lower third are dead or abnormal and discard this fraction.[19]

The water volume in the larval-rearing container is slowly increased to 15 ℓ with ultraviolet (UV) sterilized water. The added water must be within ±1°C of the water temperature in the original container so the trochophore larvae are not killed by thermal shock.[19] The containers with the newly hatched trochophore larvae are placed back into a dark, temperature-controlled room, and the larvae are maintained with no water flow or aeration until the larval shell is completely formed. The larval-rearing tanks are not aerated because this would injure the developing larvae. Therefore, very clean, UV-sterilized water is mandatory.

Development of the Larval Shell

Completion of the larval shell marks the beginning of water changes in the rearing container. Before this stage, larvae are too fragile and would be killed if retained on a filter during the water change. The larval shell must be completely formed before changing the water. After the larval shell develops, the water is changed three times a day (9 a.m., 12 p.m., 3 p.m.) by pouring the container water through a 90-μm nitex screen sitting in a low pan filled with isothermal water, and the larvae are retained on the screen.[20] Placing the screen in a tray ensures that the larvae are always covered with water. The larvae are washed thoroughly by "swishing" flowing isothermal, UV-sterilized water across the screen, making sure not to "press" the larvae against the screen. The larvae are than rinsed into a new container.[19,20] Changing the water every 8 hr would be a better rearing method but this would require working during "non-business" hours which increases labor costs.

Settlement of Larvae

The water is changed three times a day until larval development is complete and the larvae are ready for settlement. The correct larval stage for settlement can either be determined from the EAT or observation under a microscope. Settlement occurs rapidly on wavy-plate substrata if the larvae are healthy, and competent to settle and metamorphose.[19] The larvae are usually introduced into the settlement tanks after they have developed the third tubercle on the cephalic tentacle.[20]

Haliotis rufescens

Ebert and Houk[9] feel their development of a flow-through larval-rearing apparatus represents an advancement in abalone cultivation procedures. Larvae can be raised at high densities (up to 20/mℓ) with good survival and development, and the sea water in the container does not require changing as with the Japanese method.[9,22] Most mortality and abnormal development is probably due to bacteria in the culture system.[23] Bacterial growth is retarded by controlling water temperature and having continuously flowing sea water.[22] Reduction of harmful microorganisms is the most important factor in maintaining reproducible high yields, normal larval development, and growth of the larvae and juveniles.[22]

Water temperature and flow rate are the only factors monitored by the culturist. The water flow must be maintained at a low rate to ensure that the larvae are uniformly distributed and not forced against the culture screen on the bottom of the culture apparatus.[9]

Ebert and Houk[9] place free-swimming trochophore larvae after hatch out in an 8-ℓ larval-rearing apparatus with a rearing density of 5/mℓ (40,000 per apparatus). The larval-rearing apparatus consists of a container inside a container. The inner container is a 36 cm-long PVC pipe (20-cm diameter) with a 90-μm nitex screen (2.5 cm above the bottom) across the pipe. The outer container is 30 cm high and 27 cm in diameter. Water flows (200 to 300 mℓ/min) into the top of the inner container, through the nitex screen, and exits the outer container through an adjustable standpipe. Water height inside the inner container is adjusted by rotating the standpipe. The water is filtered to 3 μm and UV sterilized.

GENERAL

Larval Density

Larval-rearing density primarily depends on water flow and oxygen concentration. High larval densities may cause low survival rates but this loss might be offset by the lower costs of rearing larvae in smaller containers.[24]

Antibiotics

Adding streptomycin, an effective antibiotic against both gram-negative and gram-positive bacteria, to the larval-rearing water helps suppress bacterial growth which causes deterioration of water quality. Streptomycin helps increase survival in *H. discus* larvae. Dihydro-streptomycin sulfate (full strength) is added to the sea water in the rearing container to obtain the desired concentrations. Mortality from veliger larva (2 to 3 days after fertilization) to early juvenile can be reduced to 10 to 33% with antibiotic treatment. Survival rates of abalone raised in water containing streptomycin are always greater than the control rates. There is no significant difference in survival within the range of 50 to 250 mg/ℓ.[25]

Transporting Larvae

Kurita[26] tested the feasibility of transporting fertilized eggs and larvae from a hatchery to distant laboratories for settlement. Larvae were placed in double vinyl bags with sea water and oxygen, and the water temperature was maintained at 18°C during the trip. The trip took 4 hr by truck. The larvae were placed in a 10,000-ℓ capacity tank immediately upon arrival at the laboratory. The water temperature in the culture tank at the receiving laboratory was 3 to 4°C below the water temperature at the hatchery, but there were no adverse affects on the larval development. Fertilized eggs were also transported by this method. Embyros were transported at the morula stage (100,000 eggs in 30 ℓ of water) and had 100% hatch out after transfer to a 10,000-ℓ tank (water temperature — 15°C). The best larval density during transportation is between 5 to 10 larvae per milliliter which will produce a larval survival of 85 to 95%.[26]

ACKNOWLEDGMENTS

I would like to thank Michael Brody for the drawings of larval development.

REFERENCES

1. **Leighton, D. L.,** The influence of temperature on larval and juvenile growth in three species of southern California abalones, *Fish. Bull.,* 72 (4), 1137, 1974.
2. **Seki, T. and Kan-no, H.,** Synchronized control of early life in the abalone, *Haliotis discus hannai* Ino, Haliotidae, Gastropoda, *Bull. Tohoku Reg. Fish. Res. Lab.,* 38, 143, 1977.
3. **Kikuchi, S. and Uki, N.,** Technical study on artificial spawning of abalone, genus *Haliotis.* I. Relation between water temperature and advancing sexual maturity of *Haliotis discus hannai* Ino, *Bull. Tohoku Reg. Fish. Res. Lab.,* 33, 69, 1974.
4. **Holsinger, L. M.,** unpublished data, 1984.
5. **Ino, T.,** *Fisheries in Japan, Abalone and Oyster,* Japan Marine Products Photo Materials Assoc., Tokyo, 1980, 165.
6. **Ino, T.,** Biological studies on the propagation of Japanese abalone (genus *Haliotis), Bull. Tokai Reg. Fish. Res. Lab.,* 5, 1, 1952.
7. **Murayama, S.,** On the development of the Japanese abalone, *Haliotis gigantea, J. Coll. Agric. Tokyo Imp. Univ.,* 13, 227, 1935.

8. **Scott, K. M.**, unpublished data, 1985.

9. **Ebert, E. E. and Houk, J. L.**, Elements and innovations in the cultivation of Red abalone *Haliotis rufescens, Aquaculture,* 39, 375, 1984.

10. **Leighton, D. L., Byhower, M. J., Kelly, J. C., Hooker, G. N., and Morse, D. E.**, Acceleration of development and growth in young Green abalone *(Haliotis fulgens)* using warmed effluent sea water, *J. World Maricult. Soc.,* 12(1), 170, 1981.

11. **Leighton, D. L.**, Laboratory observations on the early growth of the abalone, *Haliotis sorenseni,* and the effect of temperature on larval development and settling success, *Fish. Bull.,* 70(2), 373, 1972.

12. **Beaudry, C. G.**, Survival and growth of the larvae of *Haliotis kamtschatkana* Jonas at different temperatures, *Natl. Shellfish Assoc.,* 3(1), 109, 1982.

13. **Koike, Y.**, Biological and ecological studies on the propagation of the ormer, *Haliotis tuberculata* Linnaeus. I. Larval development and growth of juveniles, *La Mer,* 16(3), 124, 1978.

14. **Harrison, A. J. and Grant, J. F.**, Progress in abalone research, *Tasmanian Fish. Res.,* 5(1), 1, 1971.

15. **Pena, J. B.**, Preliminary study on the induction of artificial spawning in *Haliotis coccinea canariensis* Nordsieck ('975), *Aquaculture,* 52, 35, 1986.

16. **Oba, T.**, Studies on the propagation of the abalone *Haliotis diversicolor supertexta* Lischke. II. On the development, *Bull. Jpn. Soc. Sci. Fish.,* 30, 809, 1964.

17. **Crofts, D. R.**, The development of *Haliotis tuberculata,* with special reference to organogenesis during torsion, *Philos. Trans. R. Soc. London Ser. B,* 228, 219, 1938.

18. **Crofts, D. R.**, Muscle morphogenesis in primitive gastropods and its relation to torsion, *Proc. Zool. Soc. London,* 125, 711, 1955.

19. **Kumabe, L.**, Methods used at the Oyster Research Institute, unpublished report, 1981.

20. **Hahn, K. O.**, personal observations, 1981.

21. **Hayashi, I.**, Larval shell morphology of some Japanese haliotids for the identification of their veliger larvae and early juveniles, *Venus,* 42(1), 49, 1983.

22. **Hooker, N. and Morse, D. E.**, Abalone: the emerging development of commercial cultivation in the United States, in *Crustacean and Mollusk Aquaculture in the United States,* Huner, J. V. and Brown, E. E., Eds., AVI Publishing, 1985, 365.

23. **Morse, D. E., Hooker, N., Jensen, L., and Duncan, H.**, Induction of larval abalone settling and metamorphosis by γ-aminobutyric acid and its congeners from crustose red algae. II. Applications to cultivation, seed-production and bioassays; principal causes of mortality and interference, *Proc. World Maricult. Soc.,* 10, 81, 1979.

24. **Mottet, M. G.**, A review of the fishery biology of abalones, *Washington Dept. Fish., Tech. Rep.,* 37, 1, 1978.

25. **Tanaka, Y.**, Studies on reducing mortality of larvae and juveniles in the course of the mass-production of seed abalone. I. Satisfactory result with streptomycin to reduce intensive mortality, *Bull. Tokai Reg. Fish. Res. Lab.,* 58, 155, 1969.

26. **Kurita, M., Sato, R., Nagao, N., and Ando, T.**, On the transport of abalone, *Haliotis discus* (Reeve), *Bull. Oita Pref. Fish. Exp. St.,* 10, 112, 1978. (Translated in Mottet, M. G., Summaries of Japanese papers on hatchery technology and intermediate rearing facilities for clams, scallops, and abalones, *Prog. Rep., Dept. Fish. Washington,* 203, 17, 1984.)

27. **Owen, B., DiSalvo, L. H., Ebert, E. E., and Fonck, E.**, Culture of the Californian Red abalone *Haliotis rufescens* Swainson (1822) in Chile, *Veliger,* 27(2), 101, 1984.

28. **Caldwell, M. E.**, Spawning, Early Development and Hybridization of *Haliotis kamtschatkana* Jonas, Master's thesis, University of Washington, Seattle, 1981.

29. **Tong, L. and Dutton, S.**, Induced paua spawnings, *Catch' 81,* March, 1981.

INDUCTION OF SETTLEMENT IN COMPETENT ABALONE LARVAE

Kirk O. Hahn

INTRODUCTION

Induction of settlement is probably the most critical step in the culture of abalone. It is fairly easy to have a larval survival rate, from fertilization to veligers competent to settle greater than 95%. During the transition from planktonic veliger to benthic juvenile, however, the survival rate will usually drop to approximately 10%. Some researchers do not think this low survival is very important when it is taken into account a single female can spawn a million eggs. Low survival is not a problem when annual productions are small, however, low survival does pose a problem if production is to meet the ever-increasing demand for abalone. For example, there are some laboratories in Japan which produce over 2 million juveniles a year. Higher survival allows greater efficiency which equals a lower cost per animal.

The habitat required by newly settled larvae and their ability to discriminate between substrata that may be crucial to their survival is a critical, but poorly understood aspect of most marine invertebrate life cycles. Understanding the habitat requirements of larvae of commercially important abalone species is necessary for the development of aquaculture methods and management of natural stocks.[1]

In order to facilitate settlement, and consequently increase productivity, several laboratories in Japan and the U.S. have attempted to determine the proper substratum and natural inducer of settlement. These studies can be separated into two categories; (1) settlement caused by diatoms and mucus from juvenile abalone, and (2) settlement caused by red crustose algae or γ-aminobutyric acid (GABA).[2-6]

The results of these studies will be presented and each study's method of inducing settlement will be evaluated according to the reliability and effectiveness of the procedures. An ideal method should cause rapid settlement with good survival and produce favorable conditions for the rapidly growing, newly settled juvenile.

Before discussing the methods used for settlement induction in the hatchery, it is important to understand the sequence of behavioral and developmental steps which comprise larval settlement and metamorphosis into the adult form.

SETTLEMENT BEHAVIOR AND METAMORPHOSIS

It is possible to observe certain larval structures that indicate when larval development is complete and the larva is ready or "competent" to settle[2] (see Chapter 5 — Larval Development). Competent larvae are veligers, have not lost their ability to swim or crawl, and have not yet changed shape. Seki and Kan-no[2] reported larvae are capable of crawling on the substratum after the first epipodal tentacle forms and settlement is initiated after the snout protrusions form.

The ability of veligers to begin settlement is dependent upon their larval stage. The last 11 larval stages in *H. discus hannai*, from an EAT of 750 to 1230°C-hr, were tested for ability to settle and metamorphose. Settlement was only possible in larvae that had reached an EAT of 925°C-hr (larval stage 36, snout protrusion). At this stage, veligers are able to turn the foot within the shell, right themselves, and crawl in an upright position. Metamorphosis was only possible in larvae that had reached an EAT of 1030°C-hr (larval stage 39, appearance of third tubule on the cephalic tentacles). These stages occur 74.6 and 85.5 hr after fertilization, respectively, when the water temperature is 20°C.[2]

Competent veligers repeatedly alternate swimming upward and sinking, allowing them to effectively "sample" the bottom as they are carried in water currents, until suitable (inducing) substrata are found. The veligers then exhibit a "searching behavior", alternately swimming and crawling on the substratum.[2] Initially, competent veligers spend most of their time swimming, but the duration on the substratum slowly increases with time.[3,4] This behavior ensures a highly specific induction of settlement, thus protecting against an irreversible metamorphosis (loss of the swimming cilia) into a benthic juvenile in an inappropriate (i.e., nonalgal) microhabitat. Mistakes in settlement would be common if the swimming larvae were stimulated by diffusible inducers from the algae in the agitated water.[3,4] Settlement inducers are considered to be insoluble organic materials on the substratum surface.[7,8] Larvae are capable of delaying settlement and metamorphosis if an adequate settlement substratum is not found.

Seki and Kan-no[2] conducted a detailed study of settlement and metamorphosis of *H. discus hannai*. They designed a small container made from microscope slides to allow continuous observation during the settlement process. A narrow well for a single swimming larva was created by gluing two microscope slides together with a spacer around the edge. The microscope slides were placed vertically, with the bottom spacer as the settlement substratum. A microscope was arranged horizontally to view the settlement behavior and metamorphosis of the larva.

Rotating the foot and securely clinging to the substratum indicate that larvae are ready to settle. Competent larvae exhibit two "settling" behaviors. Sometimes they swim close to the substratum, with the bottom of their foot parallel to the substratum, and settle softly when the bottom of the foot touches the surface (Figure 1 A and B). Other times the larvae collide with the substratum while the foot is inside the shell. After the collision with the substratum, velar cilia begin beating, the foot is extended from the shell and twists until it attaches to the substratum, and the larva rights itself. Larval settlement is random, larvae do not space themselves on the substratum as was previously thought to occur in abalone.[2]

It takes approximately 30 min from the initial "testing" of the substratum to settlement. The settled larva secretes a mucus from its foot to increase the strength of the attachment to the substratum. Settled larvae are so firmly attached that a stream of water from a pipette cannot remove them from the surface.[2]

After settling, the veliger crawls continuously and beats its cilia occasionally (Figure 1 C). The veliger can move very quickly, traveling up to 200 μm in 3 sec. Crawling becomes progressively slower and finally stops. At this point, the larvae do not move except to change direction and begin "testing" the surface. They show a characteristic behavior of raising the back of their shell, by contracting the juvenile retractor muscle, until the front of the shell touches the surface and the cephalic tentacles sweep the surface (Figure 1 D). This behavior can only be seen when larvae are viewed from the side.[2]

Development of sensory organs is very important before metamorphosis can proceed. The anterior region of the cephalic tentacles in *H. discus hannai*, appears to contain important chemical receptors for choosing the proper substratum.[2] The nervous system, important for chemoreception, develops at the same time as the processes on the cephalic tentacles.[9] The foot also seems to have an important role in selecting the substratum.[2]

Table 1 shows the changes in the veliger's behavior and morphology as it progresses from veliger to benthic juvenile.[2] The settlement behavior exhibited by *H. discus hannai* is characteristic of behavior found in other invertebrates during the transition to a benthic life.[2,7,10-12] There are 12 recognizable stages from the initiation of settlement until the juvenile shell grows to a diameter of 350 μm.[2] The exact timing and duration of searching by the veliger, and initiation of settlement and metamorphosis is variable between individuals. However, after initiation, metamorphosis is completed (development of the mouth, digestive tract, circulatory system, sensory organs, and adult form) in less than 24 hr.[2] Descriptions of the 12 stages follow.

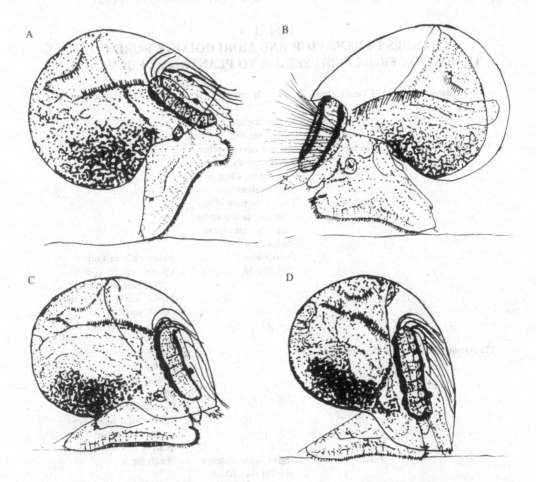

FIGURE 1. Settlement behavior and metamorphosis of *H. discus hannai*. (A, B) Searching behavior of competent larvae; (C) settlement and crawling; (D) pulling back of shell upward, while touching front edge of shell and cephalic tentacles on substratum; (E) casting off velum; (F) mantle membrane moves to shell edge; (G) peristomal shell growth; (H) settled juvenile begins feeding on diatoms (shell length — 350 μm). (From Seki, T. and Kan-no, H. *Bull. Tohoku Res. Fish. Res. Lab.*, 42, 31, 1981. With permission.) (Drawings by M. Brody)

Stage 1 — The settled larva, with assistance from the juvenile retractor muscle, casts off the 24 cilia cells of the velum and loses its ability to swim (Figure 1 E). The shedding of the velum occurs very quickly and cells are cast off either singly or in groups.

Stage 2 — A ciliated process forms near the right cephalic tentacle. The larva begins crawling again.

Stage 3 — The mantle membrane separates from the larval shell and moves to the shell edge (Figure 1 F).

Stage 4 — The juvenile shell begins to be secreted along the right edge of the larval shell (Figure 1 G). New shell forms along the left side of the larval shell opening toward the right side, and the body axis of the juvenile begins to tilt on the left side opposite the growing shell edge. The mouth has not yet developed and the juvenile is unable to eat diatoms.

Stage 5 — A pair of initial processes, which will be part of the mouth, protrude to enclose the mouth.

Stage 6 — The radula begins to form.

Stage 7 — Cephalic tentacles grow and the fifth small tubule is formed on the tentacles.

Stage 8 — The digestive tract develops.

Table 1
**CHANGES IN BEHAVIOR AND MORPHOLOGY DURING
TRANSITION FROM PEDIVELIGER TO PLANTIGRADE JUVENILE**

Stage	Classification	Behavior	Morphology
Pediveliger	Searching	Swimming and crawling on substratum	
	Settlement	Moving cilia on velum intermittently and continuous crawling on the substratum	
	Attachment	Pulling up back of larval shell and touching cephalic tentacles to substratum	
		No movement	Velum cells cast off
		Crawling begins again	Ciliated process at right of the cephalic tentacle
			Mantle moves to edge of larval shell
			Secretion of juvenile shell
Plantigrade juvenile	Metamorphosis		Processes enclose mouth
			Radula forms
			Fifth small tubule forms on cephalic tentacles
			Digestive tract develops.
			Mucus gland appears on mantle membrane
			Heart begins beating
	Feeding	Begins eating diatoms smaller than 10 μm	Mouth forms
		Vigorous feeding	Mouth 30 μm across, shell length 350 μm

(From Seki, T. and Kan-no, H., *Bull. Tohoku Reg. Fish. Res. Lab.*, 42, 31, 1981. With permission.)

Stage 9 — The mucus gland appears on the mantle membrane lining the back wall.

Stage 10 — The heart begins to beat and blood begins circulating.

Stage 11 — The mouth is formed by fusion of the right and left initial processes (stage 5) into a tube. The mouth is able to search the substratum and the radula begins to scrape the surface. The mouth is less than 10 μm across, so all food items (diatoms and bacteria) must be smaller than this size. The newly settled juveniles are able to eat some kinds of bacteria.

It is interesting *H. discus hannai* larvae, and presumably other abalone species, cannot feed immediately upon settlement and almost all of metamorphosis is fueled by yolk. The adult organs develop without energy from eating and the time sequence of the stages is constant after initiation of settlement.[2]

Stage 12 — This stage lasts much longer than the previous 11 stages and continues until 2 days after metamorphosis. At the end of stage 12, the shell diameter is approximately 350 μm in length and the mouth is 30 μm across (Figure 1 H). The juvenile still has a ciliated process on the head and an operculum which are absent in the adult. Also, the juvenile has not yet formed ctenidia.[2]

The movement of the radula and traces of feeding by juveniles, especially when the bottom of the tank is overgrown with diatoms, can be seen clearly when feeding begins. Newly settled juveniles eat vigorously after settlement.[2] At the present time, efficient culture of

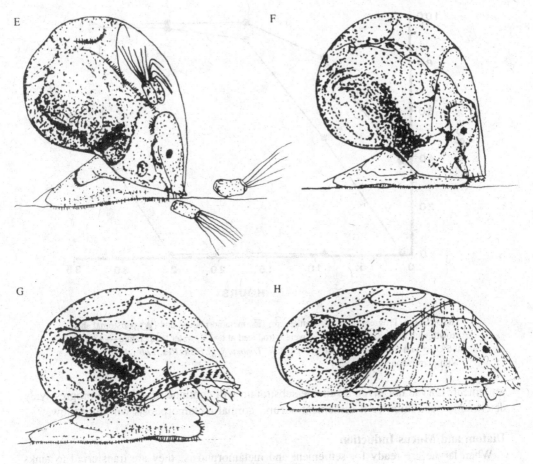

FIGURE 1 (E — H)

food organisms for newly settled juveniles is considered the most important problem in rearing juvenile abalone.[13]

Figure 2 shows the timing of the progression of settlement stages with 1230°C-hr larvae.[2] This information is useful in determining the EAT required before settlement begins naturally. Initiation of settlement is very quick but there is a short lag time after settlement before metamorphosis begins. Juvenile shell growth begins slowly, with only half of the individuals having any growth 33 hr after initiation of settlement. The mouth is completely developed in these individuals but they have not begun eating diatoms (stage 11).[2]

This information on the settlement behavior in abalone should be helpful in increasing the settlement of larvae in the hatchery, determining the biological and chemical factors connected with the selection of settlement substrata and metamorphosis, and understanding the ecological requirements for survival of veligers and juveniles in nature.

INDUCTION OF SETTLEMENT

A successful settlement induction is assured by using healthy larvae and allowing enough time for the larvae to reach the proper stage of development before they are transferred to settlement tanks and induced to settle on prepared substrata (plates or tank walls). It is important to use UV-sterilized, pollutant free, sea water during settlement induction.[14,15] Heavy metals, pollutants, chlorinated hydrocarbons, and pesticides can inhibit settlement. The precise settlement factor that larvae sense is still not clearly understood, however, suitable substratum and surface habitat conditions are known to be very important in inducing

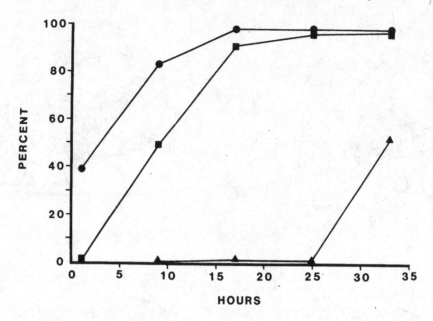

FIGURE 2 .   Progression of (●) settlement, (■) metamorphosis, and (▲) peristomal shell growth with time. Competent larvae were introduced at t = 0, water temperature — 18.2° ± 0.5°C. (From Seki, T. and Kan-no, H., *Bull. Tohoku Reg. Fish. Res. Lab.*, 42, 31, 1981. With permission.)

settlement. Larvae not given a suitable substratum will continue to swim in a progressively deteriorating condition or if settlement occurs, normal metamorphosis will not follow.[16]

Diatom and Mucus Induction

When larvae are ready for settlement and metamorphosis, they are transferred to tanks containing plates covered with a layer of benthic diatoms, microalgae, and bacteria. These organisms serve as a signal for settlement and are the food source of newly settled juveniles.[2,17] Additionally, plates can be grazed by juvenile abalone, causing the succession of a secondary diatom community and covering the plates with a layer of mucus.[16] Complete settlement and metamorphosis of competent larvae is induced by a substratum covered with encrusting diatoms and mucus secretions from grazing juvenile abalone.[5] Some laboratories do not use plates for settlement, but induce settlement directly onto the walls of the settlement tank which have previously been prepared with a layer of benthic diatoms.[6] The two methods are essentially the same, with settlement being induced by diatoms. The only differences between the two methods concern the number of larvae settled, whether the settlement tanks are indoors (plates), or outside or in a "green house" (tank walls), and future handling of the growing juveniles. A more complete description of the preparation of the settlement substratum is given in Chapter 11.

Settlement tanks used at different hatcheries vary in size and shape, but the most common are rectangular fiberglass tanks that can be stacked. The quantity of larvae introduced into a tank depends on the water quality, mortality at settlement, and food source for the newly settled juveniles. Some hatcheries use only naturally occurring diatoms while others supplement the tanks by adding a slurry of diatoms from an algal culture.[6,16] *Navicula* spp., *Cocconeis* spp., *Amphora* spp., *Nitzschia* spp., and a unicellular green algae, *Tetraselmis* spp. are typically used.[6,13,15]

H. discus hannai larvae are highly selective in choosing a suitable settlement substratum. In 1971, researchers at the Oyster Research Institute discovered *H. discus hannai* larvae use the mucus secreted by the foot of juvenile and adult abalone to select the settlement substratum before undergoing metamorphosis.[5] The Oyster Research Institute used this requirement for mucus to develop a technique to induce larvae to settle quickly and at an appropriate density on wavy plate collectors.[18] In a laboratory experiment, *H. discus hannai* larvae only settled on mucus trails from adult *H. discus hannai*, *H. discus*, *H. gigantea*, and *H. sieboldii*, while no settlement occurred on mucus trails from three intertidal gastropods (*Haloa japonica*, *Monodonta labio*, and *Cantharidus jessoensis*) or blank control plates.[5]

Larvae are able to distinguish between the different types of mucus secreted by the foot. Mucus secreted by the foot of gastropods is believed to be composed of protein and polysaccharides, and is probably insoluble in sea water.[5,19-25] The secreted mucus appears to be different depending on the activity of the juvenile. Four different types of mucus ("grazed", "crawled", "adhered", and "rubbed") were tested by Seki and Kan-no[5] for induction of settlement. Most larvae selectively settled on "grazed" mucus trails (70% after 3 hr and 100% after 22 hr). Little settlement occurred on the other types of mucus tested or blank plates, and most of the larvae remained free swimming. "Crawled" mucus caused only 16% settlement while the other treatments had less than 2% settlement after 22 hr. The settlement on the control plate was less than 1%. Metamorphosis was only observed on the "grazed" and "crawled" plates.[5]

Since settlement was not observed on blank plates placed next to mucus-covered plates, the larvae probably detect the mucus trails by contact only. Morse et al.[4] found that recognition of settlement cues on crustose red algae by red abalone (*H. rufescens*) larvae was contact dependent. Seki and Kan-no[5] theorized that the mucus trial not only serves as an inducer but also serves as a source of food (bacteria) for the larvae and juveniles during metamorphosis.[5]

Although *H. discus hannai* larvae are highly selective and can distinguish between different types of mucus, they settle on the grazing mucus trails of four different abalone species. *Haliotis* spp., therefore, are not only capable of settling in an environment inhabited by individuals of the same species but also individuals of the same genus.[5]

The selective settlement of abalone on mucus trails is similar to the gregarious settlement of barnacles, which is enhanced by the presence of individuals of the same species.[2,5,26] Crisp[7] defines "gregarious settlement" as the settlement of larvae or juvenile individuals preferentially in places where parents of the same or of a very close species grow. During the searching behavior prior to settlement, larvae are believed to detect mucus trails with their cephalic tentacles.[5]

GABA Induction

During field surveys, juvenile abalone (1 to 20 mm) of several species have been found to occur at densities several orders of magnitude higher on rocks with crustose red algae than similar nearby habitats without these algae. In the laboratory, juvenile red abalone show preferential settlement on crustose red algae, including *Lithothamnium* sp., *Lithophyllum* sp., *Hildenbrandia* sp., and closely related species. Morse et al.[4] found that *Lithothamnium* sp. and *Lithophyllum* sp. were able to induce settlement in 82% of tested larvae, while benthic diatoms (*Cyclotella* spp., *Nitzschia* spp., *Tetraselmis* spp., *Cocconeis* spp.) induced settlement in none of the larvae. This is unusual because these same diatom species are used in many laboratories to induce settlement in abalone larvae, including red abalone.[6,16] Metamorphosis and growth of newly settled abalone occurs rapidly on crustose red algae, but use of these algae to induce settlement in the laboratory is not recommended because they usually contain large quantities of predators of newly settled larvae.[15]

Crustose red algae occur over large areas of hard subtidal substrata throughout the world. These areas characteristically have few epiphytic flora and fauna, and are commonly called "barren grounds", "open space", "coralline flats", or "isoyake".[27] These habitats have been found to be very poor for adult abalone growth and survival, and are usually "reforested'' with *Laminaria* sp. to improve the habitat for abalone.[28,29] This suggests even though crustose red algae will induce settlement, it is not good for adult growth.

It is unclear how crustose red algae induce settlement of larval abalone. Intact crustose red algae does not release quantities of diffusible inducing factors. There is no chemotaxis by competent abalone larvae to the algae, no heterologous settlement, and no induction of settlement by water in which intact algae has been kept or separated by a semipermeable dialysis membrane. Direct contact with the inducing algal substratum is required.[4] Since the larval mouth does not develop until the end of metamorphosis, exploratory ingestion of the algae cannot be involved in inducing settlement.[2]

Extracts from *Lithothamnium* sp. induce rapid settlement on even clean glass. Ciliary swimming immediately stops, the larva attaches to the surface, and gliding behavior (side-to-side movement on the surface) begins, which is the same behavior caused by intact algae. Boiled and protease digested extracts also induce settlement, indicating that the inducer is not a protein.[4] Algal extracts cause an initial temporary paralysis to the larvae, but swimming and crawling begin after several hours.

Morse et al.[4] believed the abalone larvae, when they settled on crustose red algae, recognized γ-aminobutyric acid (GABA) that was covalently linked to a protein or some other macromolecular substance, or recognized phycoerythrobilin which has a backbone similar to GABA at the center of the tetrapyrrole derivative molecule. GABA was found to stimulate rapid and synchronous (99% of the larvae) induction of settlement and metamorphosis, which appeared to be normal. GABA is an amino acid, and a neurotransmitter in the human brain and other tissues of higher animals. Three amino acids similar to GABA also induce settlement in 58 to 89% of the larvae.[4]

Morse et al.[30] after further experiments, separated the settlement inducing factor(s) in algal homogenates from the pigment (phycobiliproteins). The active factor(s) has a low molecular weight, is colorless (no absorption in visible range, $A_{562} = 0$), contains no protein and is separated from the phycobiliproteins. The factor(s) is capable of inducing settlement in 80 to 90% of red abalone larvae.[27] The inducing factors in *Lithothamnium* have molecular weights of 690 and 890. Inducing factors were also found in non-natural inducing algae, *Spirulina* (mol wt = 680, 900, and 1250) and *Porphyra* (mol wt = 640 and 830 to 1230). This indicates that the true inducing factor(s) cannot be GABA (mol wt = 101), a related amino acid, or a piece of a larger protein, because the molecular weight of the factor is too high and the factor is a nonprotein molecule.[30]

Although GABA is not the true factor inducing settlement, Morse et al.[4] developed a method using a solution of GABA to artificially induce settlement. GABA can be purchased from most chemical supply companies and is very stable when stored dry at 0 to 4°C. A fresh stock solution of 10^{-2} *M* GABA in chilled filtered sea water is prepared each day of settlement induction. A final concentration of 10^{-6} to 10^{-3} *M* GABA (10^{-6} *M* is optimal) in sea water will induce competent swimming larvae of *H. rufescens* to settle and begin metamorphosis.[14] Warm water abalone species (e.g., *H. fulgens* and *H. corrugata*) require a higher water temperature (17 to 22°C) than is used with red abalone (15°C).[15]

Increasing concentrations of GABA, increases the number of larvae settling and decreases the time to settlement. For example, 10^{-3} *M* GABA causes 100% of the larvae to settle within 7 min and 10^{-6} *M* GABA induces settlement in only 26% after 38 hr.[3] The minimun concentration causing detectable settlement is 10^{-7} *M* GABA. Rapid growth of new shell becomes visible 24 to 36 hr after adding 10^{-5} to 10^{-7} *M* GABA. However, prolonged exposure to higher concentrations is toxic.[4] At concentrations higher than 10^{-6} *M*, settlement and attachment is induced but metamorphosis is inhibited and the juveniles eventually die.[31]

A desensitization or habituation caused by premature exposure of the larvae to GABA or natural inducers has been observed during settlement induction.[31-33] Both the time of addition and the concentration of the inducer are critical when using this technique.[31] This problem can be avoided by raising the larvae with the same conditions (e.g., water temperature and larval density) each time and testing samples from the culture tank for response to GABA. Larvae raised at 15°C will be competent to settle and metamorphose after 6 to 9 days, however, at 18°C they will be competent after 4 days.[4,14]

The use of GABA to induce settlement has been proposed as a sensitive bioassay for evaluating water quality at culture sites. The bioassay is based on the interference of settlement by a wide range of pollutants.[4,14,15] This procedure is supposed to be sensitive, reliable, quantifiable, and relevant to the siting of a hatchery or culture facility. This bioassay tests the effects of all pollutants in combination on the sensitive larvae.[15]

Contraindications

Akashige et al.[8] tested the effects of several neurotransmitters (GABA, L-epinephrine, L-norepinephrine chloride, acetylcholine iodide, serotonin creatinin sulfate, and L-glutamic acid) on the induction of settlement in *H. discus hannai*. Their results were in direct disagreement with results reported by Morse et al.[4] for *H. rufescens*.

Each neurotransmitter was tested for induction of settlement over a wide range of concentrations. In addition, the effect of the surface habitat of the settlement substratum (diatom/mucus covered plates and clean plates) was tested for each concentration of neurotransmitter.

Responses of larvae to the neurotransmitters were quantified by measuring the percentage of individuals exhibiting four different behaviors. Larvae were classified as fallen (larva laying, crawling, or dead on the substratum), crawling (larva in upright position with foot attached to substratum), metamorphosed (larva has lost the velum), or dead.[8]

$$\text{Rate of falling} = (\# \text{ fallen/total} \# \text{ of larvae}) \times 100$$
$$\text{Rate of crawling} = (\# \text{ crawling}/\# \text{ fallen}) \times 100$$
$$\text{Rate of metamorphosis} = (\# \text{ metamorphosed}/\# \text{ fallen}) \times 100$$
$$\text{Rate of mortality} = (\# \text{ dead}/\# \text{ fallen}) \times 100$$

Experiments were run with both positive (diatom/mucus plates) and negative (blank plates) controls. The positive control showed 91% metamorphosis with only 2% mortality after 45 hr. The negative control showed almost all of the larvae continuing to swim with less than 1% fallen after 45 hr.[8]

The effect of GABA was first tested with clean plates so there would not be any inference from organisms on the substratum.[8]

10^{-7} *M* GABA — 25% were crawling after 74 hr and there were no metamorphosed or dead larvae observed.

10^{-6} *M* GABA — 100% of the larvae fell to the substratum within 5 hr. They initially began to crawl but after 74 hr, 85% were dead and none were crawling. None of the larvae metamorphosed.

10^{-5} *M* GABA — Rapid falling of the larvae, 100% within 1 hr. However, 100% of the larvae were dead by 74 hr. None of the larvae metamorphosed.

10^{-4} *M* GABA — The same effects as 10^{-5} *M* but the larvae began to die sooner.

Metamorphosis was not induced with concentrations of GABA from 10^{-4} to 10^{-7} *M*, and there was a tendency for the falling of the larvae, temporary crawling, and death to occur earlier with each higher concentration. The ciliary action of the larvae were significantly depressed with concentrations of 10^{-4} to 10^{-6} *M* GABA.[8]

After the poor results with using clean plates, Akashige et al.[8] tested the effect of GABA when the substratum was a conditioned plate, which is normally used for settlement is Japan.

10^{-7} *M* GABA — 100% of the larvae were crawling after 22 hr and 54% were fallen after 74 hr. There was 100% metamorphosis after 74 hr with no mortality; however, no larvae metamorphosed except on the conditioned plates. About the same effects were caused by 10^{-6} M as with 10^{-7} M, except the mortality was 5% after 36 hr.

10^{-5} **M GABA** — 100% of the larvae were crawling after 5 hr and 34% were fallen and 6% were dead after 74 hr. The rate of metamorphosis was the same as for 10^{-6} and 10^{-7} M. However, larvae which had settled on the bottom of the dish, rather than on the plates, did not show any metamorphosis until after 36 hr and made up only 1% of the total number of larvae metamorphosing after 74 hr.

10^{-4} **M GABA** — 19% of the larvae were fallen after 36 hr and the number of fallen larvae remained at this level uintil 74 hr. After 5 hr, 100% of the fallen larvae were crawling but the number slowly decreased after 15 hr until it was 0% at 74 hr. Sixty percent were metamorphosed 36 hr, but mortality began at this point and increased to 100% at 74 hr.

As the concentration of GABA increased from 10^{-7} to 10^{-4} M, the number of fallen larvae decreased but fallen larvae would begin crawling earlier. In all concentrations of GABA, metamorphosis was observed after 22 hr and there were no differences in metamorphosis on conditioned plates with or without GABA.[8]

The other neurotransmitters (glutamic acid, serotonin, epinephrine, norepinephrine, and acetylcholine) tested were ineffective in inducing settlement and metamorphosis, and caused death of the larvae. These neurotransmitters probably affect the action of the cilia in the larvae.[8]

GABA appears to be inhibitory and depresses the action of the cilia, which inhibits swimming, causes the larvae to fall to the substratum, and begin crawling. Glutamic acid may cause the larvae to fall by the same biochemical pathway as GABA, since it is a precursor of GABA. This behavior could be due to the fact that larvae have both swimming and crawling capability at this stage but the neurotransmitters inhibit swimming.[8] Akashige et al.[8] beleive that since the crawling behavior observed with GABA is not the same as the normal behavior which precedes settlement and metamorphosis, the biological substance(s) that induces settlement and metamorphosis in *H. discus hannai* is not GABA or related chemicals.

EVALUATION OF SETTLEMENT METHODS

Induction of settlement in abalone larvae has been demonstrated with diatoms, mucus trials from juvenile or adult abalone, and crustose red algae and GABA. From these data, two methods have been developed to induce settlement, (1) diatoms and mucus, and (2) GABA. However, there are contraindications in the literature about the effectiveness of these methods. Akashige et al.[8] report GABA and other neurotransmitters cause death and abnormal behavior in *H. discus hannai* larvae. Seki and Kan-no[5] showed that diatoms and mucus trails from juvenile abalone induced selective settlement in *H. discus hannai* larvae. However, Morse et al.[4,14] report that none of the diatom species that are the primary food eaten during juvenile rearing had any effect in promoting settlement, development or survival, and GABA causes normal settlement and metamorphosis. The results of these two studies appear to contradict each other.

It should be concluded from these data that both methods are capable of inducing settlement and the differing results are probably not due to species differences. If one assumes induction of settlement and metamorphosis is a critical phenomenon in the life of the abalone, then the mechanism for induction should be highly conserved. Although a careful examination of the techniques used in each study to test the induction of settlement by diatoms and GABA appear to be the same, there are probably subtle differences (i.e., reagents bought from different chemical companies, age of larvae tested, water quality, length of time larvae observed, etc.) which could affect the results. Diatoms/mucus and GABA should both be considered as methods to induce settlement. However, their usefulness in a culture situation is different.

The diatom/mucus method has many advantages which makes it the preferred method in most culture situations. Settlement is highly selective and specific for the diatom and mucus-covered settlement plates without any heterologous settlement on the tank walls or bottom. Also, selective settlement ensures the newly settled larvae are in a habitat beneficial for continued development. The diatom species (*Navicula, Nitzschia*) inducing the highest rate of settlement, are the species that are the food for the newly settled larvae.[16]

The diatom/mucus settlement method probably has ecological significance and best simulates the settlement conditions found in nature. Larvae are highly selective for the mucus trails of abalone species but show no response to nonabalone gastropod mucus trails.[5] In nature, this mechanism would allow the larvae to settle in habitats which are beneficial and suitable for growth. The larvae settle "normally" when they reach the correct developmental stage and cannot be induced prematurely.[2]

Although GABA induces settlement and metamorphosis, the disadvantages of this method limit its usefulness. GABA induces nonselective, nonspecific settlement causing larvae to settle on substrata inappropriate for life (e.g., clean glass).[3,4] In a culture situation, settlement would have to be induced onto the walls and bottom of a tank, or, if settlement is onto plates, the walls and bottom of the tank would have to be lined with plates to catch the larvae settling on these surfaces. The concentration of GABA and the timing of induction must also be precise. GABA is toxic at high levels and if added too early, will cause larvae to show habituation to the chemical. The rate of settlement with GABA is variable (from 26 to 99% in the laboratory).[3,4]

ACKNOWLEDGMENTS

I would like to thank Michael Brody for the drawings of settlement behavior.

REFERENCES

1. **Shepherd, S. A. and Turner, J. A.**, Studies on southern Australian abalone, (genus *Haliotis*). VI. Habitat preference, abundance and predators of juveniles, *J. Exp. Mar. Biol. Ecol.*, 93, 285, 1985.
2. **Seki, T. and Kan-no, H.**, Observations of the settlement and metamorphosis of the veliger of the Japanese abalone, *Haliotis discus hannai* Ino, Haliotidae, Gastropoda, *Bull. Tohoku Reg. Fish. Res. Lab.*, 42, 31, 1981.
3. **Morse, D. E., Tegner, M., Duncan, H., Hooker, N., Trevelyan, G., Cameron, A.**, Induction of settling and metamorphosis of planktonic molluscan *(Haliotis)* larvae. III. Signaling by metabolites of intact algae is dependent on contact, in *Chemical Signaling in Vertebrate and Aquatic Animals*, Muller-Schwarze, D. and Silverstein, R. M., Eds., Plenum Press, New York, 1980, 67.
4. **Morse, D. E., Hooker, N., Duncan, H., Jensen, L.**, γ-Aminobutyric acid, a neurotransmitter, induces planktonic abalone larvae to settle and begin metamorphosis, *Science*, 204, 407, 1979.
5. **Seki, T. and Kan-no, H.**, Induced settlement of the Japanese abalone, *Haliotis discus hannai*, veliger by the mucous trails of the juvenile and adult abalones, *Bull. Tohoku Reg. Fish. Res. Lab.*, 43, 29, 1981.
6. **Ebert, E. E. and Houk, J. L.**, Elements and innovations in the cultivation of Red abalone *Haliotis rufescens*, *Aquaculture*, 39, 375, 1984.
7. **Crisp, D. J.**, Factors influencing the settlement of marine invertebrate larvae, in *Chemoreception in Marine Organisms*, Grant, P. T. and Mackie, A. M., Eds., Academic Press, New York, 1974, 177.
8. **Akashige, S., Seki, T., Kan-no, H., and Nomura, T.**, Effects of γ-aminobutyric acid and certain neurotransmitters of the settlement and the metamorphosis of the larvae of *Haliotis discus hannai* Ino (Gastropoda), *Bull. Tohoku Reg. Fish. Res. Lab.*, 43, 37, 1981.
9. **Crofts, D. R.**, The development of *Haliotis tuberculata*, with special reference to organogenesis during torsion, *R. Soc. London Philos. Trans. Ser. B*, 228 (552), 219, 1937.
10. **Crisp, D. J.**, Chemical factors inducing settlement in *Crassostrea virginica* Gmelin, *J. Anim. Ecol.*, 36, 329, 1967.

11. **Wilson, D. P.,** The settlement behavior of the larvae of *Sabellaria alveolata, J. Mar. Biol. Assoc. U. K.,* 48, 387, 1968.

12. **Bayne, B. L.,** The gregarious behavior of the larvae of *Ostrea edulis* L. at settlement, *J. Mar. Biol. Assoc. U. K.,* 49, 327, 1969.

13. **Ino, T.,** *Fisheries in Japan, Abalone and Oyster,* Japan Marine Products Photo Materials Assoc., Tokyo, 1980, 165.

14. **Morse, D. E., Hooker, N., Jensen, L., and Duncan, H.,** Induction of larval abalone settling and metamorphosis by γ-aminobutyric acid and its congeners from crustose red algae: II: Applications to cultivation, seed-production and bioassays; principal causes of mortality and interference, *Proc. World. Maricult. Soc.,* 10, 81, 1979.

15. **Hooker, N. and Morse, D. E.,** Abalone: the emerging development of commercial cultivation in the United States, in *Crustacean and Mollusk Aquaculture in the United States,* Huner, J. V. and Brown, E. E., Eds., AVI Publishing Co., 1985, 365.

16. **Kumabe, L.,** Methods used at the Oyster Research Institute, unpublished report, 1981.

17. **Uki, N. and Kikuchi, S.,** Food value of six benthic microalgae on growth of juvenile abalone, *Haliotis discus hannai, Bull. Tohoku Reg. Fish. Res. Lab.,* 40, 47, 1979.

18. **Seki, T.,** An advanced biological engineering system for abalone seed production, in *International Symposium on Coastal Pacific Marine Life,* Office of Sea Grant, Bellingham, Washington, 1980, 45.

19. **Lowe, E. F. and Turner, R. L.,** Aggregation and trail following of juvenile *Bursatella leachii pleii, The Veliger,* 19(2), 153, 1976.

20. **Trench, R.K.,** Further studies on the mucopolysaccharides secreted by the pedal gland of the marine slug, *Tridachia crispata* (Opistobranchia, Sacoglossa), *Bull. Mar. Sci.,* 23(2), 299, 1973.

21. **Trench, R. K., Trench, M. E., and Muscatine, L.,** Symbiotic chloroplasts: their photosynthetic products and contribution to mucus synthesis in two marine slugs, *Biol. Bull.,* 142, 335, 1972.

22. **Cook, S. B.,** A study of homing behaviour in the limpet *Siphonaria alternata, Biol. Bull.,* 141, 449, 1971.

23. **Crisp, M.,** Studies on the behaviour of *Nassarius obsoletus* (Say) (Mollusca, Gastropoda), *Biol. Bull.,* 136, 355, 1969.

24. **Peters, R. S.,** The function of the cephalic tentacles in Littorina, *The Veliger,* 7, 143, 1964.

25. **Wilson, R. A.,** An investigation into the mucus produced by *Lymnaea truncatula,* the snail host of *Fasciola hepatica, Comp. Biochem. Physiol.,* 24, 629, 1968.

26. **Knight-Jones, E. W. and Stevenson, J. P.,** Gregariousness during settlement in the barnacle *Elminius modestus* Darwin, *J. Mar. Biol. Assoc. U. K.,* 29, 281, 1950.

27. **Morse, A. N. C. and Morse, D. E.,** Recruitment and metamorphosis of *Haliotis* larvae induced by molecules uniquely available at the surfaces of crustose red algae, *J. Exp. Mar. Biol. Ecol.,* 75, 191, 1984.

28. **Kito, H., Kikuchi, S., and Uki, N.,** Seaweed as nutrition for seabed marine life technology for artificial marine forests, in *International Symposium on Coastal Pacific Marine Life,* Office of Sea Grant, Bellingham, Washington, 1980, 55.

29. **Uki, N.,** Abalone culture in Japan, in *Proceedings of the Ninth and Tenth U.S.-Japan Meetings on Aquaculture (NOAA Technical Report NMFS 16),* Sindermann, C. J., Ed., U. S. Dept. Commerce, Seattle, 1984, 83.

30. **Morse, A. N. C., Froyd, C. A., and Morse, D. E.,** Molecules from cyanobacteria and red algae that induce larval settlement and metamorphosis in the mollusc *Haliotis rufescens, Mar. Biol.,* 81, 293, 1984.

31. **Morse, D. E.,** Biochemical and genetic engineering for improved production of abalones and other valuable molluscs, *Aquaculture,* 39, 263, 1984.

32. **Morse, D. E., Hooker, N., and Duncan, H.,** GABA induces metamorphosis in *Haliotis,* V: stereochemical specificity, *Brain Res. Bull.,* 5 (Suppl. 2), 381, 1980.

33. **Morse, D. E.,** Biochemical control of larval recruitment and marine fouling, in *Marine Biodeterioration,* Costlow, J. D. and Tipper, R. C., Eds., U.S. Naval Institute, Annapolis, 1984, 134.

BIOTIC AND ABIOTIC FACTORS AFFECTING THE CULTURE OF ABALONE

Kirk O. Hahn

INTRODUCTION

Biological systems consist of functions on several different organizational levels: ecosystem mechanisms, population equilibriums, biology of individuals, physiology of organs, cell metabolism, cell organelle functions, macromolecular reactions, and synthesis and catabolism of small molecules. The biological reactions and interactions at each level are very strongly linked and automatically affect functions on all organizational levels. Natural physiological cycles are modified as environmental factors change. For example, adaptation to salinity variation is dependent on water temperature and levels of free amino acids during all stages of life from fertilized egg to adult, and nitrogen excretion is dependent upon water salinity and food composition. Changes in physiological functions also occur in stable environments due to internal circadian, lunar, annual, age, or growth cycles.[1]

This chapter will examine some factors that affect the survival and growth of abalone. Successful aquaculture requires an understanding of the physiology of the cultured animal, and the effects of biotic and abiotic factors on the animal's physiology. This knowledge will facilitate rearing the animal in the most favorable conditions for optimal growth and reproduction, and help researchers design improvements to traditional rearing methods.[1]

BIOTIC FACTORS

Oxygen Consumption

Oxygen consumption by intensively cultured abalone greatly affects oxygen concentration in the rearing water, which, in combination with reduced water exchange or recirculation, is potentially lethal. Very little is known about the oxygen consumption of abalone in general, or how it is affected by water temperature, body size, or time of day. However, these parameters were studied in *H. discus hannai*, and this information can serve as a starting point for studying other abalone species.[2-4]

The first study on oxygen consumption by abalone was conducted by Tamura[2,3] in 1939. He found oxygen consumption of *H. discus hannai* differed according to water temperature and time of day, 16.8 to 46.6 $m\ell/kg/hr$ at 13 to 14°C and 52.2 to 85.3 $m\ell/kg/hr$ at 22 to 23°C.[2,3]

Uki and Kikuchi[4] measured the oxygen consumption of *H. discus hannai* at five body weights ranging from 1.5 to 151 g and six temperatures ranging from 8.3 to 28.0°C, which resulted in separate determinations of oxygen consumption for 30 combinations of body weight and water temperature (Table 1). In addition, the effects of eating, movement, and circadian rhythms were determined by measuring oxygen consumption continuously for 24 hr at each body weight and temperature combination (Table 2).[4]

The oxygen consumption rate was calculated using the formula: oxygen consumption rate ($m\ell\ O_2$/individual/hr) = [oxygen concentration at entrance ($m\ell\ O_2/\ell$) − oxygen concentration at exit ($m\ell\ O_2/\ell$)] × [amount of water flow (ℓ/hr)/number of individuals]. The calculated relationship of oxygen consumption with body weight and water temperature is given by the formula: $R = M*w^b*A^T$, where R is the oxygen consumption ($m\ell\ O_2$/individual/hr), W is the whole body wet weight (g), T is the ambient water temperature (°C), and M, b, and A are constants. These constants for *H. discus hannai* are M = 0.0210, b = 0.8025, and A = 1.0963. This formula is only for temperatures up to 20°C. Table 3 gives the calculated

Table 1
THE OXYGEN CONSUMPTION RATE (Mℓ O₂/ INDIVIDUAL/HR) OF H. DISCUS HANNAI

Mean body weight (g)	Water temperature (°C)					
	8.3	12.0	16.1	20.0	24.0	28.0
151.0	2.65	3.56	4.95	7.70	7.64	7.73
98.0	1.83	2.48	3.49	5.42	5.45	5.85
56.0	1.08	1.69	2.22	3.37	3.55	3.78
4.8	0.164	0.228	0.290	0.504	0.427	0.446
1.5	0.043	0.061	0.082	0.184	0.152	0.149

Note: Figures show the average during the 24-hr measurement period.

(From Uki, N. and Kikuchi, S., *Bull. Tohoku Reg. Fish. Res. Lab.*, 35, 73, 1975. With permission.)

Table 2
OXYGEN CONSUMPTION RATE (mℓ O₂/individual/hr) of H. discus hannai during a 24-hr period

A. 8.3°C

Mean body weight (g)	Time											
	A. M.			P.M.						A. M.		
	6:50	8:45	10:45	12:00	2:00	4:00	6:00	8:30	10:30	12:30	2:50	4:50
151.6	1.88	2.46	2.52	2.82	3.07	3.05	3.19	2.26	3.12	2.31	2.50	2.70
98.0	1.71	1.56	1.85	1.86	1.79	1.66	2.27	2.02	1.82	1.62	1.95	1.96
56.0	1.03	1.01	1.12	1.09	1.04	1.17	1.11	1.05	1.14	—	1.08	1.06
4.0	0.169	0.164	0.152	0.154	0.154	0.152	0.194	0.200	0.174	0.150	0.151	0.157
1.5	0.044	0.041	0.044	0.038	0.044	0.044	0.054	0.049	0.039	0.042	0.036	0.037

B. 12.0°C

Mean body weight (g)	Time											
	A. M.			P. M.						A. M.		
	6:50	8:50	10:45	12:50	2:50	4:30	6:30	8:30	10:30	12:30	2:40	5:45
156.6	3.26	3.24	3.48	3.42	3.60	3.67	3.92	3.28	3.72	3.82	4.05	3.13
98.0	2.40	2.36	2.53	2.12	2.23	2.44	2.72	2.68	2.64	2.43	2.76	2.59
56.0	1.89	1.63	1.56	1.51	1.64	1.62	2.03	1.75	1.83	1.63	1.68	1.46
4.8	0.194	0.200	0.197	0.184	0.208	0.211	0.260	0.278	0.268	0.257	0.249	0.216
1.5	0.061	0.067	0.056	0.059	0.057	0.065	0.067	0.067	0.064	0.060	0.058	0.052

values of oxygen consumption from this equation for body weights from 5 to 150 g and water temperatures from 8 to 20°C. There is a linear relationship between the $\log_{10}$ (oxygen consumption), and body weight and water temperature under 20°C (Figure 1).[4] The relationship of oxygen consumption to water temperature changes between 20 and 24°C, corresponding to the highest water temperature experienced by this abalone species (Figure 2).[4,5] Q_{10} = 2.5078, and the value is independent of body weight and temperature between 8 to 20°C.[4]

Table 2 (continued)
OXYGEN CONSUMPTION RATE (mℓ O$_2$/individual/hr) of *H. discus hannai* during a 24-hr period

C. 16.1°C

Mean body weight (g)	A.M.			P.M.						A.M.	
	6:00	8:40	10:50	12:40	2:50	4:40	6:40	8:30	10:40	12:40	4:00
151.6	4.66	4.42	4.24	4.85	4.60	5.10	5.74	5.50	5.66	5.26	4.51
98.0	3.47	3.33	3.28	3.35	3.49	3.43	3.82	3.54	3.71	3.51	3.44
56.0	2.07	2.07	2.21	2.20	2.12	2.37	2.28	2.43	2.37	2.12	2.05
4.8	0.281	0.270	0.266	0.291	0.261	0.283	0.364	0.296	0.316	0.267	0.267
1.5	0.078	0.074	0.072	0.075	0.072	0.082	0.132	0.083	0.083	0.074	0.073

D. 20.0°C

Mean body weight (g)	A.M.			P.M.						A.M.		
	6:40	8:40	10:40	12:40	2:30	4:40	6:40	8:30	10:30	12:30	2:40	4.40
156.6	7.62	7.26	7.65	6.84	7.08	7.77	7.62	8.30	9.90	8.29	6.71	7.29
98.0	5.79	5.54	5.09	4.63	4.34	5.24	5.93	6.09	5.81	5.86	5.20	5.08
56.0	3.31	5.31	3.25	2.96	2.92	3.23	3.95	3.74	5.33	3.49	3.49	3.31
4.8	0.437	0.424	0.449	0.443	—	0.463	0.584	0.597	0.546	0.536	0.550	0.468
1.5	0.173	0.166	0.167	0.152	0.154	0.173	0.210	0.211	0.207	0.193	0.197	0.183

E. 24.0°C

Mean body weight (g)	A.M.			P.M.						A.M.		
	6:40	8:30	10:30	12:30	2:50	4:50	6:50	8:50	10:50	12:50	2:50	4:50
156.6	7.97	7.00	7.10	6.89	7.81	6.37	9.01	8.74	7.92	7.05	8.69	8.08
98.0	5.61	5.74	5.02	4.93	5.46	4.80	6.27	5.67	5.49	6.52	4.98	5.50
56.0	3.23	3.45	3.89	3.27	3.60	3.45	3.70	3.39	3.93	3.62	3.42	3.40
4.8	0.372	0.378	0.417	0.398	0.435	0.342	0.443	0.555	0.492	0.500	0.408	0.415
1.5	0.135	0.151	0.160	0.139	0.150	0.130	0.164	0.138	0.169	0.171	0.174	0.158

F. 28.0°C

Mean body weight (g)	A.M.			P.M.						A.M.		
	6:30	9:30	10:30	12:30	3:00	4:30	6:30	8:30	10:30	12:30	2:30	4:30
156.6	8.49	8.23	8.73	7.20	7.08	7.40	8.68	7.26	7.33	7.84	7.58	7.78
98.0	5.88	5.69	6.19	5.61	6.21	5.81	6.44	5.68	5.89	5.39	5.78	5.65
56.0	3.87	3.79	3.94	4.13	3.73	3.70	3.91	3.89	3.68	3.57	3.76	3.55
4.8	0.393	0.434	0.464	0.485	0.420	0.416	0.497	0.470	0.482	0.435	0.437	0.417
1.5	0.138	0.159	0.141	0.149	0.137	0.276	0.167	0.151	0.158	0.146	0.141	0.146

(From Uki, N. and Kikuchi, S., *Bull. Tohoku Reg. Fish. Res. Lab.*, 35, 73, 1975. With permission)

Table 3
THE OXYGEN CONSUMPTION RATE (R: mℓ O_2/kg WET WEIGHT/hr) OF *H. DISCUS HANNAI*, COMPUTED BY THE EXPERIMENTAL EQUATION, R = 0.0210 $W^{0.8025}$ * 1.0963^T * 1000/W, WHERE W IS WET WEIGHT OF WHOLE BODY IN G AND T IS WATER TEMPERATURE IN °C

Whole body wet weight (g)	Water temperature (°C)												
	8	9	10	11	12	13	14	15	16	17	18	19	20
5	31.9	35.0	38.3	42.0	46.1	50.5	55.4	60.7	66.5	72.9	80.0	87.7	96.1
10	27.8	30.5	33.4	36.6	40.2	44.0	48.3	52.9	58.0	63.6	69.7	76.5	83.8
20	24.2	26.6	29.1	32.0	35.0	38.4	42.1	46.2	50.6	55.5	60.8	66.7	73.1
30	22.4	24.5	26.9	29.5	32.3	35.4	38.9	42.6	46.7	51.2	56.1	61.5	67.5
40	21.1	23.2	25.4	27.9	30.5	33.5	36.7	40.2	44.1	48.4	53.0	58.1	63.7
60	19.5	21.4	23.5	25.7	28.2	30.9	33.9	37.2	40.7	44.7	49.0	53.7	58.8
80	18.4	20.2	22.2	24.3	26.6	29.2	32.0	35.1	38.5	42.2	46.2	50.7	55.6
100	17.6	19.3	21.2	23.3	25.5	27.9	30.6	33.6	36.8	40.4	44.3	48.5	53.2
125	16.9	18.5	20.3	22.2	24.4	26.7	29.3	32.1	35.2	38.6	42.3	46.4	50.9
150	16.3	17.9	19.6	21.5	23.5	25.8	28.3	31.0	34.0	37.3	40.8	44.8	49.1

(From Uki, N. and Kikuchi, S., *Bull. Tohoku Reg. Fish. Res. Lab.*, 35, 73, 1975. With permission.)

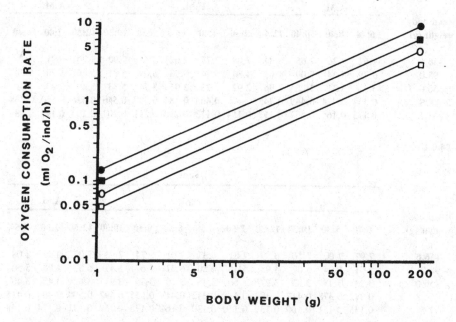

FIGURE 1.   Oxygen consumption rate vs. total wet weight of *H. discus hannai* at (□) 8.3°C, (○) 12°C, (■) 16.1°C, and (●) 20°C. (From Uki, N. and Kikuchi, S., *Bull. Tohoku Reg. Fish. Res. Lab.*, 35, 73, 1975. With permission.)

Oxygen consumption shows a circadian rhythm with the rate increasing from dusk to midnight and decreasing from midnight to midday. This indicates that the most active period and feeding period for *H. discus hannai* are mainly during the initial portion of the dark period and are not constant throughout the night.[4-6]

The general equation for the relationship between oxygen consumption and body weight in animals is R = k*W^b. The value of "b" is approximately 0.8 in most animals studied (poikilotherms, b = 0.75 and fish, b = 0.8).[7,8] The value of b found in *H. discus hannai* was very close to 0.8, so b can be safely assumed to be approximately 0.8 for all abalone species.[4]

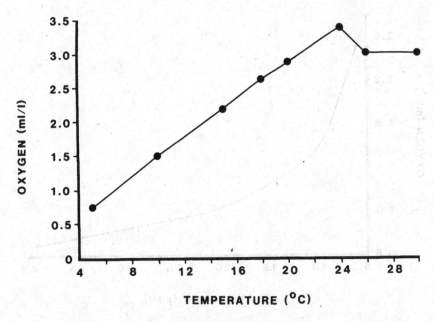

FIGURE 2. Relationship between water temperature and oxygen consumption in *H. discus hannai*. (From Sano, T. and Maniwa, R., *Bull. Tohoku Reg. Fish. Res. Lab.*, 21, 79, 1962. With permission.)

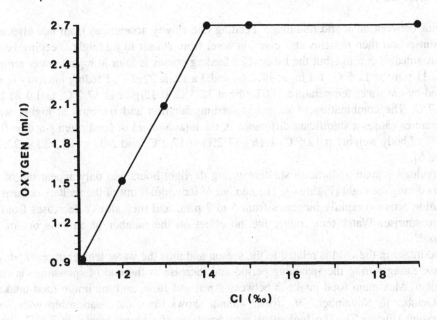

FIGURE 3. Relationship between chlorinity and oxygen consumption in *H. discus hannai*. (From Sano, T. and Maniwa, R., *Bull. Tohoku Reg. Fish. Res. Lab.*, 21, 79, 1962. With permission.)

Oxygen consumption rate is constant for chlorinity above 14‰, but decreases rapidly at chlorinities below 13‰ (Figure 3). Oxygen consumption also decreases rapidly with increasing nitrogen concentrations. The most rapid decline is at ammonia concentrations of 10μg/ℓ (Figure 4).[9]

Feeding Rate

The feeding rate of *H. discus hannai* shows a clear diurnal cycle with increased feeding

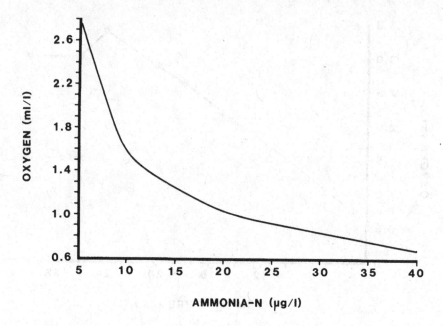

FIGURE 4.   Relationship between water ammonia concentration and oxygen consumption. (From Sano, T. and Maniwa, R., *Bull. Tohoku Reg. Fish. Res. Lab.*, 21, 79, 1962. With permission.)

behavior between dusk and midnight.[6] Feeding rate slowly accelerates from late afternoon until sunset and then remains at a constant level from sunset to midnight. Feeding begins at approximately 5 p.m., but the time when feeding ceases is later at higher water temperatures: 11 p.m. at 12.5°C, 1 a.m. at 17.2°C, and 3 a.m. at 22.7°C. Feeding intensity is also affected by the water temperature, 0.07 g/hr at 12.5°C, 0.15 g/hr at 17.2°C, and 0.21 g/hr at 22.7°C. The combination of increased feeding duration and intensity at higher water temperatures causes a significant difference in the total amount of food eaten per day: 0.43 g (2.3% of body weight) at 12.5°C, 1.16 g (7.2%) at 17.2°C, and 2.07 g (15.3%) at 22.7°C (Figure 5).[10]

Individuals remain underneath shelters during daylight hours and only appear out of the shelters during the night (Figure 6). The number of individuals out of the shelters (a measure of feeding activity) rapidly increases from 5 to 7 p.m. and then slowly decreases from 11 p.m. to sunrise. Water temperature has no effect on the number of animals out of the shelters.[10]

Food intake in the wild is related to the season and thus the water temperature. *H. discus* decrease eating during the spawning period and increase at the end of spawning, in early December. Maximum food intake is between April and June, and minimum food intake is from October to November.[5] *H. discus hannai* shows the same relationship with water temperature (Figure 7). The biological zero point for *H. discus hannai* is 7.6°C, which causes eating to almost cease below 7°C.[5,11] *H. discus hannai* eat 7.5% of their body weight in kelp per day at 10.9°C, but eat 17.6% of their body weight per day at 20.3°C. Data on growth rate and feeding rate indicates optimum water temperature for juvenile *H. discus hannai* is 15 to 20°C.[5,12] Sano and Maniwa[9] found the feeding rate of *H. discus hannai* was greatly affected by the ammonia level in the water and feeding was almost inhibited by ammonia concentrations of 70 μg/ℓ (Figure 8).

Genetics

Reseeding of juvenile abalone in Japan has been conducted for several years to maintain the fishery and repopulate depleted areas. Total production is over 10 million juveniles a

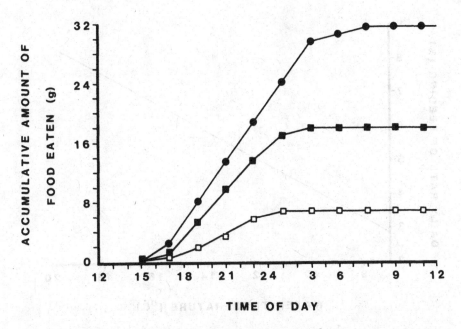

FIGURE 5. Accumulative amount of food eaten by *H. discus hannai* at (□) 12.5°C, (■) 17.2°C, and (●) 22.7°C. (From Uki, N., *Bull. Tohoku Reg. Fish. Res. Lab.*, 43, 53, 1981. With permission.)

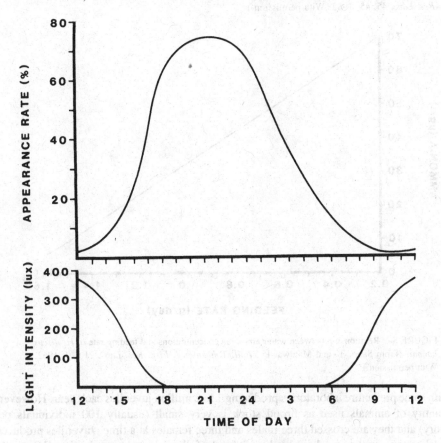

FIGURE 6. Appearance rate of *H. discus hannai* with photoperiod. (From Uki, N., *Bull. Tohoku Reg. Fish. Res. Lab.*, 43, 53, 1981. With permission.)

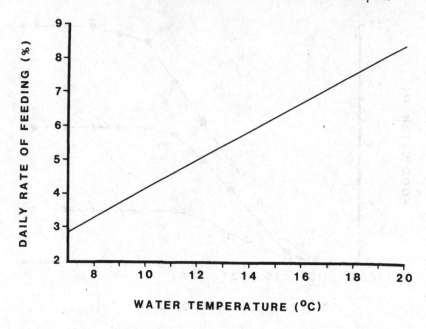

FIGURE 7.   Relationship between daily feeding rate of *H. discus hannai* and water temperature with satiation levels of food present. (From Uki, N. and Kikuchi, S., *Bull. Tohoku Reg. Fish. Res. Lab.*, 45, 45, 1982. With permission.)

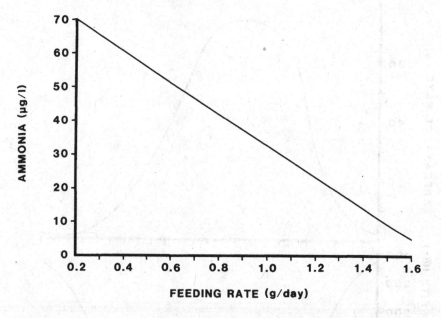

FIGURE 8.   Relationship between water ammonia concentrations and feeding rate of *H. discus hannai*. (From Sano, T. and Maniwa, R., *Bull. Tohoku Reg. Fish. Res. Lab.*, 21, 79, 1962. With permission.)

year with some prefecture laboratories producing 1 to 2 million juveniles each year. However, the quantity of animals used as brood stock is very small (usually 100 individuals per laboratory) and they are crossed three males and three females at a time. Juveniles produced at each laboratory are then used as brood stock for the following year's production.[13] Recently it was discovered that there are a large number of deficient (malnourished with reduced body weight) animals in the wild exhibiting a high level of homozygosity.[14]

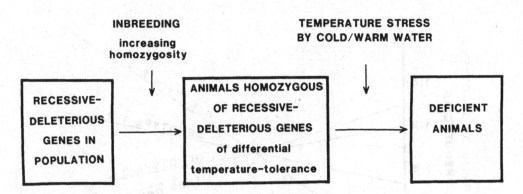

FIGURE 9. Theoretical mechanism that produces deficient animals in the wild abalone population. (From Fujino, K., Sasaki, K., and Okumura, S., *Bull. Jpn. Soc. Sci. Fish.*, 50(4), 597, 1984. With permission.)

Wild populations of *H. discus hannai* show an excessive homozygosity at multiple isozyme loci when compared to Hardy-Weinberg proportions.[14] Heterozygote deficiencies occur in many invertebrate populations, especially molluscs, and many explanations have been proposed to understand this phenomenon.[15,16] Hatchery breeding practices, however, indicate a portion of the observed excess of homozygosity may be due to genetic drift caused by low effective population size and resulting inbreeding depression. Additionally, the greater occurrence of homozygotes in deficient animals, called locally "Rho-gai" or "Yase-gai", than in normal animals suggests the deficient animals are caused by inbreeding depression resulting from a single pair of recessive-deleterious alleles. Inbreeding depression is one of the potential causes for malnutritious abalone.[14,17] Inbreeding depression in traits, such as hatching rate, growth, survival rate, and feed conversion efficiency in aquaculture populations result from a variety of breeding schemes.[17]

Genetic factors (recessive deleterious genes), breeding structure (inbreeding), and differential temperature-tolerance due to thermostability variations of enzymes are involved in the mechanisms suggested to explain the occurrence of deficient *H. discus hannai*.[17-19] Additionally, the thermostability variations of particular enzymes may contribute to the differential temperature tolerance of these animals.[17]

Figure 9 shows a suggested mechanism that could account for the presence of deficient animals in wild populations. Two steps are involved in producing deficient animals: (1) inbreeding increases homozygosity at all loci exposing recessive deleterious genes of different temperature tolerances; and (2) the habitat is periodically exposed to critical or subcritical water temperatures which stress the animals differentially depending upon the genetically determined thermostability of particular allozymes. Stress induced by water temperature will differentially affect the physiology of animals through various metabolic pathways and these differences could produce the observed deficiencies. Histological examination of the salivary gland of deficient abalones shows evidence for this mechanism. The thermostability of the enzymes could cause differential survival during critical stages in the life cycle of the animal (e.g., larval development).[17]

When wild-caught *H. discus hannai* of different size classes were analyzed by Fujino and Sasaki,[20] there was a difference in the genotypic proportions of different size classes (Figure 10). The change in the observed/expected ratio indicates unequal survival among the genotypes. These results are most striking at the Esterase M (Es*M) locus. Animals of Es*M genotype 2*3 have the lowest survival rate during the juvenile stage when compared to the whole population. However, this genotype has the highest survival overall during the life span of the animal after they reach sexual maturity. In contrast, animals of genotypes 2*2 have a higher survival rate than genotype 2*3 during the juvenile stage, but become less abundant with increasing shell length (or age). These observations strongly suggest animals

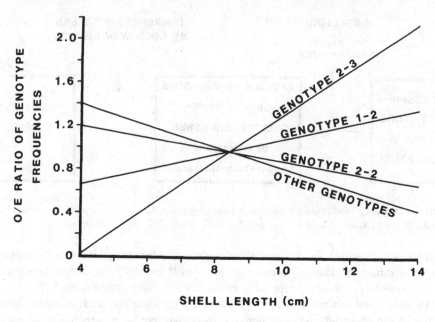

FIGURE 10.   Associations between relative survival rates and shell length of different genotypes of animals at ES*M locus in *H. discus hanna* populations from Sanriku, Iwate Prefecture, Japan. Relative survival rate is expressed as observed/expected (O/E) ratio. (From Fujino, K. and Sasaki, K., *Bull. Jpn. Soc. Sci. Fish.*, 50(1), 11, 1984.)

of genotype 2*3 are lowest in viability at the time of fertilization, but could promise the greatest survival rate up to the time of harvest. Selective breeding of abalone with genotype 2*3 for seed production could result in improving productivity per unit area or per unit effort.[20]

Several strategies have been proposed for increasing seed production by selective breeding of abalone (Figure 11).[21] The importance of removing recessive deleterious genes in animals cultured for aquaculture seed production was stressed by Fujino.[17] Abalone are inbred and selected for superior performance. Recessive and disadvantageous genes are expressed, which causes their phenotypes to be recognizable. Animals with superior qualities and performance traits (e.g., high growth rate) are selected for further crossing. Selection may have to be conducted repeatedly through several generations to obtain broodstock with the proper qualities. Strains or lines are established through repeated inbreeding for several generations. Crosses between two or more established strains will produce heterozygous animals with qualities for increasing production.[14]

Induced gynogenesis followed by diploidization could be another useful method for genetic improvement of abalone aquaculture. This method could rapidly establish inbred lines, determine sex, and eliminate recessive deleterious genes from cultured populations. The diploid number should be determined before chromosome manipulation for triploidization or gynogenesis. There is very little information on the chromosome number and morphology of most species of *Haliotis*. The diploid number of *H. discus hannai* and *H. discus* is 36 chromosomes with 20 metacentric and 16 submetacentric chromosomes, and no heteromorphic sex chromosomes.[22] However, the diploid number of *H. tuberculata* is 28 chromosomes with 16 metacentric and 12 submetacentric chromosomes.[23]

Gynogenesis is effectively induced in *H. discus hannai* by irradiating spermatozoa with (UV) light. The method used for irradiating sperm with UV light was modified from a method developed for salmon. A UV lamp is placed in the center of the ceiling of a wood box (30 cm × 66 cm × 58 cm) with a motorized rotating table, approximately 32 cm from the lamp, on the bottom. A petri dish containing 0.5 mℓ of sperm suspension, which forms

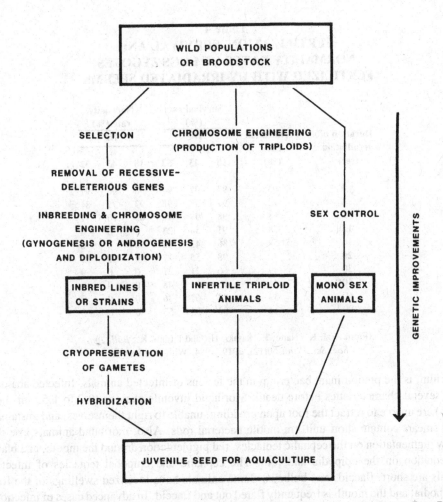

FIGURE 11. Proposed strategies for improving abalone seed production with genetic manipulations (From Fujino, K., World Conf. Aquacult., Venice, September 21 to 25, 1981.)

a layer about 0.1 mm deep, is placed on the table. This will produce an irradiation of 42 erg/mm². The haploid number of chromosomes (n = 18) appeared most abundantly in sperm irradiated for 30 sec. The low survival rate for zygotes from sperm irradiated for 12 to 20 sec. is probably due to the high frequency of dicentric and/or fragmental chromosomes (Table 4).[24]

The relationship between UV irradiation duration sperm, and fertilization rate, survival rate, and chromosome number is summarized as follows: sperm motility decreases gradually with increasing irradiation duration; survival rate decreases to a minimum of 4 to 17% at 12 to 20 of irradiation, but increases at longer lengths of irradiation; and highest survival rate of the zygotes corresponds to the modal peak in haploid number in the zygotes.[24]

The genetic material in sperm is almost completely inactivated with little effect on fertilization rate when UV irradiated at 42 erg/mm² for 30 to 60 sec. This induces gynogenesis and produces haploid individuals. These haploid individuals are deformed and nonviable, but diploidization techniques could produce viable animals of complete homozygosity.[24]

Diseases

Bacteria

Elston[25] isolated and identified a bacterium which was causing death of cultured juvenile abalone at Monterey Abalone Farms. *Vibrio alginolyticus*, a commonly occurring marine

Table 4
FERTILIZATION, SURVIVAL AND
NORMALITY RATES OF EGGS/ZYGOTES
FERTILIZED WITH UV-IRRADIATED SPERMS

Duration of irradiation (sec)	Fertilization rate (%)	Survival rate (%) Hr			Normality rate (%) Hr		
		18	43	52	18	43	52
0	93	97	97	97	96	96	97
4	72	97	95	95	93	75	84
8	58	98	98	96	63	10	7
12	58	91	54	23	1	0	0
16	53	86	43	18	0	0	0
20	43	98	55	17	0	0	0
30	49	66	21	21	0	0	0
40	41	90	61	48	0	0	0
60	39	78	62	36	0	0	0
90	14	82	53	53	0	0	0

(From Arai, K., Naito, F., Sasaki, H., and Fujino, K., *Bull. Jpn. Soc. Sci. Fish.*, 50(12), 2019, 1984. With permission.)

bacterium, is the predominant bacterium in the lesions of infected animals. Infected abalone show several characteristics before death. Moribund juvenile abalone (0.2 to 1.5 mm shell length) are unable to retract the foot upon prodding, unable to right themselves, and unstained blood smears contain short uniform motile bacterial rods. Also, moribund animals lose the yellow pigmentation on the cephalic tentacles, red pigmentation around the mouth, and black pigmentation on the epipodia and foot. The cephalic and epipodal tentacles of infected animals are short, flaccid, and bulbous. Some animals have localized swellings of the foot or epipodia, and the mouth is frequently flared out and flaccid. In advanced cases of infection, the general visceral mass shrinks and the foot retracts into the shell more than normal.[25]

The occurrence of *V. alginolyticus* infections is related to stress, i.e., high water temperatures or super-saturated oxygen. Due to the severe damage caused by neural and epithelial lesions and rapid infection, the disease is irreversible at an early stage of infection. The disease can be managed in culture systems by reducing the number of potentially pathogenic bacteria and physiochemical stresses on animals.[25]

Ebert and Houk[26] reported an occasional infection of cultures with a pathogenic bacterium, *Vibrio* sp., which causes minor to extensive larval mortality. The infection, when it occurs, usually appears on the fourth to fifth day of larval culture. A *Vibrio* sp. infection is characterized by clumps of larvae with yellow coloration on the bottom screen of the larval culture container. Healthy larvae have a green coloration. The *Vibrio* sp. infection is effectively treated with Neomycin sulfate (50 mg/ℓ).[26]

Parasites

Individuals reared under optimal conditions usually resist parasitic infection and withstand the effects of parasitism if they become infected. Parasitic infection can modify the physiology of the animal, and change the cell and tissue structure, hormonal equilibrium, water content, and mineral concentrations. A good knowledge of the parasite's biology helps in the treatment of an infection. Parasitism can be reduced by carefully controlling the rearing conditions.[1]

Pacific Trident Mariculture (Victoria, British Columbia, Canada) has had a serious disease problem for several years. This parasite is extremely pathogenic for young juvenile abalone and kills the majority of its host within 12 weeks after exposure. The abalone mortalities are caused by a new species of achlorophyllous eucaryotic protist belonging to the phylum Labyrinthomorpha and genus *Labyrinthuloides*. The parasite is readily transmitted to other abalone once its host dies. As abalone grow in size and age, the time interval increases from the initial infection to death, and the magnitude and prevalence of infection decreases. Partial resistance to infection is observed in 190-day-old (postsettlement) juveniles, and 340-day-old juveniles are completely protected from the lethal effects of the infection. This observed resistance could be from acquired immunity, since this culture population has had a history of exposure to the parasite. Disinfecting equipment and rearing tanks with chlorine (25 ppm for 40 min) prevents infection of this parasite.[27]

Only cycloheximide, out of 11 drugs tested, is capable of curing infected abalone. Cycloheximide at 1 μg/mℓ, for 5 consecutive days, is effective in curing the juveniles being raised in large grow-out tanks. However, the infection reoccurs between 14 and 19 days after treatment. Bower[27] could not determine if the parasites survived the treatment or if the tanks were reinfected from the water supply. Abalone do not show any gross signs of toxicity after 10 days of exposure to 100 μg/mℓ of cycloheximide, but 1000 μg/mℓ is lethal to all abalone after 4 to 6 days of continuous exposure. Cycloheximide also affects the viability of the diatoms in the culture tank. A concentration of 1μg/mℓ reduces the percent of living diatom cells from 77% before treatment to 13% after exposure.[27]

About 3% of *H. ruber* populations have various stages of infection with cercariae of a trematode, probably of the family Allocreadiidae. The gonads of infected animals are dull to bright orange in color with no other visible abnormalities. The infection usually causes sterilization, and determining the sex of infected individuals is impossible due to the destruction of the gametes by the parasite. The cercariae develop within sac-like sporocysts in the gonad, which even penetrate into the digestive gland tissue. The cercariae probably become sexually mature in the intestine of a host fish. *H.ruber* is not eaten by fish, therefore the parasite probably penetrates and encysts in some other marine invertebrate before being eaten by the host fish.[28]

Crofts[29] reported finding a female *H. tuberculata* with a severe trematode infection (sporocysts, rediae, and cercariae were present) in the visceral mass, mantle, and ctenidia. The outer surface of the animals was colored orange from the trematodes. The trematode could not be identified. Also, a male was found with cysts between the gonad and digestive gland. The cysts appeared to be Haplosporidian capsules.[29]

Wound Repair

Abalone lack a blood-clotting mechanism and bleed easily if cut.[30] Field observations indicate abalone die from deep cuts even though the wounds are distant from vital organs.[31] *H. rufescens* will die from cuts as small as a centimeter long.[32] A cut causes the loss of vital fluids and the animal loses the ability to hold onto the substratum, thus increasing its susceptibility to predation.[30,31] Although abalone cannot efficiently clot their blood, wound repair is possible but it takes a long time.[31]

ABIOTIC FACTORS

Oxygen Supersaturation

Although the lack of sufficient oxygen in the rearing water is the greatest worry at most culture facilities, the opposite — supersaturatation — is just as lethal. Brightly illuminated, intensive culture systems with inadequate water flow are susceptible to oxygen supersaturation from photosynthesis of kelp and diatoms in the tank. Oxygen levels can reach above 200% of saturation during periods of mechanical failure that stop water flow. Animals become stressed or behave abnormally when oxygen levels rise between 0.5 to 2.0 mg/ℓ above saturation. Prolonged exposure to these conditions causes formation of protruding subepithelial gaseous emboli. In extreme cases, emboli cause animals to float.[33]

Animals experimentally exposed to oxygen concentrations of 15.5 mg/ℓ (203% saturation) for 3 hr, lost normal oral red pigmentation and the radula became extended. After 12 hr, the animals lost the ability to firmly hold onto the substratum, were sluggish, and unable to right themselves when placed upside down. Cephalic and epipodal tentacles were not normally extended, and the mouth, foot, mantle, and epipodia were swollen. The epipodia became immobile and bloated. After 41 hr, most animals lost the black pigmentation on the side of the foot and red pigmentation around the mouth. Also the bottom of the foot became smooth and swollen. Although many animals were affected by high levels of oxygen, the effects varied between individuals. Several animals continued normal behavior and remained attached to the substratum without any obvious harm.[33]

There are numerous neural lesions in animals exposed to high concentrations of oxygen, which indicates that sensory and motor functions are quickly affected by supersaturated conditions. These lesions are correlated with observations of lethargy, loss of attachment to the substrata, and retraction of tentacles and epipodia. Emboli in the tissues may cause swelling and affect muscle function even without neural lesions.[33]

Water Temperature
Survival

Water temperature affects sperm activity. In general, sperm of Japanese abalone are active below 12 to 13°C and become less active above this temperature. However, there are variations between species, *H. sieboldii* sperm are most active at higher temperatures and less active at lower temperatures, *H. discus* sperm are less active at higher temperature and more active at lower temperatures, and *H. gigantea* sperm are intermediate between the other two species. The longest duration of sperm activity was 11.5 days in *H. discus*. High survival of *H. sieboldii* eggs occurs within the range 7.8 to 28°C, optimum survival is between 7.8 to 13.3°C.[34]

H. discus larvae are quickly killed by water temperatures at the extremes of its tolerance. Larvae die within 8 hr when raised at 32.4°C and 24 hr when raised at 5.1°C. The length of survival is a little longer at 28°C and 7.8°C. The greatest survival is found at temperatures below 13.3°C.[34] Development of *H. sorenseni* larvae is more rapid at higher temperatures but survival is reduced at above 20°C. Survival of newly settled *H. sorenseni* juveniles is also reduced at water temperatures above 20°C. However, when advanced juveniles raised at 15 to 20°C are transferred to higher temperatures, they are not adversely affected at temperatures as high as 25°C. Thermal tolerance limits are possibly more restrictive in larvae and recently settled individuals.[35]

The thermal tolerance of 1- to 2-month-old juveniles of *H. rufescens* is 10 to 19.5°C, *H. corrugata* is 10 to 23.5°C, and *H. fulgens* is 10 to 26°C. Survival is nearly 100% over a broad intermediate range of temperature but declines sharply within 2°C of the extremes.[36] The optimum temperature for *H. sorenseni* juveniles is 14 to 18°C.[35] Movement of *H. discus hannai* becomes sluggish below 7°C and the optimum water temperature is 15 to 20°C.[5,11,12]

The onset of lethal temperatures is very abrupt. The temperature difference between 100% and 0% survival is only 1°C in *H. cracherodii*. In general, no death occurs at or below 25°C, and 100% mortality occurs at or above 28°C. For example, animals that are acclimatized to 16°C, show an abrupt decline at 28°C in the time necessary before 50% of the animals die. The time quickly drops to 6 hr at 29°C. The survival rate of individuals which lose the ability to hold onto the substratum is very low (16%), even though the animals are returned to ambient sea water.[37]

Abalone show a consistent behavior to heat stress. The first sign is a loss of body turgor, followed by lifting of the anterior of the shell from the substratum until the shell is about

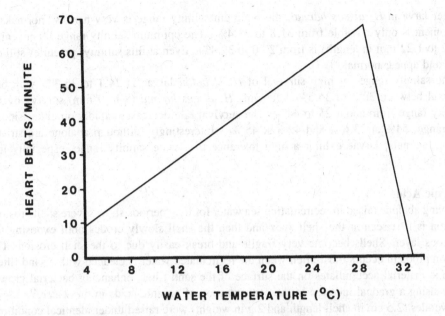

FIGURE 12. Average heart rate versus water temperature in *H. discus hannai*. (From Fujino,
K., Yamamori, K., and Okumura, S., *Bull. Jpn. Soc. Sci. Fish.*, 50(10), 1671, 1984.)

4 cm off the substratum, the epipodia spread out 1 to 2 cm beyond the shell edge, and the
tentacles no longer respond to stimuli. After these stages, animals usually can no longer
hold onto the substratum, although this can occur at any time. Males are observed spawning
at all times during the thermal stress period. Abalone experiencing thermal stress do not try
to escape the warm water.[37]

The LT_{50} (temperature where 50% of the animals die in the tested time period) for juvenile
(1 to 2 cm) *H. fulgens* is 31.5°C (48 hr).[38] The water temperature to which the abalone is
acclimatized has very little effect on the value of the lethal water temperature. The difference
in LT_{50} (96 hr) for *H. cracherodii* acclimatized at 11.5°C (26.1°C) and 16°C (27.4°C) is
1.3°C and *H. rufescens* acclimatized at 10°C (24°C) and 20°C (25°C) is only 1°C.[37]

Heart Rate

The heart rate of newly settled (0.39 mm) *H. sieboldii* is correlated to the water temperature. A normal pulse rate occurs from 8 to 27°C but there is an abrupt increase in heart
rate above 28 to 29°C. The heart rate increases to a mean rate of 131 at 30.5 to 30.8°C.
The heart rate is 20 to 30 beats per minute at 10 to 20°C and continues increasing up to
29°C but decreases at temperatures exceeding 30°C.[34]

The heart rate of juvenile *H. discus hannai* is similar to *H. sieboldii*. The heart rate is
32 beats per minute at 15°C and steadily increases up to 64 beats per minute at 29.1°C, but
quickly decreases to 9 beats per minute at 31.9°C until it is 0 beats per minute at 32.6°C.
The temperature of maximum heart rate varies between individuals. Figure 12 shows the
average relationship between heart rate and temperature.[39] The hearts of *H. sieboldii* and
H. gigantea stop beating at 3 to 4°C but the heart of *H. discus* does not cease even when
the temperature reaches 1.6°C. [5,34]

Salinity

The effect of salinity on the length of sperm life varies between species. *H. gigantea* and
H. sieboldii show a decrease in sperm life span at lower salinities but *H. discus* shows
increased sperm life span at lower salinities.[34] During development from fertilized egg to

veliger larva in *H. discus hannai,* the optimum salinity range is very narrow, normal development is only possible from 31.8 to 33.4‰. The optimum salinity range for juveniles (0.52 to 1.22 mm in length) is from 27.0 to 39.4‰. Even at this salinity, juveniles still eat food and appear normal.[40]

The salinity range for high survival of *H. sieboldii* larvae is 24.1 to 36.3‰ with best survival between 30.8 to 36.3‰[34] Juvenile *H. discus hannai* (8 to 9 mm) survive over a salinity range from about 25 to 44‰. The survival rate decreases rapidly on either side of this range, 54% at 23.6‰ and 42% at 45°‰.[5] Interestingly, although abalone are strictly marine animals, larvae exhibit a high tolerance to a wide salinity range, especially low salinities.[34]

Organic Acids

Young abalone raised in recirculating sea water for long periods show severe shell erosion. Erosion is first seen at the shell apex and then the shell slowly erodes until exposing the nacreous layer. Shells become very fragile and break easily due to the shell erosion. The erosion problem occurs most often when the recirculated water is cleaned with a sand filter. Organic material accumulates on the surface of the sand filter, enhancing bacterial growth and causing a gradual increase in the concentration of organic acids in the water.[41]

Juveniles (2.5 cm in shell length and 2 g in weight) were raised under identical conditions (200-ℓ tanks with 1ℓ/min recirculating water and coarse gravel on the tank bottom) for 314 days, except one tank had crushed oyster shells in addition to the gravel on the tank bottom. All abalone survived and the growth rates were the same in both tanks (25 μm/day). Temperature, salinity, pH, dissolved oxygen, and dissolved organic material were measured during the experiment, and there were only minor differences between tanks. The pH in the sand-filtered tank ranged from 8.0 to 8.3 and the pH of the oyster shell/sand filtered tank ranged from 8.1 to 8.4, and the pH was the same or only 0.1 lower in the sand-filtered tank at each measurement period. The organic acids produced by the animals were neutralized.[41]

Although there were not any differences in the water parameters, there was a great difference in the shells of the two groups. The sand-filtered tank contained animals with fragile, faded shells eroded down to the nacreous layer in places, especially in the oldest regions. The shells of the animals in the oyster shell/sand-filtered tank were strong and bright green in color. The organic acids dissolved the calcium carbonate in the shells of the abalone raised in the sand-filtered tank, while the crushed oyster shells neutralized the organic acids in the other tank. The build up of organic acids in the water was efficiently reduced by placing crushed oysters shells (or any source of calcium carbonate) in the filter system.[41]

Heavy Metals

The water quality at a culture site is very important to the success of the project. The presence of heavy metals is one of the most important factors to measure before selecting a site. Heavy metals have severe and disastrous effects on the health of the animals, are accumulated in the tissues of the animal, and render the meat unhealthy or unsalable for human consumption. Heavy metals can be discharged into the ocean from polluted rivers, sewage outfalls, or local industries. The pipes and materials used to construct the culture facility can also introduce heavy metals in the sea water. If there are any heavy metals present in the sea water at the culture site, these concetration levels should be tested on all life stages (larval, juvenile, and adult) for acute and chronic effects in the particular species being cultured.[1]

The lethal effects of copper toxicity on abalone was clearly demonstrated by the initial test of the cooling system at the Diablo Canyon nuclear power plant in California. Following the test, approximately 1500 dead abalone (*H. rufescens* and *H. cracherodii*) were found in the discharge area of the plant. Sea water had remained in contact with copper-nickel tubing in the condensing system during a shutdown period and copper dissociated from the tubing. The initial pulse of water from the cooling system contained 1800 μg of copper per liter. The copper concentration quickly decreased in the effluent but was still 20 μg of copper per liter after 30 days even though incoming water contained only 1 μg of copper per liter.[42]

H. cracherodii accumulates copper at a faster rate in the ctenidia than *H. rufescens*, and both species accumulate copper faster in the ctenidia than in the digestive gland. The ctenidium tissue shows a slight loss of frontal and lateral cilia but this loss probably does not seriously affect respiration. The copper LD_{50} (concentration where 50% of the animals die in the tested time period) is 65 ppb (96 hr) for adult *H. rufescens*, 114 ppb (48 hr) for larval *H. rufescens*, and 50 ppb (96 hr) for adult *H. cracherodii*. Marks[43] reported a LD_{50} of 100 ppb (72 hr) for *H. fulgens*. Mortality from copper occurs as low as 80 ppb in larvae and 50 ppb in adults of *H. rufescens* and *H. cracherodii*. Concentrations above 32 ppb in *H. cracherodii* and 56 ppb in *H. rufescens*, cause abnormality of the ctenidium tissue although death may not occur.[42] Concentrations of 40 ppb copper will kill all *H. discus hannai* larvae (0.92 to 1.7 mm) within 2 days, but concentrations of 20 and 30 ppb do not kill larvae after 6 days.[40]

The exact cause of death by copper is unknown, but could be suffocation caused by the fact that the animal secretes mucus to remove copper accumulating on the ctenidium surface. In addition, accumulation of copper in the digestive gland could cause poisoning of vital enzyme systems. Some animals build up a tolerance to chronic high levels of copper. *H. rufescens* collected near Long Beach, California had an average copper concentration of 123.5 ppm in the ctenidia.[42]

Anesthetics
Removal of Juveniles

There are many times during the culture process when juvenile abalone must be removed from the rearing substratum to maintain the proper density, transfer juveniles, or prepare settlement plates. Removal of juveniles should never be done with a brush or scraper because individuals are easily damaged by these tools.[13] Several anesthetics, alone or in combination with thermal shock, have been suggested for safe removal of juvenile abalone.

Ethyl Carbamate

A 0.1% solution of ethyl carbamic acid (common name — urethane, $NH_2CO_2C_2H_5$, mol wt = 89) has almost no effect on juvenile *H. gigantea* (11.1 to 29.8 mm in length), even after 25 min. A 0.5% solution causes all juveniles to drop off the substratum between 2 to 7 min with recovery taking from 8 to 15 min after removal from the anesthetic solution. A 1% solution causes all juveniles to fall off within 1.5 to 6 min. They recover within 10 to 14 min after being returned to fresh sea water.[44]

Recovery times are not very different after juveniles are exposed for 10 min to 0.5% and 1.0% solutions of ethyl carbamic acid (Table 5). As juveniles are exposed for periods from 30 to 180 min, the recovery time for a 1% solution is 2.3 to 5 times longer than a 0.5% solution. This trend disappears at exposures longer than 2 hr with both solutions giving almost the same recovery time. It is amazing that even after exposure for 24 hr to an ethyl carbamic solution, the abalone are still alive, although they require a long time for recovery.[44]

Magnesium Sulfate

A 5% solution of magnesium sulfate ($MgSO_4$, mol wt = 120.36) has little effect on juvenile *H. gigantea*. Higher concentrations (10 to 30%) of magnesium sulfate show some anesthesia of juveniles but give almost identical results at these concentrations. A 10% solution causes 10% of the juveniles to fall off after 12 min with 14 min required for recovery

Table 5
RECOVERY TIME AFTER
EXPOSURE TO ETHYL
CARBAMIC ACID

Length of exposure (min)	Time to recover (min) Concentration	
	0.5%	1%
10	8	13
30	10	50
60	15	50
120	20	45
180	25	60
240	20	35
300	30	30
360	50	60

(From Sagara, J. and Ninomiya, N., *Aquiculture*, 17(2), 89, 1970. With permission.)

Table 6
RECOVERY TIME AFTER EXPOSURE
TO MAGNESIUM SULFATE

Length of exposure (min)	Time to recover (min) Concentration		
	10%	20%	30%
10	28	28	28
30	32	60	65
60	30	70	45
120	85	85	205
180	145	165	—

(From Sagara, J. and Ninomiya, N., *Aquiculture*, 17(2), 89, 1970. With permission.)

after being returned to clean sea water, a 20% solution causes 50% of the juveniles to fall off after 15 min with 25 min required for recovery, and a 30% solution causes 30% of the juveniles to fall off after 12 min with 30 min required for recovery.[44]

There is a tendency for longer recovery times when juveniles are exposed for longer periods and at higher concentrations (Table 6). Exposure of juveniles to this anesthetic is only safe for 3 hr at concentrations of 10 and 20%, or 2 hr at 30%, since juveniles become weak and die if exposed for longer periods.[44]

Chloral Hydrate

A 0.1% solution of chloral hydrate [$CCl_3CH(OH)_2$, mol wt = 82] causes no juveniles to fall off or show any effects after exposure for 20 min. A 0.5% solution causes 20% of the juveniles to fall off after 1 min and 90% after 3 min exposure, with 15 min required for complete recovery after being returned to fresh sea water. A 1% solution causes 90% of the juveniles to fall off after 30 sec and 100% after 1 min. They are in a cyanotic state after 5 min exposure but recover after 15 min.[44]

Chloral hydrate has an immediate effect and anesthesia occurs quickly, however it takes juveniles a long time for complete recovery. Also, juveniles cannot be exposed for a long duration like ethyl carbamic acid or magnesium sulfate. Juveniles exposed to a 0.5% solution of chloral hydrate for 10 min require 45 min for recovery and juveniles exposed for 20 min do not completely recover, even after more than 1 hr in fresh sea water. A 1% solution is lethal and causes immediate death with even short exposure periods.[44]

Diethylbarbituratic Acid

A 1% solution of the sodium salt of diethylbarbituratic acid (common name — Barbitol, Barbitone and Veronal, $C_8H_{12}N_2O_3$, mol wt = 184.20) causes no effect or fall off after exposure for 1 hr. A 2% solution causes 20% of the juveniles to fall off after 17 min and 50% after 30 min. Most juveniles require 1.5 hr for recovery but some individuals need up to 2.5 hr for recovery. A long recovery time is a characteristic of this chemical.[44]

Ethyl p-Aminobenzoic Acid

The Oyster Research Institute in Japan routinely uses ethyl *p*-aminobenzoic acid (common name — Benzocaine, $NH_2C_6H_4COOC_2H_5$, mol wt = 165.19) to remove juveniles from the wavy plates used for settlement. A stock solution is made by dissolving 100 g of ethyl *p*-aminobenzoic acid in 1 ℓ of 95% ethyl alcohol. It is very important to use 95% ethyl alcohol and not 100% ethyl alcohol, since the latter contains benzene, which is carcinogenic. The working anesthetic solution is obtained by mixing 0.5 to 1.0 mℓ of stock solution with 1 ℓ of sea water, or more conveniently, 1 ℓ of stock solution is added to a water table containing 1.8×10^3 ℓ. The anesthetic solution, in combination with thermal shock, quickly removes (less than 1 min) juveniles from the wavy plates. The juveniles are rapidly removed from the aesthetic solution and placed in fresh sea water for recovery. Juveniles are usually allowed several hours for recovery before being placed back into culture tanks.[13]

Other Anesthetics

Prince and Ford[45] used a 0.03% (v/v) solution of diethyl carbonate [$(C_2H_5O)_2CO$, mol wt = 118.13] or an 1% (v/v) solution of ethyl alcohol (CH_3CH_2OH, mol wt = 46), mixed in sea water, for anesthesia of juvenile abalone during a field survey. Diethyl carbonic acid was not satisfactory for use in the field, but ethyl alcohol rendered juvenile abalone insensible after exposure for 10 min.[45] A 1% ethyl alcohol solution efficiently sampled rocks and boulders for juvenile abalone in the field.[46] Kurita et al.[46] used a 2% solution of potassium chloride (KCl) to anesthetize 1- to 3-mm juveniles.

Oxygen Consumption and Heart Rate With Anesthetics

Juveniles do not use much oxygen during anesthesia. The oxygen content in the water was measured before and after anesthesia with ethyl carbamic acid, magnesium sulfate, and chloral hydrate, and the oxygen content actually increased slightly between the two measurements. The effect of anesthesia on juvenile heart rate was also tested. The heart was exposed by removing the shell from 6.5 cm *H. sieboldii*. The "normal" (before anesthesia) heart rate was measured for 1 min, 5 hr after removing the shell. The heart rate (44 to 59 beats per minute) was undetectable after 10 to 15 min exposure with either 1% ethyl carbamic acid or 30% magnesium sulfate. The heart rate usually returned to normal 15 to 30 min after the juveniles were transferred to clean sea water. However, in one individual the heart rate returned to a lower than normal rate (30 beats per minute).[44]

Evaluation

Ethyl carbamic acid (0.5% or 1%) has the best anesthetic characteristics (quick action and quick recovery) and does not adversely affect abalone if they remain in this concentration for 24 hr (Table 7). Ethyl carbamic acid should not be used, however, because it is carcinogenic.[44]

Table 7
SUMMARY OF THE RESULTS ON REMOVAL OF
JUVENILE ABALONE WITH ANESTHETICS

Anesthetic	Best conc. (%)	Fall off (%)	Time to fall off (min)	Recovery time (min)
Ethyl carbamic acid	0.5	100	8	15
	1.0	100	8	15
Magnesium sulfate	20.0	40	15	25
	30.0	50	12	30
Chloral hydrate	0.5	90	3	15
	1.0	100	1	15
Sodium diethylbarbiturate	20.0	100	30	150

(From Sagara, J. and Ninomiya, N., *Aquiculture*, 17(2), 89, 1970. With permission.)

Magnesium sulfate (20 or 30%) is safe but takes a long time to work. Juveniles get weak and die if exposed for over 3 hr to a 20% solution or over 2 hr to a 30% solution.[44]

Chloral hydrate acts rapidly but takes a long time for juveniles to recover and can be lethal to juveniles. Exposure for more than 5 min is possible, however exposure for 10 min or more causes harm to the juvenile.[44]

Prince and Ford[45] found diethyl carbonic acid difficult and unpleasant to use. It is extremely volatile and evaporated rapidly out of the sea water causing the solution to lose potency. The vaporization of diethyl carbonic acid prevents its use for field studies requiring SCUBA or Hookah diving because the fumes (breathed through the Hookah or during rest breaks) could adversely affect the diver.[45]

Diethylbarbituratic acid causes a good fall-off rate but juveniles require a long recovery period. Ethyl *p*-aminobenzoic acid reliably removes juvenile abalone, and when used in combination with thermal shock, acts quickly and safely. However, both diethylbarbituratic acid and ethyl *p*-aminobenzoic acid are controlled substances that cannot be easily purchased. If these chemicals can be obtained, they are the preferred method for use at a culture facility. Ethyl alcohol is probably the preferred method for field studies requiring anesthesia or for hatchery use when diethylbarbituratic acid and ethyl *p*-aminobenzoic acid are unobtainable. It is cheap, easily obtainable, and safe to use.[44]

Desiccation

Larvae (0.45 to 0.87 mm in length) do not die after 10 min exposure to air temperatures below 31.5°C, however, exposure for 15 min causes paralysis and death. Juveniles (1.30 to 2.84 mm) can survive 30 min exposure to air temperatures below 27.0°C, but 50% die after exposure for 40 to 60 min.[40] The length of time an adult abalone can survive out of water increases with lower air temperatures, 16 hr at 27.3°C, 35 hr at 20.7°C, 94 hr at 10.6°C, and 140 hr at 5.8°C. The length of survival decreases near 0°C (e.g., 120 hr at −0.78°C). The weight loss after maximum exposure to air (at all temperatures) is 13 to 16%.[5]

The factors which determine the survival rate of abalone after removal from water are size of animal, air temperature, and length of time out of water. The optimum air temperature for survival is 6°C. The animals suffer from cold shock below 6°C, and mortality is correlated to increasing temperatures above 6°C. The higher the air temperature is above 6°C, the faster the pH falls and the sooner the animal dies.[47]

Small abalone held in air very quickly switch over to an anaerobic respiration in which CO_2 is produced but very little O_2 is consumed.[47] Larger animals (500 to 700 g) take longer to become anaerobic. The pH in abalone stays normal for a variable period depending on animal size and ambient temperature (e.g., 40 hr at 12°C).[47,48] Abalone held in air secrete a fluid which at first is colorless, and low in protein and amino acid concentration. After a time (length dependent on temperature) the fluid becomes blue and has the composition of blood. The animal apparently starts to bleed.[47] At 12°C, the fluid has the composition of blood after 30 to 40 hr, which corresponds to a decrease in pH.[47,48]

REFERENCES

1. **Ceccaldi, H. J.**, Contribution of physiology and biochemistry to progress in aquaculture, *Bull. Jpn. Soc. Sci. Fish.*, 48(8), 1011, 1982.
2. **Tamura, T.**, Influence of changes in the atmosphere on the respiration of various marine mollusks, *J. Fish.*, 43, 1, 1939. (Cited in Imai, T., *Aquaculture in Shallow Seas: Progress in Shallow Sea Culture*, Part IV, Amerind Publishing, New Delhi, 1977.)
3. **Tamura, T.**, Influence of changes in the atmosphere on the respiration of various marine mollusks, *J. Fish.*, 44, 64, 1939. (Cited in Imai, T., *Aquaculture in Shallow Seas: Progress in Shallow Sea Culture*, Part IV, Amerind Publishing, New Delhi, 1977.)
4. **Uki, N. and Kikuchi, S.**, Oxygen consumption of the abalone, *Haliotis discus hannai* in relation to body size and temperature, *Bull. Tohoku Reg. Fish. Res. Lab.*, 35, 73, 1975.
5. **Imai, T.**, *Aquaculture in Shallow Seas: Progress in Shallow Sea Culture*, Part IV, Amerind Publishing, New Delhi, 1977.
6. **Uki, N. and Kikuchi, S.**, Regulation of maturation and spawning of an abalone, *Haliotis* (Gastropoda) by external environmental factors, *Aquaculture*, 39, 247, 1984.
7. **Hemmingsen, A. M.**, Energy metabolism as related to body size and respiratory surfaces, and its evolution, *Rep. Steno Mem. Hosp. Copenhagen*, 9, 7, 1960.
8. **Winberg, G. G.**, Rate of metabolism and food requirements of fishes, *Nauche Trudy Belorusskovo Gosudarstvennovo Universiteta Imeni*, V. I. Lenina, Minsk, 1956. (Cited in Uki, N. and Kikuchi, S., Oxygen consumption of the abalone, *Haliotis discus hannai* in relation to body size and temperature, *Bull. Tohoku Reg. Fish. Res. Lab.*, 35, 73, 1975.)
9. **Sano, T. and Maniwa, R.**, Studies on the environmental factors having an influence on the growth of *Haliotis discus hannai*, *Bull. Tohoku Reg. Fish. Res. Lab.*, 21, 79, 1962.
10. **Uki, N.**, Feeding behavior of experimental populations of the abalone, *Haliotis discus hannai*, *Bull. Tohoku Reg. Fish. Res. Lab.*, 43, 53, 1981.
11. **Kikuchi, S. and Uki, N.**, Technical study on artificial spawning of abalone, genus *Haliotis* I. Relation between water temperature and advancing sexual maturity of *Haliotis discus hannai* Ino, *Bull. Tohoku Reg. Fish. Res. Lab.*, 33, 69, 1974.
12. **Sakai, S.**, Ecological studies on the abalone, *Haliotis discus hannai* Ino. I. Experimental studies on the food habit, *Bull. Jpn. Soc. Sci. Fish.*, 28(8), 767, 1962.
13. **Seki, T.**, Personal communication, 1981.
14. **Fujino, K.**, Genetic studies on the Pacific abalone. II. Excessive homozygosity in deficient animals, *Bull. Jpn. Soc. Sci Fish.*, 44(7), 767, 1978.
15. **Singh, S. M. and Green, R. H.**, Excess of allozyme heterozygosity in marine molluscs and its possible biological significance, *Malacologia*, 25(2), 569, 1984.
16. **Zouros, E. and Foltz, D. W.**, Possible explanations of heterozygosity deficiency in bivalve mollusks, *Malacologia*, 25(2), 583, 1984.
17. **Fujino, K., Sasaki, K., and Okumura, S.**, Probable involvement of thermostability variations of enzymes in the mechanisms of occurrence of deficient abalone, *Bull. Jpn. Soc. Sci. Fish.*, 50(4), 597, 1984.
18. **Okumura, S., Sasaki, K., and Fujino, K.**, Thermostability variations at multiple loci in the Pacific abalone, *Bull. Jpn. Soc. Sci. Fish.*, 47(12), 1627, 1981.
19. **Wilkins, N. P., Fujino, K., and Sasaki, K.**, Genetic studies on the Pacific abalone. IV. Thermostability difference among phosphoglucomutase variants, *Bull. Jpn. Soc. Sci. Fish.*, 46(5), 549, 1980.
20. **Fujino, K. and Sasaki, K.**, Age association of genotypic proportions of isozymes in the Pacific abalone, *Bull. Jpn. Soc. Sci. Fish.*, 50(1), 11, 1984.
21. **Fujino, K.**, Impact of Genetic Factors on Aquaculture and Stock Management, presented at World Conf. Aquacult., Venice, September 21 to 25, 1981.
22. **Arai, K., Tsubaki, H., Ishitani, Y., and Fujino, K.**, Chromosomes of *Haliotis discus hannai* Ino and *H. discus* Reeve, *Bull. Jpn. Soc. Sci. Fish.*, 48(12), 1689, 1982.

23. **Arai, K., and Wilkins, N. P.**, Chromosomes of *Haliotis tuberculata* L., *Aquaculture*, 58, 305, 1986.
24. **Arai, K., Naito, F., Sasaki, H., Fujino, K.**, Gynogenesis with ultraviolet ray irradiated sperm in the Pacific abalone, *Bull. Jpn. Soc. Sci. Fish.*, 50(12), 2019, 1984.
25. **Elston, R. and Lockwood, G. S.**, Pathogenesis of vibriosis in cultured juvenile Red abalone, *Haliotis rufescens* Swainson, *J. Fish Dis.*, 6, 111, 1983.
26. **Ebert, E. E. and Houk, J. L.**, Elements and innovations in the cultivation of Red abalone, *Haliotis rufescens*, *Aquaculture*, 39, 375, 1984.
27. **Bower, S.**, *Study of the Pathogenesis of Vibriosis in Abalone Mariculture — Abalone Disease Project: Seasonal Mortality Problems, Final Report 1984 — 1985*, Pacific Trident Mariculture, 1985, 5.
28. **Harrison, A. J. and Grant, J. F.**, Progress in abalone research, *Tasmanian Fish. Res.*, 5 (1), 1, 1971.
29. **Crofts, D. R.**, Haliotis, *Liverpool Mar. Biol. Comm. Mem. Typ. Br. Mar. Plants Anim.*, 29, 1, 1929.
30. **Cox, K. W.**, California abalones, family Haliotidae, *Calif. Fish Game Fish. Bull.*, 118, 1, 1962.
31. **Armstrong, D. A., Armstrong, J. L., Krassner, S. M., and Pauley, G. B.**, Experimental wound repair in the Black abalone, *Haliotis cracherodii*, *J. Invertebr. Pathol.*, 17, 216, 1971.
32. **Burge, R., Schultz, S., and Odemar, M.**, Draft Report on Recent Abalone Research in California with Recommendations for Management, State of Calif., The Resources Agency, Dept. Fish Game, 1975.
33. **Elston, R.**, Histopathology of oxygen intoxication in the juvenile red abalone, *Haliotis rufescens* Swainson, *J. Fish Dis.*, 6, 101, 1983.
34. **Ino, T.**, Biological studies on the propagation of Japanese abalone (genus) *Haliotis*), *Bull. Tokai Reg. Fish. Res. Lab.*, 5, 1, 1952.
35. **Leighton, D. L.**, Laboratory observations on the early growth of the abalone, *Haliotis sorenseni*, and the effect of temperature on larval development and settling success, *Fish. Bull.*, 70(2), 373, 1972.
36. **Leighton, D. L.**, The influence of temperature on larval and juvenile growth in three species of southern California abalones, *Fish. Bull.*, 72(4), 1137, 1974.
37. **Hines, A., Anderson, S., and Brisbin, M.**, Heat tolerance in the Black abalone, *Haliotis cracherodii* Leach, 1814: effects of temperature fluctuation and acclimation, *The Veliger*, 23(2), 113, 1980.
38. **Leighton, D. L., Byhower, M. J., Kelly, J. C., Hooker, G. N., and Morse, D. E.**, Acceleration of development and growth in young Green abalone *(Haliotis fulgens)* using warmed effluent seawater, *J. World Maricult. Soc.*, 12(1), 170, 1981.
39. **Fujino, K., Yamamori, K., and Okumura, S.**, Heart-rate responses of the Pacific abalone against water temperature changes, *Bull. Jpn. Soc. Sci. Fish.*, 50(10), 1671, 1984.
40. **Zongqing, N. and Wenhua, C.**, Studies on rearing conditions of abalone, *Haliotis discus hannai* Ino. II. The effects of salinity and inorganic nutrients on the development of fertilized eggs and living of larvae, *Mar. Fish. Res.*, 6, 41, 1984.
41. **Sakai, H.**, Method to check in shell quality corrosion in young abalone, *Aquiculture*, 28(2), 102, 1980. (Translated in Mottet, M. G., Summaries of Japanese papers on hatchery, technology and intermediate rearing facilities for clams, scallops, and abalones, Progr. Rep., Dept. Fish. Washington, 203, 23, 1984.
42. **Martin, M., Stephenson, M. D., Martin, J. H.**, Copper toxicity experiments in relation to abalone deaths observed in a power plant's cooling waters, *Calif. Fish Game*, 63(2), 95, 1977.
43. **Marks, G. W.**, The copper content and copper tolerance of some species of mollusks of the southern California coast, *Biol. Bull.*, 75(2), 224, 1938.
44. **Sagara, J. and Ninomiya, N.**, On the tear-off of young abalone from the attachment by four anesthetics (ethyl carbamate, magnesium sulfate, chloral hydrate and sodium diethylbarbiturate), *Aquiculture*, 17(2), 89, 1970.
45. **Prince, J. D. and Ford, W. B.**, Use of anaesthetic to standardize efficiency in sampling abalone populations (genus *Haliotis*; Mollusca: Gastropoda), *Aust. J. Mar. Freshwater Res.*, 36, 701, 1985.
46. **Kurita, M., Sato, R., Ishikawa, Y., Shiihara, H., and Ono, S.**, Artificial seed production of abalone, *Haliotis discus* (Reeve), *Bull. Oita Pref. Fish. Exp. St.*, 10, 78, 1978. (Translated in Mottet, M. G., Summaries of Japanese papers on hatchery, technology and intermediate rearing facilities for clams, scallops, and abalones, Prog. Rep. Dept. Fish. *Washington*, 203, 10, 1984.)
47. **Olley, J. and Thrower, S. J.**, Abalone — an esoteric food, in *Advances in Food Research*, Vol. 23, Chichester, C. O., Mrak, I. M., and Stewart, G. F., Eds., Academic Press, New York, 1977, 144.
48. **James, D. G. and Olley, J.**, The abalone industry in Australia, in *Fishery Products*, Kreuzer, R., Ed., Fishing News, London, 1974, 238.

NUTRITION AND GROWTH OF ABALONE

Kirk O. Hahn

INTRODUCTION

The proper nutrition and the resulting growth of cultured abalone are critical factors to successful culture of this animal. Typical growth rates of abalone are approximately 2 to 3 cm/year, therefore 2 to 5 years are required to produce a market-size abalone (Table 1).[1-17] Compounding the problem of slow growth rate, growth is also very heterogeneous. The growth rate of some individuals may be only 20 to 30% of the average at the hatchery.[6,18] Food conversion efficiency of abalone commonly ranges from 5 to 10% although it can be as high as 15%.[19] Hatchery-reared abalone often have a lower meat-to-shell ratio than individuals in the wild.[16,18,20] Also, improper nutrition can cause the animal to be more susceptible to trauma and stress.[16] All these factors can cause problems during rearing and marketing of the abalone, and make the culture of abalone expensive and time consuming.[16,18,20]

The culturist must supply optimum conditions to ensure good growth. One of the most important factors is the food source. As the animal grows, the food source must match its nutritional requirements and be in sufficient quantities. Presently, little is know about the specific nutritional requirements of abalone. The study of abalone nutrition has usually consisted of measuring the growth produced by feeding naturally occurring macroalgae and diatoms. The development of artificial foods for abalone has encouraged studies on the optimal levels of protein, lipids, and other nutrients needed for growth.

The economic success of abalone aquaculture depends, in part, on the abalone growth rate and the food conversion efficiency of the food source used during culture. It is hoped that market size (7.5 to 10 cm) abalone can be produced in 1 to 2 years with improved culture methods using the optimum food source throughout the life of the animal. The majority of the research on nutrition and growth of abalone has been conducted in Japan. The results from studies on *H. discus hannai* should be used as a starting point for determining proper food sources for the many abalone species now being cultured throughout the world.

NUTRITION

Diatoms

Diatoms are the principal food source for newly settled juveniles and are needed at sufficient levels until individuals are 7 to 8 mm, when they switch to eating macroalgae. Most culture facilities rely on naturally occurring diatom species to supply the food for juveniles (Table 2).[5,9,15,16,21]

Diatoms are unicellular algae belonging to phylum Crysophyta, division Bacillariophyta, class Diatomacea. Diatoms have siliceous cell walls that form two overlapping halves, and are golden brown from the presence of chlorophyll, carotene, and xanthophyll pigments.[9]

Diatom growth in rearing tanks or microalgal culture tanks can be enhanced by supplying phosphate, nitrogen, silica (silic acid), boron, vitamins, and trace metals to the water.[9,22] Growth of diatoms is also promoted by using natural or fluorescent lights to produce 4000 lx at the water surface for at least 12 hr/day. Light intensity and photoperiod are important factors in diatom growth. Seasonal and diurnal variations in light intensity, photoperiod, nutrients, and water temperature can cause the dominant diatom species on the rearing surface to vary throughout the year. Maintaining an adequate film of "good" diatoms for growth in the rearing tanks is a major concern in the culture process. The amount of diatoms

Table 1
AVERAGE GROWTH RATES OF
JUVENILE ABALONE

Species	Growth rate (cm/year)	Ref.
H. corrugata	2.5	1
H. cracherodii	3.0	2
H. discus hannai	2.4	3
	3.6	4
	2.0	5
H. diversicolor supertexta	2.5	6
	4.3	69
H. fulgens	2.5	1
H. iris	2.0	7
H. kamtschatkana	2.0	8
	1.5	9
	1.3	10
	2.0	11
H. laevigata	4.5	12
H. lamellosa	2.6	13
H. midae	2.7	14
H. rufescens	2.4	15
	2.5	16
H. tuberculata	1.8	17
	1.8	13

Table 2
PRINCIPAL FOOD
ITEMS FOUND ON
REARING SURFACES
FOR JUVENILE
ABALONE

Species	Ref.
Amphora spp.	70
Biddulphia sp.	9
Chaetoceros simplex	70
Cocconeis scutellum	9,26
Melanosira spp.	70
Melosira nummuloides	26
M. dubia	26
Navicula britanica	25
N. ulvacea	25
Nitzschia costerum	70
N. divergens	25
N. panduriformis	26
N. pseudohybrida	25
Platymonas sp.	25
Synedra tabulata	26
Tetraselmis suecica	71
Ulvella lens	26

in a rearing tank can vary from grazing rates of growing individuals, survival of individuals, seasonal diatom growth, and selective grazing of diatoms by individuals.[9]

Norman-Boudreau[9] found chain-forming diatom species are often the dominant species on the rearing surfaces before grazing by abalone and produce a thick mat on rearing surfaces. Grazing by juvenile abalone causes small diatom species to predominate on the surfaces. Small, motile, nonchain-forming pennate diatom species are characteristic of early diatom colonizers.[9]

Norman-Boudreau et al.[23] developed a technique to determine the diatom species eaten by newly settled juveniles. Newly metamorphosed juvenile abalone (2 days after settlement) are placed in a clean petri dish and allowed to attach to the bottom of the plate. Debris and loose diatoms are removed by rinsing (5 to 10 times) 10 μm filtered, UV-irradiated sea water over the animals. The sea water is poured out and the juveniles are immersed in a saturated EDTA (ethylenediaminetetraacetic acid) disodium salt solution for 4 to 20 hr at room temperature. The juveniles are treated until the shell is decalcified and diatoms adhering to the shell surface have fallen off. After the shell is decalcified, the EDTA solution is carefully blotted up and each animal is covered with a single drop of 6% sodium hypochlorite (bleach). Diatoms in the gut can be clearly seen with a microscope after this treatment. This technique is preferred over the squash method because diatoms that commonly adhere to the shell do not confuse the observation. Also, the first diatom species eaten by the juveniles are very small and could be obscured by shell and tissue debris in a squash preparation. One disadvantage of the EDTA/bleach technique is its destruction of other possible food items (e.g., algae, bacteria, and protozoa) that do not have an inert component, such as the siliceous frustules of diatoms.[23]

Norman-Boudreau et al.[23] found all diatoms in the 2-day-old juvenile's digestive tract were of the Order Pennales and usually smaller than 10 μm. *Navicula* sp. and *Cocconeis* sp. were the most commonly found diatoms. The juvenile abalone actively selects the diatoms during eating. The diatoms found in the gut were more uniform in size and species composition than found on the rearing surface.[23]

The active selection of diatoms by newly settled juveniles from the mixed array of diatoms on the rearing surface can be used to obtain ''good'' diatom species for monoculture. The stomachs of 2-day-old juveniles that have settled on surfaces covered with naturally occurring diatoms are removed, and the diatoms in the stomach are isolated and grown in monoculture. Good diatoms are defined as ones that produce large amounts of secretion, form sheets, and are smaller than 10 μm.[24]

Navicula spp. and *Nitzschia* spp. are the usual diatom species given to juveniles for food.[5,9,15,16,21] These diatoms occur naturally in the rearing tanks and produce good growth rates (Table 3).[25,26]

If larval settlement is reasonably successful, it becomes difficult to maintain adequate diatom films on the rearing surfaces as the animals grow and food can become limiting. Ebert and Houk[15] use a diatom-slurry method to supply sufficient food to growing juveniles. Diatoms are grown on the sides of white polyethylene buckets (2900 cm² of surface area) with ambient water, filtered to 1 μm, flowing at less than 0.5 ℓ/min. Fluorescent lamps are positioned 30 cm above the buckets to produce continuous growth. The diatom cultures are harvested after 7 days of growth. The sides of the bucket are wiped down using a latex glove and the diatoms are rinsed through a 5-μm filter and a 1-μm filter. Diatoms passing through the 1-μm filter are used for induction of larval settlement and the diatoms retained by the 1-μm filter are used for feeding juveniles.[15]

Macroalgae

Juveniles begin to eat macroalgae at about 10 mm in length and will eat from 10 to 30% of their whole-body wet weight in algae each day.[21] The high feeding rate of macroalgae

Table 3
GROWTH RATE OF JUVENILE (10 MM) *H. DISCUS HANNAI* FED MONOCULTURES OF MICROALGAE AND REARED AT 20°C

Species	Growth (μm/day)	Ref.
Navicula britanica	200	25
N. ulvacea	200	25
Nitzschia divergens	221	25
N. pseudohybrida	220	25
Platymonas sp.	23	25
Prasinocaldus marinas	126	25
Nitzschia divergens and *Navicula agnita*	270	26
Nitzschia divergens and *Amphora* sp.	260	26

Table 4
CRUDE CHEMICAL COMPOSITION OF MACROALGAE OF ORDER LAMINARIALES (DRY WEIGHT)

Species	Percent (%)					Ref.
	Protein	Lipid	Ash	Carbohydrate	Fiber	
Eisenia bicyclis	4.7	1.4	19.1			34
Laminaria religiosa	6.4	1.5	34.4			34
Undaria pinnatifida	8.7	2.3	36.8			34
Nereocystis sp.	16.1	2.9	35.5	39.8	4.6	9

is due to the high water content and relative low protein content of fresh macroalgae (Table 4).

Uki et al.[27] tested the food value of 56 species of macroalgae found along the Pacific coast of northern Japan. The growth (body weight and shell length) produced by each macroalgal species was compared to the growth produced by *Eisenia bicyclis*. Growth is expressed as *Eisenia* equivalent (growth with *E. bicyclis* is defined as 1.00).[28] The *Eisenia* equivalent reflects the combined effects of food preference and nutritive value of the macroalgae. Values of 1.00 or more, indicate the kelp is suitable as food for abalone.[28] Fourteen species of macroalgae were found to have food values equal to or greater than *E. bicyclis* (Table 5). It was interesting to note that some macroalgae produce faster shell growth than weight gain, and others produce faster weight gain than shell growth. Generally, brown algae of the order Laminariales is an optimum food for *H. discus hannai*.[27,28]

Abalone usually prefer brown algae, but there are some exceptions. Abalone species in California, depending upon location and season, eat brown algae (*Macrocystis* spp., *Nereocystis* spp., *Egregia* spp., and *Eisenia* spp.), red algae (*Gigartina* spp., *Gelidium* spp., and *Plocamium* spp.), and green algae (*Ulva* spp.).[16] New Zealand species (*H. iris* and *H. australis*) eat *Macrocystis pyrifera* when it is abundant but prefer red algae.[29] The order of food preference for *H. iris* is *Gracilaria* sp., *Glossophora* sp., *M. pyrifera*, *Lessonia variegata*, *Champia* sp., *Ulva lactuca*, and *Pterocladia* sp.[30] *Gracilaria* sp. is the best food for juvenile and adult *H. iris*, even though individuals would not normally eat it in any abundance in nature.[31] The growth rate produced by *Gracilaria* sp. is more than double the rate produced by *M. pyrifera*.[32]

Uki[28] tested the food value of eight species of *Laminaria* collected during different periods of the algal growth cycle. The experiment was conducted twice, once during the macroalgal

Table 5
**GROWTH PRODUCED BY MACROALGAE FOUND ALONG THE
PACIFIC COAST OF JAPAN (GROWTH HAS BEEN STANDARDIZED
RELATIVE TO *EISENIA BICYCLIS*)**

Species	Daily increase in SL[a] (μm)	Monthly growth rate (%)	Daily feeding rate (%)	FCE[b] (%)	Relative growth index	
					SL	BW[c]
Eisenia bicyclis	86	24	9.3	8.2	1.00	1.00
Laminaria japonica					1.10	1.17
Laminaria japonica f. *membranacea*					1.04	1.27
L. religiosa	83	32	16.5	5.6	1.44	1.47
L. angustata var. *longissima*					1.04	1.18
L. diabolica					1.18	1.66
Costaria costata	90	46	19.3	6.5	1.56	2.12
Undaria pinnatifida	90	44	17.5	6.9	1.56	2.04
Alaria crassifolia	137	46	19.3	6.5	1.41	1.14
Desmarestia ligulata	79	28	14.8	5.5	1.37	1.29
Porphyra yezoensis	91	26	5.2	14.9	0.94	1.04
Palmaria palmata	99	51	20.4	7.6	1.72	2.36
Chondria crassicaulis	90	26	15.4	5.0	1.27	1.28
Ulva pertusa	41	23	7.1	9.1	0.71	1.04
Enteromorpha sp.	62	23	10.9	6.3	0.87	1.13
Codium divaricatum	43	22	31.6	2.1	0.75	1.02

[a]   SL = shell length.
[b]   FCE = food conversion efficiency.
[c]   BW = body weight.

(From Uki, N., Sugiura, M., and Watanabe, T., *Bull. Jpn. Soc. Sci. Fish.*, 52(2), 257, 1986. With permission.)

growing season and once during the macroalgal reproductive season (October and November). The food value of each species was greater during the growth season than the reproductive season (Tables 6 and 7). Chemical analysis showed that macroalgae are higher in protein content during the growth season (Table 8).[28]

Food conversion rates of the macroalgae ranged from 4 to 7% (average 6%) on a wet-weight basis or 38 to 58% (average 50%) on a dry-weight basis.[28] Abalone fed fronds of *Laminaria religiosa* with bryozoa (*Membranipora membranacea*) on the surface grew faster than those fed the same species without bryozoa. The bryozoa probably supplies supplemental nutrition to the abalone.[28]

Artificial Food

Production levels at most abalone aquaculture facilities in Japan (>1,000,000 juveniles per year) require large amounts of food.[33] Using macroalgae as the principal food source is expensive because its use requires a large amount of labor to harvest, and electricity for frozen storage or labor for drying the macroalgae. Feeding macroalgae is very labor intensive and does not lend itself to mechanization.[33] Additionally, kelp is low in protein content (5%) and high in water content (40%).[9,34]

One of the largest costs in the culture process of abalone is the purchase and storage of kelp. In Japan, the abalone aquaculturist must either purchase macroalgae from kelp fishermen or grow it themselves. Macroalgae is not harvested from natural populations because

Table 6.
GROWTH, FEEDING RATIO, AND FOOD CONVERSION EFFICIENCY OF ABALONE FED ON 13 KINDS OF MARINE ALGAE (A) GROWTH SEASON OF ALGAE; (B) REPRODUCTION SEASON OF ALGAE

| Species | Daily increase in SL[a] (μm) | Monthly weight gain (%) | Daily feeding rate (%) | | Food conversion efficiency (%) | |
			Wet weight	Dry weight	Wet weight	Dry weight
A.						
Eisenia bicyclis	98	26.5	10.1	1.42	7.7	55
Kjellmaniella gyrata	79	25.4	13.7	1.52	5.5	50
Laminaria angustata var. *longissima*	102	31.3	22.6	2.34	4.0	39
Laminaria diabolica	116	43.9	15.7	1.66	7.6	72
Laminaria japonica	108	31.1	19.3	2.25	4.8	40
Laminaria japonica f. *membranacea*	102	33.7	16.8	1.91	5.7	50
Laminaria religiosa	101	29.7	21.6	2.06	4.0	42
Undaria pinnatifida	121	39.8	13.1	1.35	8.4	82
B.						
Eisenia bicyclis	48	23.0	8.3	1.26	8.3	54
Laminaria diabolica	45	21.9	11.8	1.67	5.6	39
Laminaria japonica	46	23.8	12.9	1.43	5.5	50
Laminaria japonica f. *membranacea*	43	21.3	9.9	1.18	6.5	55
Laminaria religiosa	42	21.9	10.1	1.19	6.6	55
Laminaria religiosa with bryozoa	56	33.7	15.3	2.65	6.3	49

[a]   SL = Shell length.

(From Unki, N., *Bull. Tohoku Reg. Fish. Res. Lab.*, 42, 19, 1981. With permission.)

Table 7
RELATIVE FOOD VALUES, *EISENIA* EQUIVALENT (1.00), OF LAMINARIALES

| Species | Growth season | | Reproduction season | |
	Shell length	Body weight	Shell length	Body weight
Kjellmaniella gyrata	0.81	0.96	—	—
Laminaria angustata var. *longissima*	1.04	1.18	—	—
L. diabolica	1.18	1.66	0.94	0.95
L. japonica	1.10	1.17	0.96	1.03
L. japonica f. *membranacea*	1.04	1.27	0.90	0.93
L. religiosa	1.03	1.12	0.88	0.95
L. religiosa with bryozoa	—	—	1.17	1.47
Undaria pinnatifida	1.23	1.50	—	—

(From Uki, N., *Bull. Tohoku Reg. Fish. Res. Lab.*, 42, 19, 1981. With permission.)

Table 8
CRUDE CHEMICAL COMPOSITION OF MARINE ALGAE LAMINARIALES (A) GROWTH SEASON OF ALGAE; (B) REPRODUCTION SEASON OF ALGAE

	Wet weight (%)					Dry weight (%)			
	Moisture	Protein	Lipid	Ash	Carbohydrates	Protein	Lipid	Ash	Carbohydrates
A									
Eisenia bicyclis	86.0	2.63	0.171	4.28	6.92	18.8	1.22	30.6	49.4
Kjellmaniella gyrata	88.9	1.84	0.098	—	—	16.6	0.88	—	—
Laminaria angustata var. longissima	89.4	1.39	0.088	4.38	4.72	13.4	0.85	41.3	44.5
L. diabolica	89.4	1.20	0.071	4.71	4.62	11.3	0.67	44.4	43.6
L. japonica	88.1	1.48	0.097	5.11	5.22	12.5	0.81	43.4	43.3
L. japonica f. membranacea	88.7	1.63	0.061	4.93	4.69	14.4	0.54	43.6	41.5
L. religiosa	90.5	0.88	0.077	4.15	4.39	9.3	0.81	43.7	46.2
Undaria pinnatifida	89.7	2.06	0.147	4.25	3.84	20.0	1.45	41.3	37.3
B									
Eisenia bicyclis	84.8	1.49	0.157	4.76	8.80	9.8	1.03	31.3	57.9
Laminaria diabolica	85.8	1.36	0.091	5.00	7.75	9.6	0.64	35.2	54.6
L. japonica	88.8	1.04	0.098	3.97	6.10	9.3	0.87	35.7	54.1
L. japonica f. membranacea	88.1	1.25	0.080	4.07	6.50	10.5	0.67	34.2	54.6
L. religiosa	88.0	1.11	0.131	4.07	6.69	9.3	1.08	34.3	55.4
L. religiosa with bryozoa	87.0	1.83	0.085	6.57	4.52	14.1	0.65	50.5	34.8

(From Uki, N., Bull. Tohoku Reg. Fish. Res. Lab., 42, 19, 1981. With permission.)

Table 9
AMINO ACID CONTENT OF ARTIFICIAL DIET AND MACROALGAE (PER 100 G DRY WEIGHT)

	Artificial diet (mg)	*Eisenia* sp. (mg)	*Laminaria* sp. (mg)
Lysine	774	114	17
Histidine	262	19	5
Arginine	845	76	17
Aspartic acid	1107	286	52
Threonine	548	76	17
Aspartic acid	1107	286	52
Threonine	548	114	17
Serine	679	114	17
Glutamic acid	2667	229	86
Proline	1202	114	17
Glycine	667	133	34
Alanine	821	95	34
Cysteine	214	38	17
Valine	690	114	17
Methionine	405	19	17
Isoleucine	583	57	17
Leucine	1250	95	17
Tyrosine	512	95	17
Phenylalanine	750	76	17

(From Anonymous, Nihon Nosan Kogyo K.K. Research Center, Jpn, Patent Public. No. 55-1586, Material No. 3, 1981.)

this would reduce the amount of macroalgae available for abalone in the wild and be counter productive to the goal of abalone aquaculture in Japan (i.e., reseeding).[5]

Presently, most abalone culture facilities must be located close to local macroalgae beds. This severely limits the number of locations that can support abalone aquaculture.[33] The development of artificial foods for abalone has removed the requirement for sources of abundant macroalgae close to the culture facility and has allowed expansion of abalone aquaculture in Japan. Some cultured abalone in Japan never eat macroalgae during the entire time they are cultured.[5] This has allowed the culture facility's production to be independent of the local macroalgae production.[5,35] Artificial food has been used to raise *H. discus, H. discus hannai,* and *H. sieboldii.*[35]

The development of an artificial feed for juvenile abalone has been a great asset to abalone seed production in Japan. An artificial food solves several of the biggest problems involved with producing large numbers of juvenile abalone. Artificial food is easy to store, less expensive than macroalgae (either dried or frozen), produces better growth in body weight and length, and allows possible mechanization of feeding. The cost in 1982 was approximately $3.50/kg.[36] Using an artificial food eliminates the need for tanks to store fresh kelp or refrigeration for frozen kelp. Artificial food can be stored for several months if placed unopened in a cool, dark location.[37] Artificial food allows juvenile abalone to be reared in high densities, since the food availability is independent of surface area.[5]

At most abalone culture facilities in Japan, artificial food is used exclusively after the juveniles (5 to 8 mm in length) are removed from the wavy plates used for rearing early juveniles with diatoms.[5,35] The artificial food must contain enough protein, essential amino acids, vitamins, and minerals to meet the nutritional requirements of the growing abalone (Table 9). Presently, there are two companies in Japan that manufacture artificial food for

Table 10
THE GROWTH OF 1
TO 2 CM ABALONE
FED DIFFERING DIETS

Diet	Growth (μm/day)
Artificial diet	83—133
Eisenia sp.	80
Undaria sp.	82
Laminaria sp.	83
Undaria (frozen)	80
Laminaria (dried)	43

(From Anonymous, Nihon Nosan Kogyo K.K. Research Center, Jpn. Patent Public. No. 55-1586, Material No. 3, 1981.)

Table 11
PERCENT CONTENT OF ABALONE FED DIFFERING DIETS

	Seaweeds		Artificial diet	
	Wet weight	Dry weight	Wet weight	Dry weight
Moisture	78.0	—	76.4	—
Crude protein	13.6	61.8	15.6	66.1
Crude fat	1.3	5.9	2.0	8.5
Crude ash	2.4	10.9	2.8	11.9
Crude fiber	0	0	0	0
NFE	4.7	21.4	3.2	13.6

(From Anonymous, Nosan Kogyo K.K. Research Center, Jpn. Patent Public. No. 55-1586, Material No. 3, 1981.)

abalone. The basic difference between the two artificial foods is that one uses fishmeal and the other uses soybean meal as the protein source, however both artificial foods are eaten by the abalone and produce excellent growth rates.[5,33]

Juvenile growth with an artificial food produced by Nihon Nosan Kogyo K.K. Research Center is 65% greater than when the animals are fed macroalgae (Table 10). The growth rate with the artificial food is directly correlated with water temperature up to 20°C. The excellent growth rate is the best feature of artificial food. Juveniles fed artificial food have a higher body weight per shell length than do juveniles fed macroalgae. In addition, the abalone meat has a relatively higher protein content and lower water content (Table 11). This indicates that the artificial food produces healthy abalone.[36]

Due to artificial food being low in water content and high in protein content, it can be given at a much lower feeding rate than macroalgae (Table 12).[36] Depending on the size of the juvenile abalone, feeding rate is from 2 to 7% of whole-body wet weight with artificial food compared with 10 to 30% with *Laminaria* spp.[35-37] The artificial food consumption per day is correlated to the water temperature and age of the animal (Tables 13 and 14). In general, at the same water temperature, daily feeding rate/body weight decreases with increasing body weight.[36]

The artificial food is scattered around the shallow net bags used for juvenile rearing, care

Table 12
PERCENT CONTENT OF ARTIFICIAL DIET AND MACROALGAE

	Artificial diet		*Eisenia* sp.		*Laminaria* sp.	
	Wet weight	Dry weight	Wet weight	Dry weight	Wet weight	Dry weight
Moisture	16.0	—	47.5	—	42.0	—
Crude protein	30.0	35.7	9.7	18.5	5.2	9.0
Crude fat	3.5	4.2	0.3	0.6	1.3	2.2
Crude fiber	2.0	2.4	3.1	5.9	3.5	6.0
Crude ash	12.0	14.3	12.9	24.6	16.3	28.1
NFE	36.5	43.5	26.5	50.5	31.7	54.7
Calcium	2.9	3.4	1.2	2.2	0.6	1.0
Phosphorus	1.4	1.6	0.2	0.3	0.1	0.2

(From Anonymous, Nihon Nosan Kogyo K. K. Research Center, Jpn. Patent Public. No. 55-1586, Material No. 3, 1981.)

Table 13
DAILY FEEDING RATE OF ARTIFICIAL DIET
(PERCENT OF ANIMAL WET WEIGHT)

Shell length (mm)	Water Temperature				
	8°C	10°C	15°C	18°C	20°C
5—10	3.2—4.4	3.5—4.9	4.5—6.3	4.7—6.5	5.2—7.2
11—15	2.1—4.0	2.3—4.5	3.0—5.7	3.1—5.9	3.4—6.6
16—20	1.6—2.8	1.8—3.1	2.2—3.9	2.3—4.1	2.6—4.5
21—25	1.3—2.1	1.4—2.3	1.8—3.0	1.9—3.1	2.1—3.4
26—30	1.1—1.7	1.2—1.9	1.5—2.4	1.6—2.5	1.7—2.8
31—40	0.8—1.4	0.9—1.6	1.1—2.0	1.2—2.1	1.3—2.3

(From Anonymous, Nihon Nosan Kogyo K.K. Research Center, Jpn. Patent Public. No. 55-1586, Material No. 3, 1981.)

Table 14
DAILY FEEDING RATE OF ARTIFICIAL DIET PER 10,000
JUVENILES (G)

Shell Length (mm)	Water Temperature				
	8°C	10°C	15°C	18°C	20°C
5—10	50—65	55—75	70—95	70—100	80—110
11—15	55—180	60—200	75—260	80—270	90—300
16—20	70—280	80—310	100—390	105—410	115—450
21—25	130—390	140—425	180—555	190—575	210—630
26—30	205—555	220—620	280—780	295—815	315—910
31—40	260—1400	300—1600	360—2000	400—2100	500—2300

(From Anonymous, Nihon Nosan Kogyo K.K. Research Center, Jpn. Patent Public. No. 55-1586, Material No. 3, 1981.)

being taken to prevent the food from gathering in corners.[5] The proper placement of the air stones will prevent this from happening. The amount of artificial food given each day must be closely regulated because excess feeding at one time will waste food and lower the water quality from disintegration of the food pellet. The food should be given according to the standard ratio, and correcting the amount by observing the actual feeding rate of the abalone. Generally, abalone are influenced by the culture environment, so the animals must be observed while feeding an artificial food, and the amount of food given must be increased or decreased to correspond to changing feeding rates.[36]

The artificial food pellet must be stable in the water for at least 24 hr at the maximum water temperature experienced at the culture facility to reduce the amount of food waste and ensure good water quality. The water temperature greatly affects the stability of artificial feeds. At 10 to 20°C (normal temperature range of water in Japan), the food pellet will keep its shape for 2 to 4 days. The food quality of the artificial food may change if it remains in the water for a long period before being eaten by the juvenile abalone, therefore it should be given to the animals once every 2 days for temperature under 15°C, and every day for temperatures above 15°C. Feeding with artificial food should be stopped during periods when the water temperature is above 24°C. Above this water temperature, the stability of the feed becomes increasingly worse and the water quality rapidly declines.[36] If artificial food must be used when the water temperature is above 24°C, the water quality should be closely monitored.[37]

The shape, size, and texture of the artificial food should be designed to produce the best attraction to the juvenile.[36,38,39] The mode of eating by abalone requires that the food pellet be flat and round. Abalone begin eating at the edge of the food, so this shape gives the maximum edge area and ease of handling for small abalone. After touching the food, abalone cover it with the anterior end of the foot and begin to eat along the edge. The radula scrapes off particles of food and sends it into the mouth. Young abalone usually eat only at night but after becoming accustomed to an artificial food, eating will also occur during the day.[36]

The artificial food produced by Nihon Nosan Kogyo contains powdered seaweed and other attractants to enhance the taste of the pellet to the juvenile. Juveniles easily and quickly change from natural seaweeds to artificial food. Juvenile abalone in Japan prefer the artificial food over *Laminaria* spp. and *Eisenia* spp. Groups of juvenile abalone were fed exclusively one of these three foods for 1 month. After this period, each group was presented pieces of all three food items and the majority of animals in each group preferred the artificial food over the macroalgae. Abalone released into the ocean for reseeding quickly switch back to eating natural macroalgae.[36]

The disk shape (10 to 15 mm in diameter and 1 to 2 mm thick) of the food pellet also helps retain the food in the net bags used to rear the juveniles.[37] Both macroalgae and artificial food produce fast growth rates, so the advantage of artificial food is the cost, ease of storage, and availability throughout the year.[35-37]

Artificial Food Composition
Protein Content
The artificial food must contain a protein content sufficient to produce an excellent growth rate. Maximum growth of *H. discus hannai* was obtained with a protein level of 25%, therefore, most artificial food is produced with an average protein level of 30% to ensure there is enough usable protein in the pellet.[36,38,40] The protein source is usually casein. The food value of casein is comparable or superior to *E. bicyclis*.[41,42] Casein has a protein efficiency ratio (weight gain/protein intake) of 2.4% and a net protein utilization [(protein gain of whole body with food + protein loss of whole body when given food with 0% protein)/protein intake] of 48%. An artificial food containing casein has a food conversion efficiency (weight gain/food ingest) and protein retention (protein gain of whole body/protein

intake) proportional to the dietary protein content. Casein has a crude protein content of 86.1%.[43]

White fishmeal, soybean meal, egg albumin, whole egg, and corn gluten meal are not good sources of protein for abalone.[42] Growth produced by these protein sources was low compared to casein; white fishmeal — 35 to 60%, soybean meal — 63%, egg albumin — 47%, whole egg — 44%, and corn gluten meal — 23%.[41,42] The inferior protein quality of white fishmeal is caused by high temperatures used during manufacture of this product. The effect of heat on the protein quality was most noticeable in larger animals.

Lipid Content

Feeding a lipid free or essential fatty-acid deficient diet results in retarded growth, and a low feed-conversion efficiency. Addition of 20:4ω6 or ω3 highly unsaturated fatty acids (HUFA) to artificial food improved weight gain and food conversion efficiency. The ω3 HUFA mixture (prepared from cuttlefish liver oil by saponification, esterification, and concentration, followed by vacuum distillation) contained 42% icosapentaenoic acid (20:5ω3) and 42% docosahexaenoic acid (22:6ω3). Abalone require ω3 and ω6 HUFA as essential fatty acids. The requirement for ω3 HUFA is approximately 1% of the 5% lipid content of the artificial food. The amount of ω3 HUFA required was found to be variable due to dietary lipid level.[34]

Attractants

An important aspect during development of an artificial food is whether the food contains a feeding attractant. Several organic molecules have been found to be feeding attractants in abalone (Table 15).[45,46]

Summary

Uki et al.[41] found that the best growth was obtained with an artificial food containing 30% casein, 20 to 30% sodium alginate, 5% lipid (mixture of soybean oil:pollock liver oil = 3:2, with 1% vitamin E[43]), 4% mineral mixture, 1.5% vitamin mixture, and 5% cellulose (Tables 16 to 18). This diet produced a daily increase in shell length of 110 μm, 0.77% daily feeding rate, 1.23% feed conversion efficiency, and 4.3% protein efficiency ratio.[41]

The daily minimum amount of food required for sustaining maximum growth rate of juvenile abalone (average weight — 2.3 g) was determined to be 2 to 3 g/100 g of total animal weight in rearing tank. This amount is almost twice the daily ingestion level. There is no relationship between abalone growth and dietary level of sodium alginate. Growth is slightly reduced in proportion to increasing dietary levels of cellulose and adding cellulose to the diet is not essential. Optimal level of mineral mixture is around 8% based upon increases in growth; however, after further tests the level was set at 4% to improve solubility of diet.[41]

GROWTH

The increase in the greater diameter of the shell (usual measure of shell length) underestimates the actual growth of new shell because the shell grows in a spiral. New shell is added mostly to the anterior of the shell with less growth on sides and posterior. The anterior of the shell shifts to the right as more shell is added to the right side. Growth rates of juveniles sharply increases after the first respiratory pore is formed (Table 19). However, the growth rate slows down after the first year as the animal grows larger and the shell length can actually decrease in older animals from shell erosion.[47]

Shepherd and Hearn[12] observed that the growth rate of female *H. laevigata* is about 25% greater than males. This finding is in opposition to common thinking that females grow

Table 15
FEEDING ATTRACTANTS[45,46]

Amino acids
  Only the L-form of the amino acids are active. The D-forms showed no activity.
  Basic amino acids
    All basic amino acids (moderate to strong)
    Hydroxylysine (strong)
    Ornithine (strong)
  Neutral amino acids
    Glycine (moderate)
    Cysteine (moderate)
    Tyrosine (moderate)
  Acidic amino acids
    Hydroxyproline (strong)
    Asparagine (moderate)
    Glutamine (moderate)
Neutral lipids
  Tristearin (strong)
Phospholipids
  Phosphatidylinositol (strong)
  Phosphatidylcholine (strong)
  Phosphatidyl-L-serine (moderate)
  Digalactosyldiacyclgycerol (strong)
  Lecithin (moderate)
  Cephalin (moderate)
  Phosphatidic acid (moderate)
  Sphinogosine (moderate)
  Cardiolipin (moderate)
Volatile nitrogenous bases
  Mono-, di-, and trimethylamine (moderate)
  Mono-, di-, and triethylamine (moderate)
  n-Mono-, n-di-, and n-tributylamine (moderate)
  Mono-, di-, and triethanolamine (moderate)
  Pyrrolidine (moderate)
Nonvolatile nitrogenous bases
  Choline (moderate)
  Ammonium acetate (moderate)
  γ-Aminobutyric acid (strong)

Table 16
MINERAL MIXTURE

NaCl	1.0 g
$MgSO_4 \cdot 7H_2O$	15.0
$NaH_2PO_4 \cdot 2H_2O$	25.0
KH_2PO_4	32.0
$Ca(H_2PO_4)_2 \cdot H_2O$	20.0
Fe-citrate	2.5
Trace element mixture[a]	1.0
Ca-lactate	3.5
Total	100.0

[a] See Table 17

(From Uki, N., Kemuyama, A., and Watanabe, T., *Bull. Jpn. Soc. Sci. Fish.*, 51(11), 1835, 1985. With permission.)

Table 17
TRACE ELEMENT MIXTURE

$ZnSO_4 \cdot 7H_2O$	35.3 g
$MnSO_4 \cdot 4H_2O$	16.2
$CuSO_4 \cdot 5H_2O$	3.1
$CoCl_2 \cdot 6H_2O$	0.1
KIO_3	0.3
Cellulose	45.0
Total	100.0

(From Uki, N., Kemuyama, A., and Watanabe, T., *Bull. Jpn. Soc. Sci. Fish.*, 51(11), 1835, 1985. With permission.)

Table 18
VITAMIN MIXTURE
(MIXUTRE IS DILUTED INTO 1 G OF CELLULOSE POWDER)

Thiamine HCl	6 mg
Riboflavin	5 mg
Pyridoxine HCl	2 mg
Niacin	40 mg
Ca pantothenate	10 mg
Inositol	200 mg
Biotin	0.6 mg
Folic acid	1.5 mg
PABA	20 mg
Menadione	4 mg
B_{12}	0.009 mg
Ascorbic acid	200 mg
Vitamin A	5000 I.U.
Vitamin D	100 I.U.

(From Uki, N., Kemuyama, A., and Watanabe, T., *Bull. Jpn. Soc. Sci. Fish.*, 51(11), 1835, 1985. With permission.)

Table 19
AGE AND SHELL LENGTH OF POST-LARVAL *HALIOTIS* AT FORMATION OF THE FIRST RESPIRATORY PORE

Species	Age (days)	Shell length (mm)	Temperature (°C)
H. corrugata	50—60	2.0—2.5	15—22
H. fulgens	30—40	1.7—2.0	16—24
H. rufescens	60—70	1.5—1.8	14—18
H. sorenseni	55—65	2.0—2.1	14—18

(From Leighton, D. L., *Fish. Bull.*, 72(4), 1137, 1974. With permission.)

Table 20
SIZE OF *HALIOTIS* AT COMPLETION OF FIRST YEAR GROWTH IN THE LABORATORY[49,50]

| Species | Shell length (mm) | |
	Mean	Range
H. corrugata	18.3	12.2—26.4
H. fulgens	29.7	23.7—33.7
H. rufescens	15.6	9.9—20.0
H. sorenseni	13.4	8.0—21.0

Table 21
DAILY SHELL ELONGATION RATE FOR GROUPS OF JUVENILE *HALIOTIS* REARED FOR 1 MONTH AT DIFFERENT TEMPERATURES (μm/DAY)

| Species | Water temperature (°C ± 1.5°C) | | | | | | |
	12	15	18	21	24	27	30
H. corrugata	41.5	43.0	54.7	63.7	58.7	48.0	14.0
H. fulgens	22.0	23.7	58.0	70.3	85.8	88.0	54.0
H. rufescens	60.9	64.5	77.3	70.0	27.3		

slower because of the energy requirements for production of oocytes. The faster growth in females was not explained but it has many ramifications for the aquaculture of this species. Production could be increased if it was possible through genetic manipulation to produce only females for grow out. Also, reseeding projects must be aware of this difference in growth. Care must be taken not to collect predominantly females from the wild population and reduce the reproductive effort of the population.[12]

Some species of abalone when cultured in the laboratory can have their growth increased dramatically by just rearing the animals at elevated constant water temperatures throughout the year (Tables 20 and 21).[5,16,22,32,48-50] The constant water temperature eliminates the slow growth during the winter in natural populations (Figures 1 to 3). Koike[22] found *H. tuberculata* grew 18 mm/year when reared at 20°C. This growth rate will produce a 45-mm juvenile in 2 years instead of 3 years required in the wild.[17] *H. fulgens* shows an enhanced growth rate when reared at water temperatures above 20°C and is an excellent species for culture with heated effluent from electric power plants or other sources of cheap, heated water.[49]

H. discus hannai has also been suggested for culture with heated effluent.[5,51] Abalone reared in warm water effluent at the Tohoku Electric Power Station were found to grow 4 to 5 times faster than those reared in ambient water.[51] *H. discus hannai* reared with ideal conditions (20°C and *ad libitum* feeding) will grow up to 6 cm/year during the first 2 years of growth. Growth with ambient water temperatures is approximately 2 cm/year (1.5 years — 3 cm, 4 years — 9 cm).[5] The use of warm water effluent for rearing *H. discus hannai* shows great promise.

Hybrids of warm water abalone species in California have shown growth rates greater than either parent species (Table 22).[52] If problems can be solved with fertilization and larval development, hybrids could significantly increase the production of abalone culture facilities.

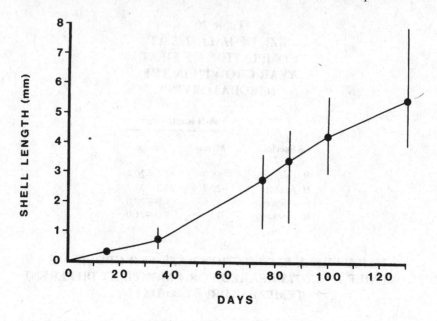

FIGURE 1. Growth of *H. sorenseni* juveniles fed naturally occurring diatoms. Vertical lines show shell length range at each sample period. (From Leighton, D. L., *Fish. Bull.*, 70(2), 373, 1972. With permission.)

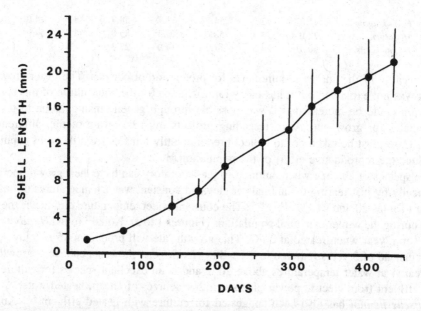

FIGURE 2. Growth of *H. tuberculata* juveniles up to 435 days after fertilization (day 0). Vertical lines show shell length range at each sample period. (From Koike, Y., *La Mer*, 16(3), 124, 1978. With permission.)

Even with optimum rearing conditions, there is great variation in the growth rates of individuals of the same age and parents, even when reared in the same tank (Table 23). The size range of individuals increases with age.[49] Momma[4] found individuals of the same size but of different ages (153, 203, and 570 days old) showed growth rates corresponding to their past growth history. Older individuals (thus slower growing) had lower growth rates than the younger individuals. Slow-growing abalone remained slow and fast-growing abalone

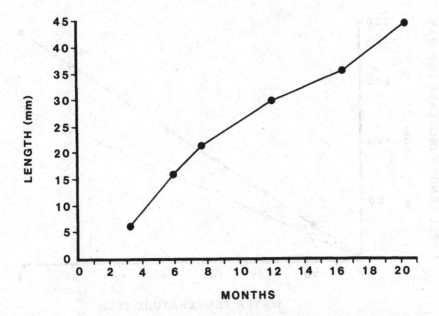

FIGURE 3. Growth of *H. fulgens* juveniles up to 20 months after settlement. Water temperature range during rearing was 18 to 24°C. (From Leighton, D. L., Byhower, M. J., Kelly, J. C., Hooker, G. N., and Morse, D. E., *J. World Maricult. Soc.*, 12(1), 170, 1981.)

Table 22
SIZE OF HYBRID *HALIOTIS* AT
COMPLETION OF FIRST YEAR
GROWTH IN THE LABORATORY

Species		Shell length (mm)	
Female	Male	Mean	Range
H. fulgens	*H. rufescens*	28.0	21.2—37.4
H. rufescens	*H. corrugata*	17.3	8.9—26.1
H. rufescens	*H. sorenseni*	24.9	15.6—33.4
H. sorenseni	*H. rufescens*	21.1	11.8—32.6

(From Leighton, D. L. and Lewis, C. A., *Int. J. Invertebr. Reprod.*, 5, 273, 1982. With permission.)

Table 23
VARIATION IN SIZE OF JUVENILE
HALIOTIS OF IDENTICAL PARENTAGE,
AGE, AND GROWING ENVIRONMENT

Species	Age (months)	Shell length (mm)		
		Mean	Range	S.D.
H. corrugata	5.0	9.2	6.8—14.1	2.42
H. fulgens	4.7	7.1	3.6—12.0	2.00
H. rufescens	3.7	4.4	2.6—6.1	1.06
H. sorenseni	3.3	4.3	3.0—5.6	0.64

(From Leighton, D. L., *Fish. Bull.*, 72(4), 1137, 1974. With permission.)

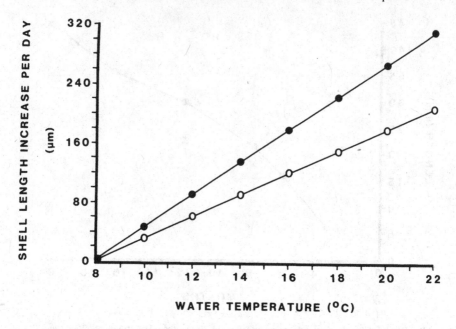

FIGURE 4.   Growth of *H. discus hannai* juveniles related to water temperature and quality of food. (○) mixed array of naturally occurring diatom species, Y = 14.75 * (T − 7.8). (●) diatom culture of *Nitzschia divergens* and *Navicula agnita*, Y = 21.95 * (T − 7.8). (From Uki, N., Grant, J. F., and Kikuchi, S., *Bull. Tohoku Reg. Fish. Res. Lab.*, 43, 59, 1981. With permission.)

remained fast when reared under identical conditions. The older group had a growth rate of 43 μm/day (range 0 to 150 μm/day), middle-aged group had a growth rate of 77 μm/day, and youngest group had a growth rate of 112 μm/day.[4]

The lowest water temperature tolerated by *H. discus hannai* is 2°C; however, the lowest temperature where growth can occur is 7.8°C.[26,53] The relationship between water temperature and growth rate of juvenile (5 to 30 mm) *H. discus hannai* is linear up to 22.5°C (approximately the maximum water temperature along the northeastern coast of Japan). The formula of the line is Y = 14.75 * (T − 7.8), where Y equals shell growth rate (μm/day) and T equals temperature (°C). Food with improved nutrition increases the slope of the line (Figure 4).[26]

The effect of water temperature on growth can be clearly seen in the growth of wild populations of *H. discus hannai* along the Pacific coast of Japan. The growth rates decrease the farther north the population is along the coast. Populations of *H. discus hannai* along Mie Prefecture require 4 to 5 years to reach 10.6 cm, Ibaraki Prefecture require 4 to 6 years to reach 11 cm, Miyagi Prefecture require 4 to 5 years to 9 cm, and Hokkaido Prefecture require 5 to 7 years to reach 7.5 cm.[54] The growth rates correlate with decreasing water temperatures and length of algae growing season.

Annual Growth Rings

Growth rates of individuals placed in the wild for reseeding or ocean ranching are hard to estimate because it is difficult to precisely age individuals. Most methods of tagging do not last for long periods or are obscured by erosion of older regions of shell. Also, there are some cases when juveniles are placed into the ocean when they are too small for tagging. Aging individuals is usually a circular argument. An average growth rate is obtained by estimating the age of individuals, and the age of animals is estimated by using the average growth rate. As was mentioned above, growth is highly variable even with identical conditions. Therefore, growth in the wild would be so variable as to make estimation of age by shell length useless.

Table 24
SHELL COLOR PRODUCED BY INGESTION OF DIFFERENT ALGAE

Species	Brown algae	Green algae	Red algae	Diatom	Ref.
H. assimilis	Mottled turquoise and white Pale brown orange		Mottled red, blue, and white		64
H. corrugata	Turquoise, white		Brown-red		64
H. cracherodii	Blue black Blue green		Reddish brown	Blue green	2, 64
H. discus hannai	Bluish green	Bluish green	Brown	Bluish green	55, 62
H. discus hannai	Green		Brown Yellowish brown		60
H. diversicolor supertexta		Green	Brown		66
H. gigantea	Green		Yellowish brown or brown		60
H. rufescens	Brown	White or greenish white		Dull brick red	65
H. rufescens	White Blue green Pale green Olive green	Pale green	Red		61, 64
H. sieboldii	Yellowish brown or brown		Yellowish brown or brown		60
H. sorenseni	Pale green Pale brown	Pale green	Red, purple		64
H. tuberculata	Green	Green	Red	Reddish brown	22

A few species of abalone can be aged by observing annual growth rings in the shell.[7,13,24,55-57] Growth rings can sometimes be seen on the outside of cleaned shells as ridges or a series of ridges marking the position of the growing edge of the shell. Growth rings on young abalone shells appear as lines on the nacre or as opaque areas in the translucent shell. If the shell is sliced in cross-section, growth rings are interruptions between the outer prismatic and inner nacreous layers.[7]

Growth rings have been used in *H. tuberculata*, *H. discus hannai*, *H. virginea*, *H. australis*, and *H. iris*.[7,13,24,55-57] Rings have not been found in *H. rufescens*, *H. cracherodii*, *H. midae*, and *H. kamtschatkana*.[7,8,14,58]

Sakai[55] found a distinct mark in the shell of *H. discus hannai* that was formed during the month of September and this ring was not formed during any other time of the year. The formation of the annual ring is a result of temporal cessation of shell growth at the period of gonad maturation and spawning. This conclusion is supported by the fact that no annual ring is seen in abalone that are immature.[55] Shell growth of most abalone species in Japan is discontinued during the spawning season and the shell increases in thickness during this period. This causes the formation of an annual ring and makes it possible to determine the age.[21]

Age can also be determined by heating the shell to 400°C, which causes the shell to fall apart in sections corresponding to each year's growth.[55,59] The annual rings can also be seen after cleaning the shell and holding it up to a red light in a dark room.[59]

Shell color is determined by the food eaten by the abalone and the shell will show banding from the succession of the dominant algal species during the year (Table 24).[2,22,55,60-66] The age of the animal can sometimes be estimated by observing the color-banding pattern in the shell.[63,67,68] Some researchers have suggested varying the algal species given to culture abalone to produce color bands as a biological tag on reseeded juveniles. This method has limited use because the bands would only be visible until shell erosion of the older regions obscures the colors.

REFERENCES

1. **Burge, R., Schultz, S., and Odemar, M.,** Draft Report on Recent Abalone Research in California with Recommendations for Management, State of Calif., The Resources Agency, Dept. Fish and Game, 1975.
2. **Leighton, D. and Boolootian, A.,** Diet and growth in the black abalone, *Haliotis cracherodii, Ecology,* 44(2), 227, 1963.
3. **Kawamura, K., Hayashi, T., Sato, M., and Takano, M.,** Ecological observations on the abalone, *Haliotis discus hannai* Ino, in the closed fishing grounds along the coast of Otaru, Hokkaido, Japan, *Sci. Rep. Hokkaido Fish. Exp. Lab.,*12, 33, 1970.
4. **Momma, H.,** Studies on the variation of the abalone. I. On the growth of the different aged young abalone, *Aquiculture,* 28(3), 142, 1980. (Translated in Mottet, M. G., Summaries of Japanese papers on hatchery, technology and intermediate rearing facilities for clams, scallops, and abalones, *Progress Rep., Dept. Fish. Washington,* 203, 19, 1984.)
5. **Kumabe, L.,** Methods used at the Oyster Research Institute, unpublished report, 1981.
6. **Oba, T., Sato, H., Tanaka K., and Toyama, T.,** Studies on the propagation of an abalone, *Haliotis diversicolor supertexta.* III. On the size of the one-year-old specimen, *Bull. Jpn. Soc. Sci. Fish.,* 34(6), 457, 1968.
7. **Poore, G. C. B.,** Ecology of New Zealand abalones, *Haliotis* species (Mollusca: Gastropoda). III. Growth, *N. Z. J. Mar. Freshwater Res.,* 7(1 and 2), 67, 1973.
8. **Quayle, D. B.,** Growth, morphometry and breeding in the British Columbia abalone *(Haliotis kamtschatkana* Jonas), *Fish. Res. Board Can. Tech. Rep.,* 279, 1, 1971.
9. **Norman-Boudreau, K.,** Study of Nutrition in Abalone Mariculture, Pacific Trident Mariculture, Final Report 1984 — 1985, 1985.
10. **Caldwell, M. E.,** Spawning, Early Development and Hybridization of *Haliotis kamtschatkana* Jonas, Master's thesis, University of Washington, Seattle, 1981.
11. **Olsen, S.,** Abalone and scallop culture in Puget Sound, *J. Shellfish Res.,* 3(1), 113, 1983.
12. **Shepherd, S. A. and Hearn, W. S.,** Studies on southern Australian abalone (genus *Haliotis).* IV. Growth of *H. laevigata* and *H. ruber, Aust. J. Mar. Freshwater Res.,* 34, 461, 1983.
13. **Forster, G. R.,** The growth of *Haliotis tuberculata:* results of tagging experiments in Guernsey 1963 — 65, *J. Mar. Biol. Assoc. U. K.,* 47, 287, 1967.
14. **Newman, G. G.,** Growth of the South African abalone *Haliotis midae, Div. Sea Fish. Invest. Rep.,* 67, 1, 1968.
15. **Ebert, E. E. and Houk, J. L.,** Elements and innovations in the cultivation of Red abalone *Haliotis rufescens, Aquaculture,* 39, 375, 1984.
16. **Hooker, N. and Morse, D. E.,** Abalone: the emerging development of commercial cultivation in the United States, in *Crustacean and Mollusk Aquaculture in the United States,* Huner, J. V. and Brown, E. E., Eds., AVI Publishing Co., 1985, 365.
17. **Clavier, J. and Richard, O.,** Growth of juvenile *Haliotis tuberculata* (Mollusca: Gastropoda) in their natural environment, *J. Mar. Biol. Assoc. U. K.,* 66, 497, 1986.
18. **Morse, D. E.,** Prospects for the California abalone resource: recent development of new technologies for aquaculture and cost-effective seeding for restoration and enhancement of commercial and recreational fisheries,in *Ocean Studies,* Hansch, S., Ed., Calif. Coastal Comm., San Francisco, 1984, 165.
19. **Kikuchi, S., Sakurai, Y., Sasaki, M., and Ito, T.,** Food values of certain marine algae for the growth of the young abalone, *H. discus hannai, Bull. Tohoku Reg. Fish. Res. Lab.,* 27, 93, 1967.
20. **Morse, D. E.,** Biochemical and genetic engineering for improved production of abalones and other valuable molluscs, *Aquaculture,* 39, 263, 1984.
21. **Ino, T.,** *Fisheries in Japan, Abalone and Oyster,* Japan Marine Products Photo Materials Assoc., Tokyo, 1980, 165.
22. **Koike, Y.,** Biological and ecological studies on the propagation of the ormer, *Haliotis tuberculata* Linnaeus. I. Larval development and growth of juveniles, *La Mer,* 16(3), 124, 1978.
23. **Norman-Boudreau, K., Burns, D., Cooke, C. A., and Austin, A.,** A simple technique for detection of feeding in newly metamorphosed abalone, *Aquaculture,* 51, 313, 1986.

155

24. **Tong, L.,** personal communication, 1984.
25. **Uki, N. and Kikuchi, S.,** Food value of six benthic microalgae on growth of juvenile abalone, *Haliotis discus hannai, Bull. Tohoku Reg. Fish. Res. Lab.,* 40, 47, 1979.
26. **Uki, N., Grant, J. F., and Kikuchi, S.,** Juvenile growth of the abalone, *Haliotis discus hannai,* fed certain benthic micro algae related to temperature, *Bull. Tohoku Reg. Fish. Res. Lab.,* 43, 59, 1981.
27. **Uki, N., Sugiura, M., and Watanabe, T.,** Dietary value of seaweeds occurring on the Pacific coast of Tohoku for growth of the abalone *Haliotis discus hannai, Bull. Jpn. Soc. Sci. Fish.,* 52 (2), 257, 1986.
28. **Uki, N.,** Food value of marine algae of order Laminariales for growth of the abalone, *Haliotis discus hannai, Bull. Tohoku Reg. Fish. Res. Lab.,* 42, 19, 1981.
29. **Poore, G. C. B.,** Ecology of New Zealand abalones, *Haliotis* species (Mollusca, Gastropoda). I. Feeding, *N. Z. J. Mar. Freshwater Res.,* 6(1 and 2), 11, 1972.
30. **Dutton, S. and Tong, L.,** Food preferences of paua, *Catch' 81,* March 15, 1981.
31. **Tong, L. J.,** Paua research shows progress, *Catch' 83,* September 18, 1983.
32. **Tong, L. J.,** The potential for aquaculture of paua in New Zealand, in *Proc. Paua Fish. Workshop,* Fish. Res. Div. Pub. No. 41, Akroyd, J. M., Murray, T. E., and Tayler, J. L., Eds., N. Z. Min. Agric. Fish., Wellington, 1982, 36.
33. **Hahn, K.,** personal observation, 1981.
34. **Uki, N., Sugiura, M., and Watanabe, T.,** Requirement of essential fatty acids in the abalone *Haliotis discus hannai, Bull. Jpn. Soc. Sci. Fish.,* 52(6), 1013, 1986.
35. Anonymous, The Production of Abalone Seeds Using "Nosan's Abalone Feed", Nihon Nosan Kogyo K.K. Research Center, Jpn. Patent Publ. No. 55-1586, Material No. 2, 1981.
36. Anonymous, The Feeding Method of "Nosan's Abalone Feed", Nihon Nosan Kogyo K.K. Research Center, Jpn. Patent Publ. No. 55-1586, Material No. 3, 1981.
37. Anonymous, Nosan's Abalone Feed (for Young Abalone Breeding), Nihon Nosan Kogyo K.K. Research Center, Jpn. Patent Publ. No. 55-1586, Material No. 1, 1981.
38. **Ogino, C. and Ohta, E.,** Studies on the nutrition of abalone. I. Feeding trials of abalone, *Haliotis discus* Reeve, with artificial diets, *Bull. Jpn. Soc. Sci. Fish.,* 29(7), 691, 1963.
39. **Sagara, J. and Sakai, K.,** Feeding experiment of juvenile abalones with four artificial diets, *Bull. Tokai Reg. Fish. Res. Lab.,* 77, 1, 1974.
40. **Ogino, C. and Kato, N.,** Studies on the nutrition of abalone. II. Protein requirements for growth of abalone, *Haliotis discus, Bull. Jpn. Soc. Sci. Fish.,* 30(6), 523, 1964.
41. **Uki, N., Kemuyama, A., and Watanabe, T.,** Development of semipurified test diets for abalone, *Bull. Jpn. Soc. Sci. Fish.,* 51(11), 1825, 1985.
42. **Uki, N., Kemuyama, A., and Watanabe, T.,** Nutritional evaluation of several protein sources in diets for abalone *Haliotis discus hannai, Bull. Jpn. Soc. Sci. Fish.,* 51(11), 1835, 1985.
43. **Uki. N., Kemuyama, A., and Watanabe, T.,** Optimum protein level in diets for abalone, *Bull. Jpn. Soc. Sci. Fish.,* 52(6), 1005, 1986.
44. **Uki, N. and Watanabe, T.,** Effect of heat-treatment of dietary protein sources on their protein quality for abalone, *Bull. Jpn. Soc. Sci. Fish.,* 52(7) 1199, 1986.
45. **Harada, K. and Kawasaki, O.,** The attractive effect of seaweeds based on the behavioral responses of young herbivorous abalone *Haliotis discus, Bull. Jpn. Soc. Sci. Fish.,* 48(5), 617, 1982.
46. **Sakata, K. and Ina, K.,** Digalactosyldiacylglycerols and phosphatidylcholines isolated from a brown alga as effective phagostimulants for a young abalone, *Bull. Jpn. Soc. Sci. Fish.,* 51(4), 659, 1985.
47. **Mottet, M. G.,** A review of the fishery biology of abalones, *Washington Dep. Fish., Tech. Rep.* 37, 1, 1978.
48. **Leighton, D. L.,** Laboratory observations on the early growth of the abalone, *Haliotis sorenseni,* and the effect of temperature on larval development and settling success, *Fish. Bull.,* 70(2), 373, 1972.
49. **Leighton, D. L.,** The influence of temperature on larval and juvenile growth in three species of southern California abalones, *Fish. Bull.,* 72(4), 1137, 1974.
50. **Leighton, D. L., Byhower, M. J., Kelly, J. C., Hooker, G. N., and Morse, D. E.,** Acceleration of development and growth in young Green abalone *(Haliotis Fulgens)* using warmed effluent seawater, *J. World Maricult. Soc.,* 12(1), 170, 1981.
51. **Shaw, W. N.,** The culture of molluscs in Japan, *Aquacult. Mag.,* November — December, 43, 1982.
52. **Leighton, D. L. and Lewis, C. A.,** Experimental hybridization in abalones, *Int. J. Invertebr. Reprod.,* 5, 273, 1982.
53. **Sakai, S.,** Ecological studies on the abalone, *Haliotis discus hannai* Ino. IV. Studies on the growth, *Bull. Jpn. Soc. Sci. Fish.,* 28(9), 899, 1962.
54. **Imai, T.,** *Aquaculture in Shallow Seas: Progress in Shallow Sea Culture,* Part IV, Amerind Publishing Co., New Delhi, 1977.
55. **Sakai, S.,** On the shell formation of the annual ring on the shell of the abalone, *Haliotis discus* var. *hannai* Ino, *Tohoku J. Agric. Res.,* 11, 239, 1960.

56. **Cochard, J.,** Recherches sur les Facteurs Determinant la Sexualite et la Reproduction Chez *Haliotis tuberculata* L., Ph. D. thesis, Universite de Bretagne Occidentale, 1980.

57. **Flassch, J. and Aveline, C.,** Production de Jeunes Ormeaux a la Station Experimentale d'Argenton, *Pub. C.N.E.X.O., Rapp. Sci. Tech.,* 50, 1, 1984.

58. **Boolootian, R. A., Farmanfarmaian, A., and Giese, A. C.,** On the reproductive cycle and breeding habits of two western species of *Haliotis, Biol. Bull.,* 122, 183, 1962.

59. **Uki, N.,** Personal communication, 1981.

60. **Ino, T.,** Biological studies on the propagation of Japanese abalone (genus *Haliotis*), *Bull. Tokai Reg. Fish. Res. Lab.,* 5, 1, 1952.

61. **Leighton, D. L.,** Observations on the effect of diet on shell coloration in the red abalone, *Haliotis rufescens* Swainson, *The Veliger,* 4(1), 29, 1961.

62. **Sakai, S.,** Ecological studies on the abalone, *Haliotis discus hannai* Ino. I. Experimental studies on the food habit, *Bull. Jpn. Soc. Sci. Fish.,* 28(8), 766, 1962.

63. **Olsen, D.,** Banding patterns of *Haliotis rufescens* as indicators of botanical and animal succession, *Biol. Bull.,* 134, 139, 1968.

64. **Olsen, D. A.,** Banding patterns in *Haliotis.* II. Some behavioral considerations and the effect of diet on shell coloration for *Haliotis rufescens, Haliotis corrugata, Haliotis sorenseni,* and *Haliotis assimilis, The Veliger,* 11(2), 135, 1968.

65. **Shibui, T.,** Studies on the transplantation of Red abalone and its growth and development, *Bull. Jpn. Soc. Sci. Fish.,* 37(12), 1168, 1971.

66. **Chiu, C. C.,** Biological Studies on the Propagation and Larval Rearing of Abalone, *Haliotis diversicolor,* Master's thesis, National Taiwan University, Taipei, 1981.

67. **Sakai, S.,** Ecological studies on the abalone, *Haliotis discus hannai* Ino. II. Mutuality among the colored shell area, growth of the abalone and algal vegetation, *Bull. Jpn. Soc. Sci. Fish.,* 28(8), 780, 1962.

68. **Sakai, S.,** Ecological studies on the abalone, *Haliotis discus hannai* Ino. IV. Studies on the growth, *Bull. Jpn. Soc. Sci. Fish.,* 28(9), 899, 1962.

69. **Peon, S. C.,** Studies on the Age and Growth of Small Abalone in Hua Lien, Master's thesis, National Taiwan University, Taipei, 1980.

70. **Saito, Y.,** Information on the culture of phytoplankton for aquaculture needs, *N.O.A.A. Tech. Rep., NMFS,* 442, 1, 1982.

71. **Koike, Y., Flassch, J., and Mazurier, J.,** Biological and ecological studies on the propagation of the ormer, *Haliotis tuberculata* Linnaeus. II. Influence of food and density on the growth of juveniles, *La Mer,* 17(1), 43, 1979.

ABALONE SEEDING

Mia J. Tegner and Robert A. Butler

INTRODUCTION: THE RATIONALE FOR ABALONE SEEDING

Abalones, prosobranch gastropods of the family Haliotidae, are highly prized shellfish especially in the Orient and the U.S. While abalones occur in many parts of the world, the species large enough to support commercial fisheries are found along temperate coasts. The major producing countries are Mexico, Japan, Australia, South Africa, and the U.S.[1] The natural history of abalones facilitates overfishing, and this has been a problem in every producing country. Abalones live in shallow water near stands of their algal food, in relatively predictable and accessible locations. Once larvae settle to the bottom, the juveniles may move on scales of tens of meters and the adults are virtually stationary. Furthermore, abalones have unpredictable recruitment,[2] slow growth rates, and sublegal sizes are subject to substantial mortality from bar cuts.[3] In combination, these factors make abalones particularly susceptible to overexploitation.

The recent history of the California abalone fishery (Figure 1) illustrates the pressure on these animals. Traditionally the most commercially important species has been the red (*Haliotis rufescens*), followed by the pink (*H. corrugata*), the green (*H. fulgens*), and the black (*H. cracherodii*). From 1951 to 1968, total abalone landings were more than 1800 metric tons per year. Despite increased harvest of the less desirable greens and blacks in the early 1970's, total landings have since declined to about 600 t/year. The evidence associating the decline with a large increase in fishing pressure includes the development of more practical diving gear, the introduction of faster boats making the offshore islands more accessible, a 300% increase in the commercial fleet between the early 1950's and 1974, and a similar increase in the number of sport divers. Other factors, such as the range expansion of the sea otter (*Enhydra lutris*) which is a voracious abalone predator, encroachment by sea urchins, environmental degradation of some coastal sites, and closures in some areas on the offshore islands, acted to concentrate the increased effort in smaller areas.[34] To reduce the pressure on a declining resource, California instituted limited entry to the commercial fishery, reduced the sport take, and closed a stretch of the coastline to all abalone fishing to give remaining stocks a chance to recover. While such efforts may halt the decline of abalone stocks, recovery will be slow.

The Japanese pioneered the use of hatchery-reared juvenile abalones to enhance native stocks. As in the U.S., abalones are an expensive food, not a low-priced dietary staple. Despite this status, culture from egg to adult is not economically feasible because of the slow growth rates.[5] The techniques for seed production were developed in the 1960's[6] and by 1971 there were 16 laboratories producing 2 to 3 million seed annually.[7] Hatchery development continued at a rapid rate and 11,658,000 seed were produced in 1979, almost all of which were delivered to fishing cooperatives for release in the sea.[5] The success of this effort is hard to judge. A number of Japanese fishery biologists have reported experimental results ranging from less than 1 to 80% seed survival.[8,9] However, despite the massive nationwide seeding effort, Kafuku and Ikenoue[5] report that "the number released is too low to determine whether there is any effect on (the total national) catches". These authors document a slow decline in the national harvest from 1970 to 1979.

Seeding is an appealing approach to abalone resource enhancement for several reasons: (1) by analogy with the well-known successes of salmon and trout, (2) no restrictions are put on existing levels of abalone fishing effort, (3) seeding is potentially much more rapid than natural stock recovery, and (4) seeding could eventually be self sustaining as the seed

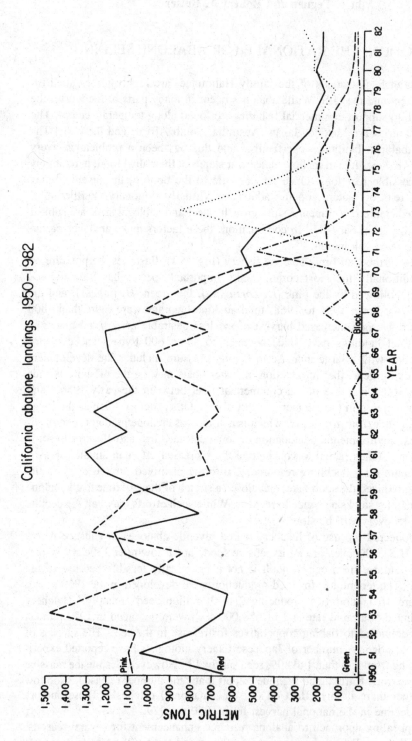

FIGURE 1. California abalone landings 1950 to 1982 for the major commercial species. (Calif. Dept. Fish Game).

reproduce before harvest. Interest is also growing in sea floor farming, the leasing of state-owned bottom for cultivation of a seed "crop" by private enterprise. Given the reduced state of the common property abalone resource in California, the enthusiasm for seeding in Japan, and the availability of hatchery-reared abalones, the University of California Sea Grant College Program and the California Department of Fish and Game sponsored studies to determine the scientific and economic feasibilities of seeding as an approach to stock enhancement in southern California. Here we discuss general considerations for abalone seeding, some of the techniques developed, experimental results from this research, and future directions for abalone seeding in California, and compare these with the situation in Japan.

HABITAT SELECTION

Survival and growth rates of hatchery-reared abalones planted in the ocean are likely to be higher in areas that are not only good quality abalone habitat generally but also provide for the specific needs of the seed employed. There are eight species of abalones in California, each with its own characteristic natural history.[10-12] Although there is considerable overlap, these species vary in their preferred temperature range, depth distribution, physical habitat, and food preferences (Table 1).

The depth and latitudinal distributions of the different species appear to be most closely related to temperature.[13] Of the major commercial species, greens and pinks are found along the warm, southern Channel Islands and in shallow waters along the mainland of southern California. Pinks tend to be found somewhat deeper than greens and are also more abundant in habitats of intermediate temperatures. Red abalones are typical of shallow habitats in cooler waters such as the northern Channel Islands and San Nicolas Island, and the mainland coast north and west of Santa Barbara. Reds are also found at depths below pink abalones in areas off San Diego and the Palos Verdes Peninsula in Los Angeles County where upwelling tends to keep temperatures lower. Leighton[13] showed that eggs and early larvae are the abalone life history stages most sensitive to temperature; older larvae, juveniles and adults survive a wider range of temperatures. Because of the larval sensitivity, field distributions of juveniles and adults correspond to the thermal tolerances of each species' larvae. Although seeding bypasses the larval stage, water temperatures are still an important issue. At cool temperatures (12 to 15°C) reds grow faster than pinks which grow faster than greens, at intermediate temperatures (18 to 21°C) these differences converge, and at warm temperatures (>21°C) greens grow faster than pinks which grow faster than reds.[13]

General habitat requirements for abalones, in addition to the proper temperature/depth combination, include good water circulation to remove wastes and sediments, rocky substrates of the appropriate size and configuration to provide protection from predators, and good supply of drift algal food. Native abalones are not found in bays or estuaries, preferring the enhanced circulation of the open coast.[10] Influxes of fresh water or sand can cause substantial mortality,[14] so seeding sites should be located away from input sources, even dry ones. Within these general water circulation requirements are species-specific differences. For example, Tutshulte[11] found that because green abalones choose more cryptic microhabitats than pink abalones, they can extend their distribution onto more wave-beaten promontories than pinks which are restricted to more sheltered sections of the shallow subtidal and deeper water.

Finding the appropriate rocky habitat may be the most difficult aspect of seeding site selection; the necessary temperature/depth, water circulation, and food requirements are much more widespread. Abalones are subject to a wide variety of predators[9,10,12,15] and the nature of the physical habitat plays a critical role in avoiding predation. Small- and intermediate-sized abalones, which have little to no refuge in shell thickness, must rely on

Table 1
COMPARISON OF THE NATURAL HISTORY OF THE THREE MAJOR COMMERCIAL ABALONE SPECIES

	Red abalones (*Haliotis rufescens*)	Pink abalones (*H. corrugata*)	Green abalones (*H. fulgens*)	Ref.
Range	Sunset Bay, Oregon to Bahia Tortugas, Baja California	Point Conception, California to Bahia Tortugas, Baja California	Point Conception, California to Bahia Magdalena, Baja California	10
Depth range	Low intertidal (north of Point Conception) to more than 180 m; major concentrations shallower than 11 m in northern California, and 8 — 25 m in southern California	Low intertidal to 60 m; major concentrations 6 to 24 m	Low intertidal to 18 m; major concentrations 1.5 to 8 m	10, 11
Relative thermal preference	Coolest	Intermediate	Warmest	
Optimum temperatures (T_o) for growth of juveniles	15—21°C	18—24°C	21—27°C	13
Relative juvenile growth rates at their T_o	Intermediate	Slowest	Fastest	13
Relative juvenile growth rates at Santa Catalina Island		Faster	Slower	11
Habitat preferences	Shallow, high energy environments in north; deeper (cool) water in south	Occupies quieter water, either more sheltered or deeper than green abalones	Areas exposed to high-wave action and currents	10, 11
Laboratory food preferences	*Macrocystis, Egregia Laminaria, Eisenia*	*Macrocystis, Egregia Eisenia, Laminaria Pterygophora*	*Egregia, Macrocystis Eisenia, Laminaria* Various red algae	25
Food preferences at Santa Catalina Island		*Eisenia, Plocamium*	Various red algae	11

inaccessibility to predators in order to survive. Thicker-shelled adult abalones appear to be less vulnerable to certain predators; e.g., we have found no red abalone shells above 125 mm with octopus drill holes. There is no refuge in size against such predators as sea otters and bat rays (*Myliobatis californica*) however, so even adult abalones tend to concentrate in habitats with some degree of protection. Abalones require approximately planar surfaces to attach. We have found that small (<30 mm) and large (>100 mm) abalones live in quite distinct microhabitats (Table 2). The open surfaces necessary for 150- to 200-mm adults offer little protection to juveniles of one tenth the size; conversely, the habitats where juveniles are commonly found will generally not accommodate adults. Intermediate sizes occupy some microhabitats characteristic of each of the other groups. Shepherd[15] reports similar size-specific microhabitats for Australian species.

The importance of the different juvenile microhabitats varies considerably between areas. For example, a large pipe laid on a pavement bottom near Point Conception supports a substantial red abalone population in the absence of turnable rocks because sea urchins (Figure 2) provide protection for small- and intermediate-sized abalones.[16] Adult red urchins (*Strongylocentrotus franciscanus*) are supported 1 to 2 cm above the substrate by the oral

Table 2
HABITAT SELECTION BY SMALL, INTERMEDIATE AND LARGE ABALONES

	Small abalones (<30 mm)	Intermediate abalones (30 — 100 mm)	Large abalones (> 100 mm)
Habitats where found	Small rocks Sea urchins Medium rocks *Macrocystis* holdfasts Large rocks	Ledges Medium rocks Large rocks Sea urchins Small rocks Crevices	Ledges Large rocks Crevices Exposed faces
Habitats where absent	Ledges Crevices Exposed faces	*Macrocystis* holdfasts Exposed faces	Small rocks Medium rocks Sea urchins *Macrocystis* holdfasts

Habitat Descriptions

Small rocks	< 0.1 m² bottom surface area
Medium rocks	0.1 — 0.25 m² bottom surface area
Large rocks	> 0.25 m² bottom surface area
Sea urchins	Area protected by the spine canopies of *Strongylocentrotus franciscanus* and *S. purpuratus*
Ledges	Undercut surfaces in reefs, typically with narrow openings and parallel to the bottom
Crevices	Narrow vertical cracks or openings in reefs
Exposed faces	Habitats which offer no apparent protection from predators

Note: Habitats are ranked in order of importance. Data are pooled for red, pink and green abalones from many different sites.

FIGURE 2.   The spine canopy of adult red sea urchins, *Strongylocentrotus franciscanus,* provides shelter for juvenile abalones and sea urchins. (From Tegner, M.J. and Dayton, P.K., *Science,* 196, 324, 1977. With permission.)

spines; juvenile sea urchins and abalones up to 50 mm clustered underneath derive protection from predators and a share of the drift algae snared by the adult urchin.[17,18] A site on San Miguel Island with a large number of turnable rocks had juvenile abalone densities of 1.4/m² in boulder fields and 1.0/m² on urchin-dominated pavement. In contrast, Day[19] found no juvenile abalones under urchins on reefs near Santa Barbara; all were under rocks. Thus, while abalones do show some flexibility in habitat selection, seed survival rates are likely to be higher in sites which provide ample habitat for all size classes, a characteristic of the most productive natural beds.

There are important species-specific considerations as well. Whether by preferential settlement or differential survival, juvenile and adult red abalones are considerably more abundant on rocks set in sand than on rocks set on a rocky substrate within the same area. In shallow green abalone habitat, which is characterized by much higher water movement, turnable rocks are less common and the sand is much coarser. Adult green abalones are typically found in crevices or cavities at the base of rocky outcrops or within narrow, undercut ledges. Juvenile greens are found under rocks, in narrow crevices, or under sea urchins within the same area. Pink abalone habitat is similar to that of greens in shallow water and converges with red abalone habitat in deeper water.

An abundant supply of the appropriate food is the final consideration for site selection. In the San Miguel Island example described above, abalones under rocks average 30 mm and those under urchins, 19 mm. Because 30-mm abalones fit easily under adult red urchins, this segregation implies a change in abalone feeding behavior with size. Sea urchins maintain patches of encrusting coralline algae, an important food source for small abalones,[20] free from overgrowth by other organisms.[21] Juveniles also feed on other encrusting algae such as diatoms. As abalones grow larger, they depend increasingly on drift algae which tend to accumulate at the sand/rock interface.

Very small (≤5 mm) abalones are found almost exclusively on encrusting coralline algae,[22-24] which act as inducers of larval settlement and metamorphosis,[23] as food sources,[20,23] and because the settled abalones quickly acquire pink coloration, as a refuge from predators.[22,23] Abalones of this size class also appear to require clean, relatively sediment-free crustose coralline surfaces. The most productive juvenile abalone habitat we have found, the side of the San Diego sewage outfall very distant from the release site, seemed to be distinguished from the large number of other crustose coralline habitats in the area by the lack of sediments. A layer of cobbles (8 to 15 cm diameter) rested against the pipe and on top of large boulders protecting the structure (Figure 3). Crustose coralline algae covered the exposed side of the pipe (this large expanse may have been acting as a larval trap) and much of the cobble surface area; probably as a result of water movement through the rock pile, these surfaces were very clean. A sample of 24 cobbles contained 33 red abalones (1 to 22 mm in size), a density of 224 juvenile *H. rufescens* per square meter of bottom surface area.[23] Sea urchins at the pipe-cobble interface also sheltered juvenile abalones, up to ten small red abalones under individual *S. franciscanus* and *S. purpuratus*. This structure has been an important nursery habitat for several years. Similarly, Saito[24] reported that small abalones disappeared when their crustose coralline substratum became covered with sediments and epiphytes.

Leighton's[25] laboratory food preference studies indicate that red, pink, and green abalones all prefer species of Laminarian algae (Table 1). He concluded that minor differences in preference may relate to algal abundance at the depths where the abalones are found. Tutshulte[11] obtained somewhat diferent results in field studies. His data show that greens eat *Macrocystis* and *Eisenia* in the proportion they occur in the drift, select for foliose red algae and avoid *Egregia*. This selection for red algae suggests that the decrease in red algal drift (along with the decrease in temperature) with depth may contribute to the lower depth limit of green abalones. Pink abalones ate *Macrocystis* and *Dictyopteris/Pachydictyon* in the proportions in which they appeared in the drift, selected for *Eisenia* and *Plocamium* and

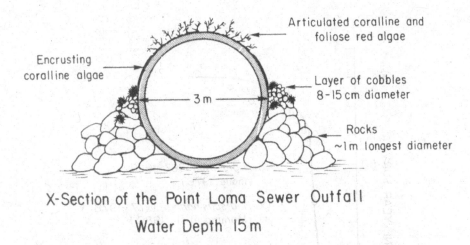

X-Section of the Point Loma Sewer Outfall

Water Depth 15m

FIGURE 3.   Cross-section of the San Diego sewer outfall structure within the Point Loma kelp forest distant from the actual release site, a highly productive nursery for juvenile abalones. Encrusting coralline algae line the sides of the pipe and much of the surface of the cobbles which overlie the larger rocks. Apparently because of their position well above the sea floor, water movement keeps these cobbles and adjacent sections of the structure virtually silt free. Sea urchins contribute to the quality of this habitat for juvenile abalones by maintaining areas of the encrusting coralline algae, specific inducers of settlement and metamorphosis as well as food for the newly settled abalones and protection from predators, free from overgrowth by other organisms.[20-23]

avoided *Egregia*.[11] Abalones in culture grow faster on a mixture of macroalgae[26] and this is likely to be the case in the field as well.

Abalones are predominantly drift feeders which rarely graze on attached plants. Thus, feeding will be facilitated in habitats where drift algae is likely to be found such as sites adjacent to sand channels, near the sand/rock interface of reefs, and groups of sea urchins. Shepherd[15] showed that moderate levels of water movement enhance drift transport and capture; abalones in very calm or very rough seas feed less successfully.

CHOICE OF SEED

The choice of abalone species to use for seeding should be based on both the nature of the site available and the characteristics of each species (Table 1). Of the important California species, only reds and greens have been produced in large numbers by the hatcheries and California law precludes the introduction of nonindigenous species. At their respective optimal temperatures, juvenile greens grow faster than small red abalones[13] but reds (which are commercially harvested at a larger size and older age) bring a higher price to fishermen. We have no evidence that either species is more or less susceptible to predation. Red abalones have been used more for seeding because they are the traditional market species, they have been available longer, and their habitat specifications are more widely distributed.

The optimal size for seeding is a critical but unresolved issue. Presumably the larger (and thicker shelled) an abalone is, the less susceptible it is to predation. The goal of raising juvenile abalones in a hatchery is to ease the animals through the larval and early settlement stages when mortality rates are highest. But maintaining abalones in a hatchery is expensive and growth rates of animals cultured in ambient temperature water may be reduced compared with abalones in the wild.[9,27] It is necessary to determine the seed size which optimizes both survivorship and cost. There are some data available from Japan. Working with a range of sizes of *H. gigantea*, Inoue[28] reported that 1-year survival rates increased rapidly with seed size and began to level off at about 60% survival for seed 40 mm at the time of planting (Figure 4). Momma and colleagues[29] found 1-year survival rates for *H. discus hannai* of

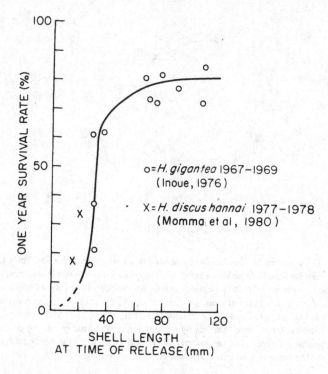

FIGURE 4. Japanese studies of size-specific seed survival. One-year seed survival rates as a function of seed size at the time of release. (From Inoue, M., *Suisan Zoyoshoku Deeta Bukku*, Suisan Shuppan, 1976, 19, and Momma, H., Kobayashi, T., Kato, Y., Sasaki, T., Sakamoto, T., and Murata, H., *Suisan Zoshoku*, 28, 59, 1980. Translations by Mottet.)

17% for 15-mm seed and 33% for 21-mm seed. Inoue[28] stressed that recapture rates vary with species, size at planting, and depth and habitat of the plant site. We have not been able to repeat this size-specific pattern of survival in California.

The actual size at which the Japanese plant abalones appears to be variable. Mottet[1] reports that the seed are usually released after 1 year of rearing at sizes of 20 to 30 mm. Saito[8] indicates that the seed should be more than 20 mm if released in the spring and 30 mm in the fall. Kafuku and Ikenoue[5] report seed release at 30 to 50 mm. To get seed into a more favorable size category before planting, some fishing cooperatives take delivery of the seed at 20 to 30 mm and then rear the abalones in intermediate culture (e.g., floating net cages in bays) for 1 year until the seed are 40 to 60 mm.[30]

To evaluate the success of resource-enhancement efforts, it is necessary to be able to distinguish between hatchery-reared and native abalones. Furthermore, California requires that abalones harvested by sea-floor farmers be positively identifiable at the time of collection, a requirement which probably precludes anything but simple visual markers. Individual tags are highly labor intensive and the shells of seed-sized abalones are too thin for tags wired through the respiratory pores. We have had considerable success notching the shells of abalones 25 mm and larger.[9] A Dremel-Moto Tool equipped with a grinding disc is used to grind a V-shaped notch about 5 mm deep into the growing edge of the shell just posterior to the respiratory pore line (Figure 5). If the mantle is pushed away from the grinder, notching does not appear to injure the animal and the notch is filled in within a few days. As long as the shell is not heavily encrusted or eroded, the repaired notch remains visible and is an unambiguous indicator of both the origin of the animal and its size at planting (Figure 5). While this method is still labor intensive, a practiced worker can notch about 200 animals per hour.

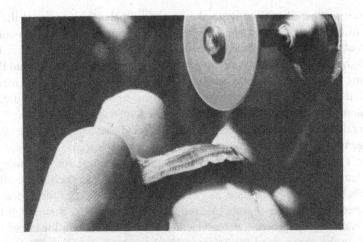

FIGURE 5. Notches ground into the growing edge of seed at the time of planting provide an unambiguous indicator of both the origin of the seed and the size at planting. (Top) A Dremel-Moto Tool equipped with a grinding disc is used to make the notches while the abalone's mantle is gently pushed out of the way. (Bottom) Growth after planting. The top animal grew from 39 mm at time of planting to 70 mm. Note shell color change of new growth. Bottom sell grew from 50 to 72 mm. An octopus drill hole is circled.

The simplest method of tagging large numbers of animals involves manipulating the diet of the seed in the hatchery. Red abalones fed diatoms produce a turquoise-colored shell, brown or green algae result in white or very pale tints of green or blue, and red algae lead to a dark red shell.[31] While fancy banding patterns are possible, we have found that the pale shells resulting from diets of diatoms and brown algae in the hatchery change markedly after planting because a field diet almost always contains some red algae (Figure 5). Native juvenile abalones, because of their relationship to encrusting coralline algae,[20] are almost invariably coralline colored. This biological tag has the disadvantage that the oldest part of the shell is generally eroded by the time abalones reach harvest size. We estimate that the seed must be a minimum of about 25 mm at the time of planting for this biological tag to be useful at harvest. Seed abalones frequently exhibit a change in shell shape after planting which is probably due to an increase in growth rates. A central California hatchery using ambient-temperature seawater reports that red abalones averaged 2 mm/month for the first year in tanks and that the growth rate declined to about 1 mm/month for larger animals (50 to 80 mm).[27] In the ocean, seed (some from the above hatchery) up to 70 mm at the time of planting averaged 30 mm growth per year[9] and the new growth was noticeably flatter than the portions of the shell laid down in the hatchery. However, this is not an unambiguous indicator of hatchery origin; native abalones subjected to major fluctuations in food availability show similar shape changes.

Another possible approach to seed identification is the use of hybrids. Abalones hybridize easily and hybrids are found at a low but variable frequency in nature.[32] If large-scale production of hybrids proves feasible, and if the hybrid characters are easily distinguishable, this may be one of the easiest methods of seed identification. An important advantage of working with hybrids may be the effects of heterosis on growth rates.[33]

Seed abalones should be inspected before leaving the hatchery. In some cases when seed have been in recirculating seawater for long periods, shell erosion may be substantial. Erosion exposes the nacreous layer, making the shells shiny and fragile (Figure 6). The problem begins at the apical spire and spreads toward the growth edge of the shell. In severe cases, the respiratory pores expand and join, leaving large holes in the shell. Often resembling fishing lures, such seed have a poor probability of survival in the ocean. Erosion results from the production of organic acids in culture systems where water is recirculated through sand filters. Calcium carbonate dissolves out of the abalone shells to buffer the effects of the acids.[34]

SEEDING METHODS

The Japanese have found that abalones can be transported long distances out of water as long as they are kept cool and moist.[1] We have successfully transported seed in water when the time between hatchery and the ocean is short but such containers are heavy and delays can cause problems. A more practical and safer method is to ship the abalones in styrofoam ice chests (Figure 7). Depending on the size of the seed, 100 to 500 abalones are placed in a large plastic bag along with a sponge saturated with seawater. Care should be taken not to allow standing water in the bag which can become hazardous to the seed from waste accumulation and oxygen depletion. The bags are filled with oxygen, tied off, and placed into the chests where insulation (e.g., newspaper) shields the abalones from direct contact with the ice. These containers are lightweight and acceptable for commercial transport. Ideally the abalones should be unpacked within 24 hr, but this system will work for longer periods. In one test we opened a chest 42 hr after packing and the temperature inside was still cold. The abalones required several minutes to warm up and become active but all recovered. If the chests become warm before they are opened, oxygen depletion may become critical.

FIGURE 6. Examples of shell erosion caused by poor pH conditions in hatcheries with recirculating seawater (left) compared with native red abalones of similar size (right). Erosion has exposed the shiny nacreous layer and increased the size of the respiratory pores, leading to fusion of the pores in the upper and lower eroded shells.

Shipments should be timed so that the abalones can be placed into ambient-temperature seawater as soon as possible. On-board vessel the chests are unpacked, the seed allowed to warm before being placed in mesh bags, and the bags are hung either in bait tanks with flowing seawater or over the side. Several hours of acclimation to receiving water temperatures appear to ease the transition to a benthic existence. While ideally the abalones should not be placed into intermediate holding facilities, it is important to make contingency plans to place the seed in flowing seawater if unexpected bad weather delays planting after a seed

FIGURE 7.   Transport of seed abalones. (Top) Styrofoam ice chests provide thermal insulation and may be shipped by commercial carriers. (Bottom) Seed abalones packed for transport inside a plastic bag. Note attachment to each other.

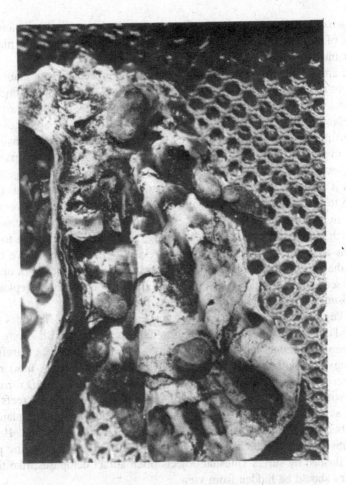

FIGURE 8. Oyster shell used as a seed substrate during planting. If the oyster shell is placed into hatchery tanks the day before transport, the seed will attach overnight. The oyster shell with its attached seed can then be planted as a unit with minimal stress to the abalones.

shipment has been received. On a larger time scale, seeding should be timed seasonally to minimize temperature differences between the hatchery and receiving waters, and to ensure that adequate algal food will be available.

The objectives of seeding are to place the abalones into natural environments in such a way as to maximize protection from predators and to minimize physiological stress to the seed. To achieve these objectives, we have tested two methods of seeding and compared artificial and natural habitats. If a number of abalones are placed in a bag, they will attach to the only available hard substrate — each other. It is a poor idea to plant such clumps because the abalones may not move to safer attachment sites until after dark and, in the meantime, are quite vulnerable to predators. Separating the seed so that they can attach individually probably stresses the abalones. This problem can be circumvented by planting abalones preattached to small substrates. If the substrates [we used oyster shell (Figure 8), and laminated pieces of corrugated plastic] are placed into hatchery tanks the day before the seed are to be packed, the abalones will attach during the night. These substrates can then be packed and handled like clumps of seed. The advantages are the reduction in handling, since the abalones neither have to be scraped off aquarium walls nor broken apart during seeding, and a substantially lower planting effort. One disadvantage of this method is the relatively high planting density; predators may locate a bonanza. This method is especially valuable for small (<25 mm) seed but it becomes cumbersome as the abalones get larger. An advantage of individual seeding is better spacing of seed. However, this planting method is quite labor intensive.

When seed are not attached to a substrate, the easiest procedure is to plant in rock piles. Choose piles of turnable rocks which offer adequate habitat for small abalones to disperse. Very small rocks can be moved by predators and should be avoided. Remove some rocks to expose a planting site. The site should have clean rock surfaces; abalones do not attach well to encrusting animals or silty surfaces and a poor hold on the substrate increases susceptibility to predation. In some cases we have used steel brushes to prepare surfaces for attachment, especially for larger animals. Check for predators and scavengers, and collect these animals if they are accessible. If the predators cannot be removed, abandon the site. Drop a few (5 to 10) abalones onto rock substrate and allow them to turn over on their own. Do not press an abalone to make it attach; excessive handling appears to cause the release of mucus which attracts predators and scavengers. If an abalone is unable to attach on its own but seems alive, tuck it into a protective crevice. When the seed have begun to turn over, replace the rocks over the planting area taking care not to set any rock on the abalones and to get the seed out of sight of visual predators. It is also important to replace rocks right side up; exposure of the underside fauna attracts predators.

Removing large sea urchins often exposes good seeding habitat. Abalones can be added to walls behind or crevices underneath urchins and the urchins replaced. Along with urchins, crabs are common in ledge and crevice habitats; such areas should be carefully searched before planting. Kojima[35] reports that *H. discus discus* seed (8 to 18 mm) released from cages were recovered several days later under the spines of *Anthocidaris crasspina*.

Substrates with preattached abalones are placed into boulder piles or reefs in much the same manner as individual seed. Adequate habitat for dispersal from the plant site is even more critical because of the number of seed (often 30 to 40) per substrate. If placed under ledges, the substrates should be braced by rocks to prevent them from being pulled out by a predator or flushed by surge. Unusual objects often attract octopuses; like the individual seed, substrates should be hidden from view.

To protect abalones during seeding and minimize transplant stress, the Japanese use a system of wave plates inside a cage.[1] Wave plates are corrugated pieces of polyethylene, vinyl chloride, or fiberglass which are used as growth substrates in hatcheries. Wave plates covered with seed are put into a cage with a mesh size just larger than the size of the abalones. Placed on the bottom, these cages will protect the abalones by day and the seed will move off at night. If the wave plates are transparent, all the abalones will move out within 2 to 3 days.[1]

We conducted one small-scale artificial habitat experiment using concrete cinder blocks which measured 10 × 40 × 20 cm with two holes 4 × 15 × 20 cm. These holes were thought to offer considerable protection to small abalones and a good surface for attachment. The blocks were placed on the bottom 3 weeks before planting to leach and to acquire a microflora. Abalones were planted into the cinder blocks, natural boulders, and urchin-lined crevices in the same general area. When the experiment was evaluated 3 months later, the cinder-block habitat had both the highest number of mortalities (as indicated by shell collections) and the lowest number of live animals. There were fewer shells and many more live seed in the habitats typical of where native juveniles are found, rock piles and urchin crevices, presumably as a result of long-term natural selection. Thus, given the expense of artificial habitats and the apparent advantages and availability of natural habitats, we have not pursued this further.

The Japanese widely use artificial structures to augment natural habitats.[36] Fishermen pile rocks of appropriate sizes in areas of soft sediment or pavement bottom. For culture of abalones, rocks are layered to a height of 0.4 to 0.6 m. In areas where the rock piles are not stable, they are contained in artificial structures ranging from synthetic fiber nets to concrete cribs.[36] Concrete blocks engraved with slits are also used as nursery grounds to protect seed from attack. Seed reared in the two types of nursery grounds are said to have the same survivorship as seed 1 cm larger in natural habitat.[30] Various concrete-block structures are provided for larger abalones as they move away from protected crevices and attach to relatively flat surfaces. Such habitats are often accompanied by longline culture of seaweeds to provide food for the grazers.[30,33]

PREDATORS

Native abalone populations are subject to intense predation pressure by a variety of predators;[9,10,12,15] in addition to this pressure, seed must cope with the stresses of transplanting. Despite extreme care, handling stresses abalones, both seed and transplanted native animals, causing the secretion of mucus which is highly attractive to predators. Divers' activity also attracts predators.

Predator problems can be divided into three nonmutually exclusive categories: planting operation opportunists, scavengers on stressed abalones, and perpetual predators. A number of fishes are attracted to seeding operations and will take dropped or unprotected seed. These include sheephead (*Semicossyphus pulcher*), Garibaldi (*Hypsypops rubicundus*), various surf perches (Embiotocidae), and others. However, these fishes do not seem to be a concern once the abalones enter protective habitat.

Scavengers are a much more serious problem, attacking live but weakened abalones of all sizes in their protective habitat.[9] These include Kellet's whelk (*Kelletia kelletii*), *Octopus* spp., and several species of starfish (Table 3). Although primarily nocturnal, we have observed octopuses entering seeding sites within 15 min of planting during the day. *Kelletia* are very abundant in many habitats although this may not be apparent because they tend to bury in the sand after a meal and are more active at night. These whelks insert their proboscises under the shells of weakened abalones and drill into the flesh. Large *Kelletia* completely engulf small seed. Starfish are the final important scavengers; different species are involved with changing latitude. Seeded abalones appear to recover from transplant stresses and lose their susceptibility to scavengers after a few days to a week. The normal reflex reaction, where an abalone draws its shell tightly to the substrate in response to disturbance, has been observed as soon as one day after planting. This reaction is important in reducing susceptibility to predation; e.g., an octopus might have to drill a tightly attached abalone instead of just pulling it off.[9]

Other predators will continue to prey on the abalones after they have recovered from transplant stress (Table 3). In southern California, the important predators on juvenile abalones are octopuses and crustaceans. More than half the native shells found during a study at Santa Catalina Island[11] and 38% of the shells from a site on the Palos Verdes Peninsula[9] had octopus drill holes. Since octopuses also feed by wrenching abalones off the substrate, these numbers may be low estimates of the role of octopus predation in these habitats. About a third of the Palos Verdes shells had the chipped edges characteristic of crustacean predation.[9] Bat rays feed primarily on adult abalones, usually breaking the shells into several large pieces. Cabezon (*Scorpaenichthys marmoratus*) take small and intermediate-sized abalones but are not abundant. These fish feed by ingesting whole abalones and later regurgitating shells with a characteristic acid-etched appearance.[12] A predaceous gastropod, *Ceratostoma nutalli*, is also an occasional predator of small- and intermediate-sized abalones. The crustaceans vary considerably between habitats and spiny lobsters (*Panulirus interruptus*) and sheep crabs (*Loxorhyncus grandis*) also undergo seasonal onshore-offshore migrations. Starfish are generally not considered to be predators of healthy adult abalones.[10,12] *Astrometis sertulifera*, which frequents rock piles and other habitats characteristic of juvenile abalones

Table 3
CALIFORNIA ABALONE PREDATORS (ABUNDANCES VARY CONSIDERABLY BETWEEN HABITATS AS WELL AS LATITUDINALLY AND TEMPORALLY)

Predators	Comments	Ref.
Mammals		
Enhydra lutris sea otter	Major predator of midsized and adult abalones; presently found in central California	12
Fishes		
Myliobatis californica bat ray	Primarily a predator of semi-exposed adult abalones; attracted to transplant operations	
Scorpaenichthys marmoratus Cabezon	Swallows abalones to 125 mm whole; regurgitates shells with a characteristic acid-etched appearance	10,12
Molluscs		
Octopus spp. Several species especially *O. bimaculatus, O. bimaculoides*	Major predator on small and midsized abalones, both stressed and adapted. Also preys on stressed adult transplants. Characteristic bevelled drill hole but not all molluscan prey are drilled	9,10,59
Kelletia keletii Kellet's whelk	Major scavenger on stressed transplants; ignores adapted abalones in field and lab. Difficult to clear from plant sites because often buried in sand	9
Ceratostoma nuttalli Nuttall's hornmouth	Occasional predator on juvenile abalones; characteristic oval drill hole, larger and straight-sided compared with octopus drill holes	9
Crustaceans		
Panulirus interruptus Spiny lobster	Nocturnally foraging generalist; moving abalones are most accessible but spiny lobsters have been observed attacking abalones in cryptic locations	9
Cancer antennarius *C. productus* Rock crab Red crab	*C. antennarius* is an important predator of all sizes of abalones. Common in protected abalone habitat. *C. productus* found in kelp forest in some areas and may also prey on abalones, especially smaller sizes	10
Loxorhynchus grandis Sheep crab	Seasonally important in some areas. Fed on abalones to 80 mm in field and aquaria experiments	9
Taliepus nuttalli Southern kelp crab	Considered herbivorous but consumed abalones up to 80 mm in aquaria experiments; commonly found in crevices with abalones	60
Asteroids		
Astrometis sertulifera Fragile rainbow sea star	May be an important predator of juvenile abalones in the field because it lives in similar, protected habitat. Consumes juvenile abalones to 80 mm in aquaria experiments. Also a scavenger on stressed seed	
Pisaster giganteus Great sea star	Major scavenger on stressed transplants; ignores adapted abalones	9
P. ochraceus Ocher sea star	Scavenger on stressed transplants; ignores adapted animals	9
Pycnopodia helianthoides Sunflower star	Scavenger on stressed transplants; may also feed on juvenile abalones.	

FIGURE 9.   Shell remains characteristic of various predators. (A) A native red abalone consumed by a spiny lobster in an aquarium experiment. The anterior portion of the shell is broken off and there are three V-shaped indentations along the growth margin of the shell. (B) Native red abalone shell recovered from the field with similar indentations which probably resulted from lobster predation. (C) Black abalone with the beveled hole (circled) characteristic of an octopus drill.[59] (D) Black abalone with much of the periostracum digested by a cabezon. (E) A hatchery-reared red abalone with an octopus drill hole. (F) Native green abalone consumed by a predaceous gastropod, *Ceratostoma nuttali*, in an aquarium experiment. The drill hole is much larger and less beveled than that made by an octopus.

and feeds on small and mid-sized abalones in aquaria, appears to be an exception. Juvenile *Pycnopodia helianthoides*, also common in cryptic juvenile abalone habitat, may be another exception. Examples of the kinds of shell damage caused by different predators are shown in Figure 9. Starfish leave shells unmarked, but as discussed above for octopuses, such a shell is not an unambiguous indicator of starfish predation.

Japanese fishermen apparently do not control for predators; once the abalones are planted on the designated growing grounds, the seed are left alone to grow to market size.[37] However, it should be noted that the Japanese have intensive fisheries for octopuses, crustaceans, and gastropods as well as fishes. Many of these animals are minimally if at all exploited in California.

Predator control at the time of seeding appears to reduce short-term mortality. Predators were not manipulated during our initial seeding experiment with 10,000 red abalones and 173 shells were recovered within the first month. We collected predators and scavengers before and during planting in a second seeding experiment at the same site with the same number of animals. The seed used in the second experiment were smaller than the first and the probability of total shell destruction by predators or loss from the habitat was higher with smaller seed. Nevertheless, our observations of predator and scavenger activity and a reduction to 36 shells recovered within the first month strongly suggested that predator control efforts would significantly decrease initial mortality rates. This hypothesis was supported by a series of four native adult green abalone transplants to one location.[9] No predators were controlled in the small scale, preliminary plant; the area covered and cumulative effect (as measured by a decrease in catch per unit effort) of predator collections increased with each subsequent transplant. Short-term mortality was reduced substantially in each successive transplant.[9]

After preliminary data suggested that long-term survival of 20 to 35 mm seed was very poor, we conducted an experiment with larger (40 to 80 mm), thicker-shelled seed and added long-term predator control to the collections at the time of seeding.[9] Most of these abalones, all notched (Figure 5) to indicate size at planting, had shells too thick for predators to destroy. After intensive efforts to collect predators before and immediately after planting, monthly visits were made to continue predator control, monitor the abalones, and collect shells. Shells were examined for evidence of cause of death (e.g., octopus drill holes or the chipped edges characteristic of crustacean predation, Figure 9) and assessed for growth since planting. After 1 year, the entire seeding site was destructively sampled. Shells with no growth, an indication of death at or soon after planting, were found up to 17 months later; thus, shells remained intact at the study site for a long period of time. Forty-four percent of the seed were eventually recovered as shells. Almost half of the shells showed no growth and can be considered planting mortalities. The other half showed growth up to 40 mm, presumably an indication that these animals had adapted to the new habitat before they were eaten. The value of the monthly predator collections is not clear. Octopuses were taken on 11 of the 12 visits, suggesting continual migration into the area. Survival, as measured by recovery of live animals after 1 year, was no better than other experiments with no predator control after planting. However, more than half the animals were unaccounted for and some probably migrated out of the area. Presumably many more seed would have died if the predators were not removed, but the data suggest that nearly continual predator control would be required to substantially improve survivorship.[9]

SEED BEHAVIOR

The intensity of predation pressure apparent in southern California puts a premium on seed behavior which minimizes susceptibility to predation. Abalones raised in a hatchery are exposed to substrates, lighting conditions, and feeding schedules very different from what they will encounter in the ocean, and obviously there are no predators or other animals with which they interact. During planting most seed display behavior expected of juvenile abalones; when dropped into rock piles during the daytime, they turn over, move into a dark crevice, and attach. A small percent of the seed will continue moving, sometimes across open areas of reef. These animals are highly susceptible to diurnal fish predators. Many seed begin dispersing after dark of the first night.

Abalone behavior changes with size and this change may be important to survival. To

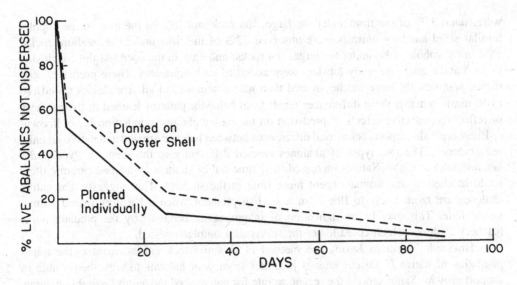

FIGURE 10 . Comparative movement of 13-mm red abalones planted on oyster shell and abalones planted individually from their respective seeding sites.

compare dispersal behavior as a function of seed size and species, about 200 abalones of each type were planted into mapped habitats during different seeding experiments and recensused at various intervals. Abalones which were not found live in the mapped sites or dead were considered to have dispersed to new habitat. These results suggest that dispersal rate decreases with increasing size for both red and green abalones, e.g., after 1 or 2 days, 38% of the 13-mm reds, 23% of the 45-mm reds, and 18% of the 71-mm reds had dispersed from their planting sites. Since abalones are more vulnerable to predation when they are not firmly attached to the substrate, differential dispersal rates as well as changing shell thickness may contribute to the size-specific variations in survivorship reported by the Japanese (Figure 4). There appeared to be little difference in either dispersal rate (Figure 10) or short-term mortality (4.0 vs. 3.6%) between 13-mm red abalones planted on oyster shell and seed of the same size planted individually. Aquaria experiments with native red abalones also indicate a reduction in nocturnal foraging movement with increased abalone size. Juvenile abalones (≤80 mm) regularly leave their protective habitats to forage in the open. Essentially stationary, larger red abalones wait in place for drift to be carried to them. Thus, in terms of size-specific mobility patterns, native and hatchery-reared abalones behave in a qualitatively similar manner.

In an effort to understand the poor survival of seed in the ocean, we compared habitat selection by similar sized hatchery-reared and native green abalones in aquaria. The abalones were presented three rock habitats [cobble (0.01 to 0.02 m^2 bottom surface area), medium rocks (0.04 to 0.05 m^2), and large, flat rocks (0.18 m^2)] with differing degrees of exposure to predators, and their location monitored daily. Food was anchored in the open so the abalones had to move in order to feed. Similar to the dispersal experiments described above, there was a change in habitat selection with abalone size. There was also a highly significant difference between hatchery-reared and native greens of the same size. Generally the hatchery abalones changed habitat more frequently, were more widely dispersed among the three rock habitats, and were more often found out in the open than the natives. For example, 100% of the observations of 10 to 20 mm native greens were in the cobble. The disappearance of food and trails in the sediment indicated that the animals were moving. In contrast, 10 to 20 mm hatchery abalones were found 44% of the time in the cobble, 18% in the medium rocks, 33% under the ledge-like flat rock, and 3% in the open. Larger (40 to 50 mm) natives were found 81% of the time under the large, flat rock and 19% of the time in the cobble. Similar sized hatchery animals were observed 52% of the time under the medium rocks, 15% in the cobble, 17% under the large, flat rocks, and 15% in the open. At the conclusion of the habitat analysis, spiny lobsters were added to each aquarium. These predators consumed significantly more hatchey-reared than native abalones in all size classes tested (10 to 50 mm). Whether these differences result from behavior patterns learned in the hatchery or reflect the selective effects of predation on the native abalone population is not known.

Breitburg[38] also reports behavioral differences between hatchery-reared and native juvenile red abalones. The two types of abalones reacted differently to the scent of *Pycnopodia helianthoides* at night. Natives increased their time out of shelters (V-shaped ceramic tiles) while hatchery-reared animals spent more time in the shelters. This suggests that native abalones are more likely to flee from a starfish predator when a hatchery-reared animal would hide. This may be an indication of inappropriate behavior by the predator-naive hatchery abalones which could lower the survival of outplanted seed.

In Hokkaido, northern Japan, one method of abalone stock enhancement is the transplantation of native *H. discus hannai* juveniles from poor habitats to areas better able to support growth. Saito[8] reports the recapture rate for native seed moved to favorable habitats at 20 to 25% and the recapture rate for hatchery-reared seed at 5 to 10% in similar areas. These results offer further evidence of differences between hatchery-reared and native abalones which affect their survival in the ocean.

ASSESSING RESULTS

There are two approaches to evaluating seeding success: (1) collection of shells to assess the causes, timing, and rates of mortality, and (2) monitoring live animals for growth and survival rates. Each approach has advantages and disadvantages. Both are time and labor intensive, but given the high cost of seed and the number of variables involved, are necessary to evaluate progress.

Shell collections offer the first indication of short-term survival. When seeding experiments are known to have failed, the results are obvious. The California Department of Fish and Game (DFG) planted 14,000 red abalones in Orange County in 1975. Apparently killed by thermal stress, hundreds of shells with no growth were found 1 year later.[39] R. Schmitt,[40] a researcher at U.C. Santa Barbara who conducted several test plants on Naples Reef, received a shipment of 9945 poor quality seed (20 to 40 mm); 431 shells were found 3 weeks after planting and only two live seed were found 4 months later. The DFG planted 3066 red abalones (40 to 80 mm) into an area which turned out to have a predator problem. Within 1 month, positively identifiable (intact or notched fragments) shells indicated a mortality of 14% and there were many smaller pieces. Less than 1% of the animals were observed live at this time.[41] Furthermore, shells stay in the area and are easier to locate than live animals. Hines and Pearse[12] found that marked shells dispersed only 2 to 3 m in 3 months with no net directionality. Schmitt seeded 1194 marked shells into cryptic sites in his study area in the same manner as live seed. Fifty percent were recovered in four person-dives 2 weeks later and 71% were eventually found without destructive sampling.[40] In the experiment with large seed described previously,[9] only 8% of the shells found in the year-long study were recovered by destructive sampling. Even if abalones are consumed in cryptic locations such as octopus middens, most shells appear to flush into the open.

The usefulness of this method varies with seed size. The smaller the abalone, the more likely certain predators will fragment the shell beyond recognition. Spiny lobsters, for example, grind very small abalones into shiny sand. Predators such as cabezon may transport seed out of the area. Finally some shells may be buried or large waves may scour the bottom of all small debris. Nevertheless, shell collections provide the best indicator of short-term survival, and in combination with field observations, should identify unanticipated predator problems. If shell collections are continued periodically, evidence of growth after planting will allow a separation of mortalities associated with planting (no growth) from those of seed which adapted to the new environment before they died.[9]

Censuses of live seed provide a more direct indication of seed survival than a low percent of recovered shells, but censuses are very much more time consuming and risk disturbing to the animals. Breitburg[38] found that 28 mm hatchery-reared red abalones showed significantly reduced shelter fidelity when shelters were disturbed. Shelter disturbance also resulted in increased daytime use of aquarium walls. Thus, disturbing the animals may make them more susceptible to predation. Furthermore, juvenile abalones are very mobile; one red seed moved 90 m in 30 days[40] and we have observed red seed more than 150 m from the seeding site. Red abalones tend to be cryptic, not visible to a diver without turning rocks, until they reach a size of 125 mm or larger. Cryptic habitat selection and the rapidly decreasing density resulting from dispersal make adequate census taking extremely difficult after a short period of time.

Growth of the seed after planting reflects the capacity of the habitat to support abalones. In the experiment with large red abalone seed described previously, growth averaged 30 mm per year and was as high as 53 mm per year.[9] This sort of growth suggests that neither water quality nor food availability caused the poor observed survival, and further supports the evidence from shell collections that intense predation pressure was limiting seeding success. A portion of the abalones used in this experiment were 2.5 years old and averaged 45 mm at the time of seeding. The growth observed after these animals were planted in the ocean indicates that the capacity for rapid growth was not lost under hatchery conditions.

Final analysis must be based on the recovery of live, harvest-size abalones which can be attributed to seeding.

DISCUSSION

In five large-scale seeding experiments with red and green abalones conducted on the Palos Verdes Peninsula and on Santa Cruz Island, we observed poor long-term survival of hatchery-reared animals in our study sites. In each case the estimated 1- or 2-year survival rate based on the recovery of live seed was 2.8% or less. Seed sizes ranged from 10 to 80 mm but we have no evidence that survivorship increases with seed size. The DFG had similar results with a plant of 20,000 greens on Santa Catalina Island.[42] These survival rates were based on censuses of the planting sites and their immediate surroundings; they do not account for possible larger scale migration because of the problems of sampling rare, cryptic abalones. All of the sites were known to support abalone populations in the past. Some had significant native abalone populations at the time of seeding, others did not. All sites had plenty of food available and good habitat for juveniles. We are unaware of any environmental fluctuations (such as unusual temperatures) which could have affected the abalones with the exception of one site which was inundated with sand during the severe winter of 1983. Schmitt and Connell[43] attributed the failure of a red abalone plant near Santa Barbara to storm-induced food shortage leading to starvation.

Shell collections suggest that mortality related to the stresses of transplanting is not the major source of the poor observed survival. Intensive shell searches within a month of planting small seed (10 to 35 mm) generally accounted for 2 to 3% of the animals. The lack of difference in short-term mortality between 13-mm abalones planted individually and the presumably lesser-stressed animals planted on oyster shell supports this hypothesis. The best supporting data came from the experiment with large, thick-shelled seed where shells were collected monthly and the entire planting habitat was destructively sampled after 1 year.[9] In this case, 44% of the animals are known to have died, 21% near the time of planting, and 23% after growing up to 40 mm. We recovered 1% live and were unable to account for 55% of the planted seed. The observation of a live animal 10 m from the site and shells up to 100 m away suggests that emigration may have been important in this study. Censuses of live animals further support the hypothesis that planting mortality is not the explanation for poor long-term survival. We have observed live seed at up to 20% of the planting density 4 months later without sampling for dispersal. Furthermore, studies of native red abalones on the Palos Verdes Peninsula suggest that population growth is limited by predation on juveniles, not recruitment rates.[9] Thus we conclude that predation on adapted animals and emigration probably are the major reasons for the poor seed survival observed in our study sites.

Estimates of the mortality rate of native abalone populations, made on different species in different locations by a variety of methods, vary considerably.[2] Studies from southern and central California suggest that natural finite mortality rates are quite high. Tutshulte[11] estimated average annual finite mortality rates for sublegal pink and green abalones at Santa Catalina Island at 0.5 and 0.6, respectively. Smith[44] calculates an annual rate of 0.6 for blacks at Palos Verdes although this value also includes minimal fishing mortality. In central California, three methods resulted in finite mortality rate estimates of 0.3 to 1.0 for one population of red and flat (*H. walallensis*) abalones.[12] Intuitively, there should be a decrease in susceptibility to many predators with increasing size; e.g., we have not observed red abalones larger than 125 mm with octopus drill holes. However, this does not apply to sea otters, the major predators in the central California study, or to bat rays. Tutshulte[11] did not find evidence for a decreasing mortality rate to age seven but noted that the rate must decline after age seven to account for the observed number of larger animals in the population.

Department of Fish and Game tagging studies of several populations of red, pink, and green abalones in southern California indicate that sport fishing size is reached after about 8 years and commercial size after a minimum of 10 years for pinks and greens, and 15 years for red abalones.[3,45] If we assume that abalones are seeded at 1 year of age, and that the mortality rate remains constant until the abalones enter the sport fishery 7 years later, we can estimate survival rates under different levels of natural mortality. If the annual finite mortality rate is 0.2, 21% will be alive 7 years later. If the mortality rate is 0.3, perhaps a minimum estimate for southern California, survival to sport legal size will be 8%. These calculations do not take into account additional losses from transplant stress, inappropriate behavior patterns, or bar cutting of sublegals.

Other studies, generally from higher latitudes, estimate natural annual finite mortality rates in the range 0.05 to 0.32.[2, 44, 46-49] It may be inappropriate to assume that there is a mortality rate characteristic of the genus as a whole. Predation pressure increases from high to low latitudes[50] and there are latitudinal clines in life history attributes such as recruitment, growth rates, and size and age distributions of species with long ranges.[51, 52] A significant reduction in predation may make seeding more feasible in higher latitudes.

We are unaware of any estimate of natural mortality rates for Japanese abalones but there are several related factors which favor the success of seeding in that country. The Japanese harvest abalones at small sizes, thus reducing the time of exposure to predators, e.g., in Chiba Prefecture, 90% of *H. discus discus* seed (mode 30 mm) reach market size of 120 mm in 3 years.[53] The Japanese have a very high per capita rate of fish consumption which includes a diverse array of marine products.[54] Fisheries for finfishes, octopuses, crustaceans, and gastropods doubtlessly reduce the natural mortality rate of abalone populations. Various institutional factors also play a role. Coastal fishing areas are under the control of local fishermen organized into cooperatives. Thus, there is strong incentive to manage, modify, and enhance the habitat. Government subsidies in the form of guaranteed low-interest loans, insurance, extension services, educational support, and applied research aid aquaculture and fishing. Finally the government subsidizes the hatcheries and the seed are sold below cost.[54, 55]

The high price of abalones and the depressed commercial catch has stimulated the interest of private enterprise in California. Three approaches have been explored: land-based culture, containers in the ocean, and sea-floor farming. Of the three approaches, sea-floor farming potentially has the advantages of considerably lower capital and operating costs. Oyster culture set a precedent in California for the productive use of water-bottom leases.[56] California law deals with general aspects of aquaculture and specific requirements for farming abalones are written into lease agreements; both are administered by the DFG. The DFG inspects the proposed sites before leases are granted and sites with large existing abalone populations are not approved. The DFG monitors the number, sizes, and general characteristics of seed (including obvious color or pattern differences which will allow the seed to be distinguished from natives) at the time of purchase from hatcheries. Harvesting, also supervised by the DFG, may take abalones below commercial legal size and is not restricted by other fishing regulations such as seasons or quotas. Harvesting at sublegal size allows the abalones to be taken while growth rates are still high, reduces the exposure to predators, and precludes take by other fishermen. The DFG stipulates that no more than 80% of the number planted may be harvested; the remainder are to be left for the benefit of the resource.

Starting in 1981, a number of sea-floor farming leases were granted. Only one group has actually planted seed on lease sites and it has not produced any marketable abalones to date. Presently the only cultured abalones being sold are from one company using containers in the ocean and another company employing land-based methods. The California Abalone Association, an organization of commercial fishermen, has planted seed at several sites on the Channel Islands not covered by lease agreements to augment the common property resource.

It should be noted that oyster farming as practiced in California[56] has important advantages over abalone farming. First, oysters are sessile; they cannot emigrate. Second, oysters are harvested after 1 to 2 years in culture and various off-bottom culture techniques have minimized loss to predators.[56] The success of abalone farming will depend upon learning how to contain the seed within reasonable areas and ensuring adequate recovery rates by reducing mortality rates and/or harvesting at smaller sizes.

There are a number of potential directions for future seeding research. One unknown issue concerns the relationship between seed parental stock and the receiving habitat. If there are physiological races adapted to environmental gradients, seed from brood stock originating from one end of an abalone's range may be at a disadvantage planted at the other end. Uki and Kikuchi[57] suggest genetic manipulation to obtain the most suitable seed for effective restocking. Moving from questions of seed stock to the receiving habitat, the Japanese make extensive use of prepared rock beds, and generally through other fisheries, control many predator populations.[36,54] These sorts of intensive habitat modification and management efforts may be beyond the means of state agencies charged with enhancing common property resources but may be reasonable for private enterprise on sea-floor leases.

We have had poor success attempting to adapt the Japanese approach of seeding roughly year-old abalones to California. The fundamental differences between the two systems may well preclude the Japanese method. D. Morse[58] has suggested the use of miniseed as an alternate approach. Very small (approximately 3 to 5 mm) abalones are inexpensive; large numbers could be planted at very low cost using some sort of transfer substrates. Because these abalones are very young, this may eliminate hatchery-acquired behavior problems. The mortality rate would doubtlessly be very high, but at the small unit cost, it may be possible to get an economically efficient number to survive by swamping the system. Miniseed would preclude positive visual identification unless hybrids are used; evaluation of seeding success would require comparison with control sites. An extension of this approach would be to consider the release of advanced-stage larvae which are competent to settle.

The poor survival rates we have observed suggest that seeding will not be a panacea for the depressed California abalone fishery. Despite this record, seeding appears to have benefits in addition to the few animals which may survive to enter the fishery. Tutshulte[11] reported that abalones settle gregariously, that an increase in the number of benthic abalones increased the number of larval recruits. Preliminary data suggest that our seeding experiments have attracted increased settlement of natives. Second, abalones become sexually mature between 3 and 5 years of age[11] and begin contributing gametes to the wild stock. While these benefits may assist in the long-term recovery of abalone stocks, the present record for seeding in California underscores the importance of careful conservation of existing stocks.

ACKNOWLEDGMENTS

Many people, too many to list here, contributed to the field work and laboratory studies described; to all we are grateful. We thank P. Breen, P. Dayton, and D. Parker for commenting on the manuscript. This work is the result of research sponsored in part by the Department of Fish and Game, in part by NOAA, National Sea Grant College Program, Department of Commerce, under grant number NA80AA-D-00120 and in part by the California State Resources Agency, project numbers R/F-47 and R/F-73. The U.S. Government is authorized to produce and distribute reprints for governmental purposes.

REFERENCES

1. **Mottet, M. G.,** *A Review of the Fisheries Biology of Abalones,* State of Washington, Dep. Fish. Tech. Rep. 37, 1, 1978.
2. **Breen, P. A.,** Management of the British Columbia fishery for northern abalone *(Haliotis kamtschatkana),* in *Proceedings of the North Pacific Workshop on Invertebrate Stock Assessment and Management,* Jamieson, G. S., and Bourne, N. F., Eds., *Canadian Spec. Public. Fish. Aquatic Sci.,* 92, 300, 1986.
3. **Burge, R., Schultz, S., and Odemar, M.,** Draft Report on Recent Abalone Research in California with Recommendations for Management, presented to the Calif. Fish Game Comm., San Diego, January 17, 1975.
4. **Cicin-Sain, B., Moore, J. E., and Wyner, A. J.,** *Management Approaches for Marine Fisheries: The Case of the California Abalone,* IMR Reference 77-101, Institute of Marine Resources, University of California, La Jolla, 1977, chap. 1.
5. **Kafuku, T. and Ikenoue, H.,** *Modern Methods of Aquaculture in Japan,* Kodansha, Tokyo, 1983, chap. 18.
6. **Kan-no, H. and Hayashi, T.,** The present status of shellfish culture in Japan, in *Proceedings of the First U.S.-Japan Meeting on Aquaculture at Tokyo, Japan, October 18—19, 1971,* Shaw, W. N., Ed., NOAA Technical Report NMFS CIRC-388, 23, 1974.
7. **Shaw, W. N.,** Shellfish culture in Japan, in *Proceedings of the First U.S.-Japan Meeting on Aquaculture at Tokyo, Japan, October 18—19, 1971,* Shaw, W. N., Ed., NOAA Technical Report NMFS CIRC-388, 107, 1974.
8. **Saito, K.,** Ocean ranching of abalones and scallops in northern Japan, *Aquaculture,* 39, 361, 1984.
9. **Tegner, M. J. and Butler, R. A.,** The survival and mortality of seeded and native red abalones, *Haliotis rufescens,* on the Palos Verdes Peninsula, *Calif. Fish Game,* 73, 150, 1985.
10. **Cox, K. W.,** California abalones, family Haliotidae, *Calif. Dep. Fish Game Fish Bull.,* 118, 1, 1962.
11. **Tutshulte, T. C.,** The Comparative Ecology of Three Sympatric Abalones, Ph.D. thesis, University of California, San Diego, 1976.
12. **Hines, A. H. and Pearse, J. S.,** Abalones, shells, and sea otters: dynamics of prey populations in central California, *Ecology,* 63, 1547, 1982.
13. **Leighton, D. L.,** The influence of temperature on larval and juvenile growth in three species of southern California abalones, *U.S Natl. Mar. Fish. Serv. Fish. Bull.,* 72, 1137, 1974.
14. **Bonnot, P.,** Abalones in California, *Calif. Fish Game,* 16, 15, 1930.
15. **Shepherd, S. A.,** Studies on southern Australian abalone (genus *Haliotis*). I. Ecology of five sympatric species, *Aust. J. Mar. Freshwater Res.,* 24, 217, 1973.

16. **Curtis, M.**, personal communication, 1984.
17. **Tegner, M. J. and Dayton, P. K.**, Sea urchin recruitment patterns and implications of commercial fishing, *Science*, 196, 324, 1977.
18. **Tegner, M. J. and Levin, L. A.**, Spiny lobsters and sea urchins: analysis of a predator-prey interaction, *J. Exp. Mar. Biol. Ecol.*, 73, 125, 1983.
19. **Day, R.**, unpublished data, 1978.
20. **Morse, A.N.C. and Morse, D. E.**, Recruitment and metamorphosis of *Haliotis* larvae induced by molecules uniquely available at the surfaces of crustose red algae, *J. Exp. Mar. Biol. Ecol.*, 75, 191, 1984.
21. **Breitburg, D. L.**, Residual effects of grazing: inhibition of competitor recruitment by encrusting coralline algae, *Ecology*, 65, 1136, 1984.
22. **Shepherd, S. A. and Turner, J. A.**, Studies on southern Australian abalone (genus *Haliotis*). VI. Habitat preference, abundance and predators of juveniles, *J. Exp. Mar. Biol. Ecol.*, 93, 285, 1985.
23. **Morse, D. E., Tegner, M., Duncan, H., Hooker, N., Trevelyan, G., and Cameron, A.**, Induction of settling and metamorphosis of planktonic molluscan *(Haliotis)* larvae. III. Signaling by metabolites of intact algae is dependent on contact, in *Chemical Signals*, Muller-Schwarze, D. and Silverstein, R. M., Eds., Plenum Publishing Co., New York, 67, 1980.
24. **Saito, K.**, The appearance and growth of 0-year-old Ezo abalone, *Bull. Jpn. Soc. Sci. Fish.*, 47, 1393, 1981.
25. **Leighton, D. L.**, Studies of food preference in algivorous invertebrates of southern California kelp beds, *Pac. Sci.*, 20, 104, 1966.
26. **Owen, B., DiSalvo, L. H., Ebert, E. E., and Fonck, E.**, Culture of the California red abalone *Haliotis rufescens* Swainson (1822) in Chile, *Veliger*, 27, 101, 1984.
27. **Ebert, E. E. and Houk, J. L.**, Elements and innovations in the cultivation of red abalone *Haliotis rufescens*, *Aquaculture*, 39, 375, 1984.
28. **Inoue, M.**, Awabi (Abalone), in *Suisan Zoyoshoku Deeta Bukku* (Fisheries Propagation Data Book), Suisan Shuppan, 1976, 19. (Translated by Mottet, M. G., in *A Review of the Fisheries Biology of Abalones*, State of Washington, Dept. Fish., Olympia, 1978, chap. 2.)
29. **Momma, H., Kobayashi, T., Kato, Y., Sasaki, T., Sakamoto, T., and Murata, H.**, Results of abalone culture in shallow water on a rocky coast. I. Survival of laboratory cultured seed abalone *(Haliotis discus hannai)* when transplanted into concrete-beam cribs filled with rocks, *Suisan Zoshoku (The Aquiculture)*, 28, 59, 1980. (Translated by Mottet, M. G., in *Summaries of Japanese Papers on Hatchery Technology and Intermediate Rearing Facilities for Clams, Scallops, and Abalones*, State of Washington, Dept. Fish., Progr. Rep. No. 203, 19, 1984.)
30. **Uki, N.**, Abalone culture in Japan, in *Proceedings of the Ninth and Tenth U.S.-Japan Meetings on Aquaculture*, Sinderman, C. J., Ed., NOAA Technical Report NMFS 16, 83, 1984.
31. **Olsen, D.**, Banding patterns of *Haliotis rufescens* as indicators of botanical and animal succession, *Biol. Bull.*, 134, 139, 1968.
32. **Owen, B., McLean, J. H., and Meyer, R. J.**, Hybridization in the eastern Pacific abalones *(Haliotis)*, *Bull. L. A. Co. Mus. Nat. Hist. Sci.*, 9, 1, 1971.
33. **Leighton, D. L. and Lewis, C. A.**, Experimental hybridization in abalones, *Int. J. Invertebr. Reprod.*, 5, 273, 1982.
34. **Sakai, H.**, A method to prevent erosion in the shells of young abalone, *Suisan Zoshoku (The Aquiculture)*, 28, 102, 1980. (Translated by Mottet, M. G., in *Summaries of Japanese Papers on Hatchery Technology and Intermediate Rearing Facilities for Clams, Scallops, and Abalones*, State of Washington, Dept. Fish., Progr. Rep. 203, 23, 1984.)
35. **Kojima, H.**, Mortality of young Japanese black abalone *Haliotis discus discus* after transplantation, *Bull. Jpn. Soc. Sci. Fish.*, 47, 151, 1981.
36. **Mottet, M. G.**, *Enhancement of the Marine Environment for Fisheries and Aquaculture in Japan*, State of Washington, Dept. Fish., Tech. Rep. 69, 1, 1981.
37. **Shaw, W.N.**, The culture of molluscs in Japan. III. Abalone culture, *Aquacult. Mag.*, 9, 43, 1982.
38. **Breitburg, D. L.**, Laboratory behavior of hatchery-reared and native juvenile red abalone *(Haliotis rufescens)* : implications for outplanting success, unpublished manuscript, 1984.
39. **Burge, R.**, personal communication, 1977.
40. **Tegner, M. J., Connell, J. H., Day, R. W., Schmitt, R. J., Schroeter, S., and Richards, J.**, Experimental abalone enhancement program, in *California Sea Grant College Program 1978-1980 Biennial Report*, Institute of Marine Resources, University of California, La Jolla, 1981, 114.
41. **Henderson, K.**, unpublished data, 1983.
42. **Haaker, P.**, unpublished data, 1984.
43. **Schmitt, R. J. and Connell, J. H.**, Field evaluation of an abalone enhancement program, in *California Sea Grant College Program 1980-1982 Biennial Report*, Institute of Marine Resources, University of California, La Jolla, 1984, 172.

44. **Smith, S. V.,** Production of calcium carbonate on the mainland shelf of southern California, *Limnol. Oceanogr.*, 17, 28, 1972.

45. **Parker, D.,** unpublished data, 1985.

46. **Beinssen, K. and Powell, D.,** Measurement of natural mortality in a population of blacklip abalone *Notohaliotis ruber, Int. Counc. Explor. Sea, Rapp. P. V. Reun.*, 175, 23, 1979.

47. **Sainsbury, K. J.,** Population dynamics and fishery management of the paua, *Haliotis iris.* I. Population structure, growth, reproduction, and mortality, *N. Z. J. Mar. Freshwater Res.*, 16, 147, 1982.

48. **Shepherd, S.A., Kirkwood, G. P., and Sandland, R. L.,** Studies on southern Australian abalone (genus *Haliotis*). III. Mortality of two exploited species, *Aust. J. Mar. Freshwater Res.*, 33, 265, 1982.

49. **Fournier, D. A. and Breen, P. A.,** Estimation of abalone mortality rates with growth analysis, *Trans. Am. Fish. Soc.*, 112, 403, 1983.

50. **Vermeij, G. J.,** *Biogeography and Adaptation, Patterns of Marine Life,* Harvard University Press, Cambridge, Mass., 1978, chap. 2.

51. **Frank, P. W.,** Latitudinal variation in the life history features of the black turban snail *Tegula funebralis* (Prosobranchia: Trochidae), *Mar. Biol.*, 31, 181, 1975.

52. **Fawcett, M. H.,** The Consequences of Latitudinal Variation in Predation for some Marine Intertidal Herbivores, Ph.D. thesis, University of California, Santa Barbara, 1979.

53. **Sakamoto, J., Tanaka, K., and Kobayashi, K.,** On the utilization of the marine environment for the intermediate culture of laboratory reared abalone seed, Chiba Ken Suisan Skikenjo Kenkyu Hokoku, *Chiba Prefect. Fish. Res. Lab. Rep.*,40, 123, 1982. (Translated by Mottet, M. G., in *Summaries of Japanese Papers on Hatchery Technology and Intermediate Rearing Facilities for Clams, Scallops, and Abalones,* State of Washington Dept. Fish. Progr. Rep. 203, 26, 1984.

54. **Mottet, M. G.,** Factors Leading to the Success of Japanese Aquaculture with an Emphasis on Northern Japan, State of Washington, Dept. Fish. Tech. Rep. 52, 1, 1980.

55. **McCormick, T. B. and Hahn, K. O.,** Japanese abalone culture practices and estimated costs of juvenile production in the USA, *J. World Maricult. Soc.*, 14, 149, 1983.

56. **Conte, F. S. and Depuy, J. L.,** The California oyster industry, in *Proc. Nor. Am. Oyster Workshop, March 6-8, 1981,* Chew, K. K., Ed., World Maricult. Soc. Spec. Public. No., 1, 43, 1982.

57. **Uki, N. and Kikuchi, S.,** Regulation of maturation and spawning of an abalone, *Haliotis* (Gastropoda) by external environmental factors, *Aquaculture*, 39, 261, 1984.

58. **Morse, D. E.,** personal communication, 1977.

59. **Pilson, M. E. Q., and Taylor, P. B.,** Hole drilling by octopus, *Science*, 134, 1366, 1961.

60. **Morris, R. H., Abbott, D. P., and Haderlie, E.C.,** *Intertidal Invertebrates of California,* Stanford University Press, Stanford, California, 1980, 597.

Culture Techniques

ABALONE AQUACULTURE IN JAPAN

Kirk O. Hahn

INTRODUCTION

An abalone fishery has existed in Japan since early history. The earliest reference to ama abalone divers is in a document from the reign of Emperor Suinin around 30 A.D. The tradition of ama divers being exclusively female began in the sixth century when many men were taken to serve on war ships (Figure 1). Women left behind had to take care of themselves and became self reliant. The government tribute was paid with awabi by the ama divers. From this beginning, it has become traditional for women to dive for abalone.[1]

Japan traded with foreign countries only through the port of Nagasaki prior to the beginning of the Meiji Era (1865 to 1912). Dried abalone was the most important export, constituting about 80% of marine exports. During the Meiji Era, there was a strong desire in Japan to generate more money from exports by increasing the abalone harvest. This desire led to the initiation of biological studies of abalone in Japan by Uchimura (1881), Matsuhara (1882), and Kishigami (1894).[2]

Japan has an unique system of controlling the fisheries along the coasts. Fishing cooperatives own ocean areas along the coast in much the same way as a person owns farmland. The fishing cooperative has complete control over their area and manage all aspects of the abalone resource (e.g., harvest, predator removal, habitat restoration, reseeding). Members of each cooperative can fish only in their own area; therefore, each area must support a sustainable harvest.

The total annual harvest of abalone in Japan is approximately 5.7×10^6 kg, which represents 57×10^6 individuals or 15% of the total abalone population.[3] The levels of abalone harvest in Japan are relatively stable, although the demand is still high. Abalone imports ($\sim 2 \times 10^6$ kg/yr) from other countries have steadily increased to meet the high demand.[4] After the initiation of international fishing regulations in foreign seas, Japan began investigating aquaculture as a means for increasing harvests in their own coastal waters.[2,5]

Research on hatchery methods for abalone aquaculture began at the Tokai Regional Fisheries Laboratory in 1959.[6] The present abalone aquaculture techniques are based on the pioneering work of Murayama[7] on *Haliotis gigantea* and Ino[8] on *H. discus*. Modern methods of seed production and seaweed reforestation have greatly improved the harvests in Japan (Table 1).[5,9]

SEED PRODUCTION

Aquaculture Laboratories

Abalone seed production has been developed on both the national and prefecture level in Japan. There are 43 national and prefecture laboratories, and one research foundation (Oyster Research Institute — ORI) producing abalone seed.[10] The majority of abalone seed produced is *H. discus hannai*, but small quantities of seed are also produced of *H. discus*, *H. gigantea*, and *H. diversicolor supertexta*. Abalone seed production is a major project by the government of Japan to maintain abundant abalone populations, and thus ensure continuation of the abalone fishery. Total seed production in 1978 was 10,729,000.[3] The production levels at the major facilities were: 2,693,000 (*H. discus hannai*) at the ORI, 1,242,000 (*H. discus hannai*) at the Hokkaido Prefecture Laboratory, and 1,214,000 (*H. discus*, *H. gigantea*, and *H. diversicolor supertexta*) at the Nagasaki Prefecture Laboratory.[3] Abalone seed production increased in 1979 to 11,658,000.[11] The production levels in 1980 at the major facilities were: 1,500,000 at the Hokkaido Prefecture Laboratory, 1,000,000 at the Iwate Prefecture

FIGURE 1.   Wood-block print showing an Ama diver.

Laboratory, 2,000,000 at the ORI, 1,000,000 at the Miyagi Prefecture Laboratory, and 500,000 at the Chiba Prefecture Laboratory. Seed production has increased every year. The Miyagi Prefecture Laboratory increased production to 4,000,000 after 1980.[3,5] The Iwate Prefecture Laboratory also expanded and increased production to 4,000,000; with a projected annual production greater than 10,000,000 seed per year.[3,5] The Iwate Prefecture Laboratory cost 300,000,000 ¥ to build, and requires 50,000,000 ¥ to run the facility and 300,000 kg of *Laminaria* sp. for food each year.[3]

Abalone seed produced at the laboratories are sold (at a subsidized cost) to fishing cooperatives for planting into the open ocean (no containers or cages are used). In 1980 the cost of juveniles from prefecture laboratories was 20 ¥ per 20 mm seed or 10 ¥/g.[11] The fishing cooperatives manage their reseeded areas and harvest the abalone when they reach market size.[12] Each cooperative finances the cost of abalone seed by charging a fee to each member of the cooperative depending on their harvest [e.g., 1.5% of catch in northern Iwate Prefecture and 20% of catch in southern Miyagi Prefecture (few members in the cooperative)].[3]

Hatchery Location

The ideal location for a hatchery is on the shore of a small protected bay along the open rocky coast near the center of the cultured abalone species' distribution.[4] The site should have clean water with good water movement along the coast, a nearby kelp forest, and

Table 1
STEPS FOR IMPROVING AN ABALONE FISHERY WITH AQUACULTURE OF JUVENILES FOR RESEEDING

	Techniques of artificial seed production	Spawning:	Conditioning of brood stock for spawning throughout the year
		Rearing:	Mass culture system of juveniles
Development of abalone fisheries			
	Methods to improve the rocky coastal environment	Food:	Seawood afforestation
		Habitat:	Availability of food Protection from predators and competitors for juvenile and adult abalone

From Kan-no, H., *Proc. First Int. Conf. Aqua Cult.*, 195, 1975. With permission.

abudant and cheap electrical power.[12] The grow-out area should be near the tip of a headland with a good current. As long as the necessary electrical power and some means of transporting the abalone seed are available, the more remote the laboratory is from human populations, the better the location is for abalone aquaculture.[4]

After finding a hatchery site, the culturist must select the species best suited for the location. The qualities to consider are taste, toughness of the meat, selling price, adaptability to environmental conditions (all life stages), availability of adults for brood stock, and ease of inducing spawning. The next step is conditioning of the brood stock. The sex ratio of the brood stock should be about 70 to 80% females. Depending on the desired amount of seed production, a large number of comparatively small-sized mature individuals is better than a small number of large-sized individuals. This will help reduce costs of maintaining the brood stock in good rearing conditions (food supply, rearing density, water circulation, and containers) to ensure reliable spawning.[4]

Hatchery Techniques
Miyagi Prefecture Laboratory

The Miyagi Prefecture Laboratory exclusively cultures *H. discus hannai*. The annual seed production is about 2 to 4 million and the abalone seed is sold for 1.5 ¥/mm. The laboratory is a very large facility with a total area of over 24,000 m^2. There are 60 tanks (1 m × 20 m × 1.5 m) for raising juveniles up to 30 mm (Figure 2). There are 68 baskets for rearing abalone seed and two buckets with inorganic nutrients to enhance diatom growth in each tank. The laboratory has three pumps which can pump 200,000 ℓ/hr. Each culture tank receives 6,000 to 8,000 ℓ/hr of filtered water. The facility has seven large filters (100,000 ℓ/hr) for filtration of the water to the rearing tanks. At all times, two filters are cleaning the water to the rearing tanks, four are backwashing, and one is in reserve. In addition, there are two smaller filters (30,000 ℓ/hr) which are used for the larval and spawning water in the hatchery. The sand in the filters is changed every 2 years.[3]

Conditioning tanks for the brood stock receive 800 ℓ/hr at 20°C with an artificial 12/12 photoperiod (1 a.m. to 1 p.m. light, 1 p.m. to 1 a.m. dark). Fully mature individuals are air dried for 1 hr before being placed in the spawning tank. Each spawning tank receives 200 mℓ/min of ultraviolet (UV)-irradiated sea water. The larval rearing tanks have densities of 300,000 per 15 ℓ of water.[3] The settlement tanks are 280 cm × 170 cm × 50 cm and

FIGURE 2.   Miyagi Prefecture Laboratory. Foreground — rearing tanks for juvenile abalone. Background — filters for rearing and hatchery water.

receive UV-sterilized water at 600 ℓ/hr.[4] Diatom-covered wavy plates with mucus from juvenile abalone are used to induce settlement of competent larvae. During settlement induction, 3000 larvae per wavy plate are introduced into the settlement tank.[3]

Shizuoka Prefecture Laboratory

Hatchery production begins in November. The brood stock is air dried for 1 hr; then each individual is placed in a separate tank with UV-irradiated water for induction of spawning. Males begin spawning first and released sperm is added to each tank containing a female, stimulating the females to spawn. Fertilization is immediate and uncontrolled. The eggs are siphoned out, washed, and placed into incubation tanks. Hatch out occurs in 12 to 18 hr. The floating larvae are poured into large (about 12,000 ℓ) concrete tanks. After 4 days, settlement occurs on plates placed in the tanks. Algae are grown directly on the settlement plates as a food source. The abalone grow on the plates for 6 months until May/June. By this time they are 0.5 to 1.5 cm long and are transferred to floating cages for further grow out.[13]

Sado Island Laboratory

In 1981, the Sado Island Laboratory started culturing *H. discus*. The laboratory is using the same techniques as developed for *H. discus hannai* at the Miyagi Prefecture Laboratory. Due to the severely cold winters in this region of Japan, rearing tanks are inside five buildings (26 tanks in each building) which act like hot houses. There are slight modifications to the standard methods developed for *H. discus hannai*. The settlement plates for the larvae are more rectangular than those used in the Tohoku region. This is probably due to the shallowness of the rearing tanks.[3]

Intermediate Culture

Intermediate culture of abalone (growing abalone from 30 mm to a larger size before

FIGURE 3. Typical apparatus for intermediate rearing of juvenile abalone with a long line. Netting retains the abalone and kelp placed in the apparatus for food.

release or to harvest size) has been proposed for reducing the high mortality of abalone seed placed in the wild. Cages used for intermediate culture can either be placed underneath oyster rafts or hung from long lines. A typical plastic intermediate culture structure is shown in Figure 3. It consists of four flat shelves surrounded by netting. This cage can be used for abalone from 20 to 40 mm. The cost of the cage is 10,000 ¥. The size of the abalone determines the stocking densities of the cages: 30 mm — 2,300, 40 mm — 1,300, 50 mm — 800, and 60 mm — 550. The survival for seed is 5 to 6% if stocked into the open ocean at 30 mm, but increases to between 50 and 60% if stocked at 45 mm into the plastic cages.[3]

Live Storage Facilities

Due to a short fishing season and the desire for live abalone in Japan, fishing cooperatives build facilities to store live abalone until they are sold at market. Along the coast of Miyagi Prefecture, the annual harvest occurs during a 3-day period (6 a.m to 9 a.m.) in December. Fishermen stand in boats, looking in water boxes, and use 7- to 10-m-long bamboo poles with a large metal hook on the end to catch the shell edge (not the animal) and quickly pull the abalone to the surface. A good fisherman can catch 120 kg in 3 hr using a hook.[3]

The storage facility near Kensennuma has 65 tanks and each tank can store 1000 kg of abalone. The abalone are fed *Undaria* sp. and *Laminaria* sp. During the summer, two baskets of kelp are required for each tank, which equals a total of 600 to 800 kg/day for the facility. The price of abalone slowly increases during the year as the abalone supply decreases and the cost of storing the abalone increases. The cost in March is 6500 ¥/kg but increases to 8000 ¥/kg by October (just before the harvest).[3]

RESEEDING

Release of Seed

The minimum size for release of seed is 15 mm, but 30 mm is the preferred size. Small abalone suffer high mortality from crab and fin fish predation. Before reseeding, juveniles are placed on half-round pipes that are planted directly into the field without the necessity of removing the juveniles. The release site should have a wild abalone population present, stones 5 to 30 cm in diameter on the ocean bottom, gentle wave action (not sufficient to move stones), good seaweed populations, shallow water (<10 m deep), and not be close to freshwater run off.[13]

Abalone reach market size (11 cm) in about 3.5 years after release. Typical recapture rates are between 0.5 and 10%, with as high as 20% returns from one area of Japan.[13] Abalone seed (20 mm) released in Hokkaido took 3 years to reach commercial size (80 mm) and the recovery rate was 5 to 10%. Reseeded animals in Hokkaido increased 58 times in weight during the 3 years from release (20 mm) to harvest (80 mm). However, the total biomass of abalone seed only increased from 2.9 to 5.8 times (total weight of harvested abalone vs. total weight of abalone seed). Thus, a release of 40,000 seed produced only an increase of 150 to 270 kg.[14] If a typical prefecture laboratory produced 1,000,000 seed, only 5,000 to 100,000 abalone would reach market size and be harvested. This is an increase of only 1,000 to 25,000 kg (3 to 4 abalone per kilogram) to the fishery.[13]

Predation is the major factor for loss of seed. Artificial habitats and improved release techniques are needed for better seed survival. Migration of the seed away from the release site also reduces the harvest to the fishing cooperative. Some of the major factors affecting migration after release are water depth, currents, seaweed distribution pattern, and topography. The last two factors are the most important. Interestingly, the larger the seed, the farther they migrate from the point of release. (This is thought to be the reason why seed >34 mm have the lowest recapture rates).[14]

Artificial habitats protect the abalone seed and help reduce predation. Rock-filled cement bins are easy to build and inexpensive. A typical habitat is 2 m × 2.5 m × 1 m, with two stacked concrete beams (25 cm²) as walls. The center of the cement bin has an area (1.4 m × 0.4 m) that holds a metal mesh cage. This area provides temporary protection for newly released seed. The bin is filled with 30- to 50-cm rocks and weighs about 4000 kg. The open sides of the bin allow water movement through the rocks. A bin can hold 1000 to 5000 abalone.[15]

The most significant factor affecting the abalone population is the natural production of kelp. High kelp production means there will be a large harvest of abalone. The variation in the Kurile current, called "Oyashio", causes an increase in kelp production which in turn causes an increase in abalone production.[16]

Reforestation

The distribution of abalone in Japan corresponds to the distribution of the algal order Laminariales, except in the coldest regions of eastern Hokkaido. The main foods for abalone in cold areas are *Laminaria* spp. and *Undaria pinnatifida*, and in warm areas are *Eisenia bicyclis*, *Ecklonia cava* and *U. pinnatifida*. *H. discus hannai*, the most important abalone species in Japan, grows best on *Laminaria* spp. The annual net production of algae in the fertile bed is about 10 kg (wet weight) per square meter and in a ordinary bed about 4 kg (wet weight) per square meter. Besides abalone, there are many herbivores (e.g., sea urchins, limpets) that eat macroalgae. The usual total biomass density of all herbivores is about 100 to 200 g (wet weight) per square meter and, in a good abalone habitat, abalone comprise more than 60% of the biomass.

Some areas along the northeast corner of Honshu lack macrolagae on the sea bed, particularly the species *Laminaria* spp. and *U. pinnatifida*. In these areas the sea bed is almost a desert, with no large kelp and only a dense growth of crustose coralline red algae covering all the rock surfaces. These areas are called "isoyake" (sea desert) or "pink rock".[5,17]

Isoyake areas have been observed for over 70 years and are caused by many factors: (1) change of environmental conditions (ice along coast, change in ocean currents, very low tide during a severe winter, or influx of fresh water due to floods), (2) biological effects (over grazing by herbivores and disease), and (3) pollution.[17]

The sea-water temperature along the northeast coast of Honshu is highly variable due to the meeting of the Oyashio current (cold current from north Pacific) and the Kuroshio current (warm current from central Pacific). The exact location of the meeting changes from year to year and has a dramatic effect on the sea life along the coast. Cold water is an important factor in the establishment of *Laminaria japonica*. The optimum temperature for growth is 10°C, when the grazing activity of sea urchins and abalone is very low. Once the kelp forest is established, the algae are capable of withstanding grazing by herbivores. *Laminaria* spp. density will stabilize if there are no changes in the environmental conditions or the density of grazers does not increase.[17]

High water temperature is the strongest factor accelerating the formation of sea deserts. Grazers remain active during the winter when young kelp plants begin growing. The grazers quickly eat all the young algae fronds and only crustose algae remain.[5,17] However, isoyake areas are caused by more than environmental factors, otherwise the habitat would recover after the environmental conditions returned to normal. Sea deserts can remain for several years or decades without succession of other algal communities.[17]

Isoyake areas typically have high densities of small and starving sea urchins and abalone on the sea bed. The herbivore biomass in these areas is over 300 g (wet weight) per square meter.[5] Large kelp (*Laminaria* spp., *E. bicylis* and *U. pinnatifida*) are only seen in the high intertidal zone where they escape grazing by the sea urchins and abalone. The large sea urchin populations eat any young kelp fronds before they have a chance to grow. In a typical isoyake area, the density of sea urchins is 15/m² and abalone (*H. discus hannai*) is 5/m². However, approximately 90% of the sea urchins and 95% of the abalone are under the legal harvest size limit and are of poor quality due to the lack of meat or good gonad condition. This makes these areas unsuitable for a fishery.[5,17]

Isoyake areas recover once kelp is given to the herbivores and a source of young plants is supplied to reforest the sea desert areas. Kelp reforestation was developed to rehabilitate these sea deserts. The reforestation methods were designed for ease of implementation by abalone fishermen. A long-line method was chosen since kelp fishermen had already perfected these techniques and spores released from the cultured *Laminaria* spp. help re-establish the kelp population during the next growing season.[17] Long-line ropes are seeded with either *L. religiosa* or *L. japonica* spores and placed in the area to be reforested. As the kelp grows, the rope gets heavier and finally lays on the sea floor. The mature kelp supplies food to the herbivores, reduces the grazing pressure on the rocky substrata, and spreads spores for establishing a new kelp forest. The total biomass is also reduced by harvesting the herbivores in the area. After reforestation of a previously denuded area, 6 to 8 kg (wet weight) per square meter of *Laminaria* spp. is produced the next year. Once the isoyake area has recovered, abalone seed are introduced. The plant communities in the reforested area change with increasing grazing pressure. *Laminaria* and *Eisenia* communities are dominant with mild to moderate grazing pressure; *Sargassum, Neodilsea*, and *Dictyota* communities become dominant with heavier grazing pressure; and Corallinaceae community is dominant with the heaviest grazing pressure.[5]

A long-line method of reforestation with *L. japonica* was tested in Miyagi Prefecture. Long lines were set at intervals of 2 to 3 m over the isoyake area. This method produced 33,300 kg of kelp with 3825 m of line of 1972 and 41,500 kg from 4200 m of line in 1973. The increased food from the cultured kelp produced well-developed gonads in the sea urchins and the increase in body weight of abalone was four times greater than individuals in nonreforested areas. In addition to reforestation, herbivore removal was conducted to get the habitat in ecological balance. The abalone density was reduced from $2.1/m^2$ to $0.58/m^2$.[17]

After long lines of *L. japonica* were started, *Sargassum horneri* began growing in the area and grew to 10 m by the summer of the next year. In portions of the reforestation area, many small thalli of *L. japonica* appeared in the late spring.[17] Two years after beginning long-line culture of *L. japonica*, a dense algal community was established with *L. japonica, U. pinnatifida, Alaria crassifolia, Desmarestia liguilata* and *Sargassum* spp. *U. pinnatifida* dominated the side slopes of the bay while *L. japonica* composed 80% of the algae on the flat bottom.[17]

The results achieved at Enoshima Island in Miyagi Prefecture are an example of the kind of improvement obtained with these techniques. A typical sea desert has no commercial value, but 1 to 2 × 10^3 kg of abalone per hectare was harvested after reforestation.[5] The abalone growth rate in the reforested area was 20 mm/year and 70 g/year.[18] The total production of the reforested area was 3100 kg of abalone and 6600 kg of sea urchins. Thus, 100,000 kg of kelp produced 10,000 kg of commercially important herbivores.[17]

E. bicyclis reforestation techniques are being developed to improve sea deserts in the warm water regions in Japan. Continuous production of kelp is being increased by the introduction of the perennial *E. bicyclis* into *Laminaria* spp. regions. *Laminaria* spp. is nonproductive in winter, while *E. bicyclis* is productive throughout the year and has a life span of 7 to 8 years.[5]

ABALONE PEARLS

Although it is not common knowledge, abalone are capable of producing very beautiful pearls. The nacre of the abalone pearl has a faint bluish tint and produces delicate rainbow tints. Abalone pearls are usually large. The largest abalone pearl was found in 1914 and weighed 17.22 g. Natural abalone pearls were very special to the Japanese. The oldest abalone pearls in Japan are on the statue of the Buddhist Goddess of Mercy in the Sangatsu-do Temple in Nara. Also, the Manyoshu (the oldest collection of poetry in Japan) contains five poems which refer to abalone pearls.

Since abalone pearls are so beautiful, artificial induction has been attempted from ancient times in Japan. Uno in 1957 obtained pearls in 70% of animals implanted with a semispherical bead (16 to 22 mm) nucleus. Smaller nuclei (9 mm) produce semispherical pearls in about 3 months. Perfectly round pearls are obtained by inserting small nuclei (3 to 5 mm) beads into the body cavity and are produced in 50% of the animals. Care must be taken during the operation not to injure the intestinal tract. A nucleus has a 644-μm-thick nacre layer after 175 days.[4]

Kaneichi Komatsu built an abalone culture laboratory at Oura in Miyagi Prefecture to produce semispherical pearls from *H. discus hannai*. So far, he has been unable to produce high quality pearls. The outer nacre layers on the nucleus are dark brown from the deposition of conchiolin. The Ojika Fishery Association is also experimenting with the culture of abalone pearls. Their method consists of inserting a bead nucleus between the mantle lining and the inner surface of the shell.[4]

UNUSUAL PROPERTIES OF JAPANESE ABALONE

Antiviral and Antibacterial Factors

Abalone juice has strong antiviral and antibacterial action when injected into mice, and presumably in other mammals. The juice from canned abalone reduces the rate and severity of paralysis, and death from poliomyelitis. Abalone juice also inhibits the growth of *Staphylococcus aureus*. The antiviral and antibacterial activities of untreated abalone juice can be separated by ion exchange chromatography. The antibacterial effect is not changed by heating to 95°C for 45 min or dialyzed against water overnight. Therefore, it is not an ordinary antibiotic.[19,20]

Toxicity of Abalone Guts

In regions where dried abalone is produced, the guts removed from the animals are also used as food. The abalone guts are usually salted but fishermen often eat them raw or boiled. "Uro-Zuke" (salted abalone guts) is produced in small quantities, barely enough to meet local demand.[4]

In the prefectures which produce dried abalone and salted abalone guts, there often occur cases of poisoning by abalone guts. The first case of poisoning by abalone guts was reported by Takenaka in 1899. The fishermen of Iwate Prefecture report abalone poisoning occurs mostly in early summer, indicating the toxicity of abalone varies with the season. An unusual observation was made by Toba in 1928 when he reported that a cat poisoned by abalone guts had lost its ears. Animals fed abalone viscera become photosensitized by a photodynamic agent in the abalone liver. Toxicity is characterized by hypersensitivity to light and exposure to sunlight after feeding works in combination with the poison.[21,22]

When rats were fed abalone guts, blood congested in the ears causing inflammation and the animal scratched the area vigorously. Any area exposed to sunlight and not covered by hair (e.g., nose and limbs) also had the same inflammation. In the worst cases of poisoning, the limbs of the animals were paralyzed and death occurred within 1 hr. The toxic substance, pyrophenophorbide, is heat stable, and unaffected by $-20°C$ storage for 10 months. The substance is a dark greenish brown pigment and exhibits intense crimson luminescence when exposed to sunlight.[21,22]

The amount of toxicity varies in the different abalone species. The toxicity in *H. discus hannai* varies from February to June with the strongest period during April and May. The toxicity in *H. sieboldii* is strongest in April and disappears in May. The toxicity in *H. diversicolor supertexta* is strongest in March and April.[21,22]

REFERENCES

1. **Miyamoto, T.,** *Ama; The Women Sea Divers in Japan,* Chunichi, Tokyo, 1962, 1.
2. **Imai, T.,** *Aquaculture in Shallow Seas: Progress in Shallow Sea Culture,* Part IV, Amerind Publishing Co., New Delhi, 1977.
3. **Hahn, K. O.,** personal observation, 1981.
4. **Ino, T.,** *Fisheries in Japan, Abalone and Oyster,* Japan Marine Products Photo Materials Assoc., Tokyo, 1980, 165.
5. **Uki, N.,** Abalone culture in Japan, in *Proc. Ninth and Tenth U.S.-Japan Meetings on Aquaculture (NOAA Tech. Rep. NMFS 16),* Sindermann, C. J., Ed., U.S., Dept. Commerce, Seattle, 1984, 83.
6. **Cuthbertson, A.,** *The Abalone Culture Handbook,* Tasmanian Fisheries Development Authority, Hobart, Australia, 1985, 1.
7. **Murayama, S.,** On the development of the Japanese abalone, *Haliotis gigantea, J. Coll. Agric. Tokyo Imp. Univ.,* 13, 227, 1935.
8. **Ino, T.,** Biologicial studies on the propagation of Japanese abalone (genus *Haliotis*), *Bull. Tokai Reg. Fish. Res. Lab.,* 5, 1, 1952.

9. **Kan-no, H.,** Recent advances in abalone culture in Japan, Proc. First Int. Conf. Aquacult. Nutr., 1975, 195.

10. **Cowan, L.,** Hatcheries — the basis of modern Japanese mariculture, *Aust. Fish.*, March, 20, 1981.

11. **Kafuku, T. and Ikenoue, H.,** Abalone *(Haliotis (Nordotis) discus)* culture, in *Modern Methods of Aquaculture in Japan: Developments in Aquaculture and Fisheries Science*, Vol. 11, Kafuku, T. and Ikenoue, H., Eds., Kodansha, Tokyo, 1983, 172.

12. **Shaw, W. N.,** The culture of molluscs in Japan, *Aquaculture Magazine*, November-December, 43, 1982.

13. **Smith, P.,** Abalone culture in Japan, *Catch '81*, September, 11, 1981.

14. **Miyamoto, T., Saito, K., Ito, M., and Mizutori, Y.,** On the releasing of the cultured seeds of Ezo abalone, *Haliotis discus hannai*, in the northern coast of Shiribeshi, Hokkaido, *J. Hokkaido Fish. Exp. St.*, 39, 169, 1982. (Translated in Mottet, M. G., Summaries of Japanese papers on hatchery, technology and intermediate rearing facilities for clams, scallops, and abalones, *Prog. Rep., Dept. Fish. Washington*, 203, 18, 1984.)

15. **Momma, H., Kobayashi, K., Kato, T., Sasaki, Y., Sakamoto, T., and Murata, H.,** On the artificial propagation method of abalone and its effects on rocky shore. I. Remaining ratio of the artificial seed abalone *(Haliotis discus hannai* Ino) on the latticed artificial reefs, *Aquiculture*, 28(2), 59, 1980. (Translated in Mottet, M. G., Summaries of Japanese papers on hatchery, technology and intermediate rearing facilities for clams, scallops, and abalones, *Prog. Rep. Dept. Fish. Washington*, 203, 19, 1984.)

16. **Sakai, S.,** Ecological studies on the abalone, *Haliotis discus hanni* Ino. III. Study on the mechanism of production of the abalone in the region of Onagawa Bay, *Bull. Jpn. Soc. Sci. Fish.*, 28(9), 891, 1962.

17. **Kito, H., Kikuchi, S., and Uki, N.,** Seaweed as nutrition for seabed marine life technology for artificial marine forests, in *International Symposium on Coastal Pacific Marine Life*, Office of Sea Grant, Bellingham, Washington, 1980, 55.

18. **Kikuchi, S.,** Marine forest establishing experiment for sea urchin and abalone, in *The Farming Fisheries*, Ohshima, Y., Ed., Japanese Fisheries Agency, Tokyo, 1976, 292.

19. **Li, C. P.,** Antimicrobial effect of abalone juice, *Proc. Soc. Exp. Biol. Med.*, 103, 522, 1960.

20. **Li, C. P., Prescott, B., and Jahnes, W. G.,** Antiviral activity of a fraction of abalone juice, *Proc. Soc. Exp. Biol. Med.*, 109, 534, 1962.

21. **Hashimoto, Y. and Tsutsumi, J.,** Isolation of a photodynamic agent from the liver of abalone, *Haliotis discus hannai*, *J. Food Sanit. Soc. Jpn. Bull. Jpn. Soc. Sci. Fish.*, 27(9), 859, 1961.

22. **Tsutsumi, J. and Hashimoto, Y.,** Isolation of pyrophenophorbide as a photodynamic pigment from the liver of abalone, *Haliotis discus hannai*, *Agric. Biol. Chem.*, 28(7), 467, 1964.

JAPANESE ABALONE CULTURE TECHNIQUES OF THE OYSTER RESEARCH INSTITUTE

Kirk O. Hahn

INTRODUCTION

The Oyster Research Institute (ORI) was originally founded by Takeo Imai to develop techniques for oyster aquaculture. However, the emphasis of ORI has slowly changed to developing techniques for abalone aquaculture, and it is now one of the leading abalone aquaculture laboratories in Japan. ORI is also unique by being a privately owned, nonprofit company, in contrast to other laboratories culturing abalone established by the prefecture or national government.

ORI is supported by donations from abalone fishing associations that enable ORI to conduct research on abalone culture. Juvenile abalone, a byproduct of the research, cannot be sold (due to ORI's nonprofit status) are are distributed to the abalone fishing associations in proportion to the amount each association donated. The value of each 30-mm juvenile abalone is 90 ¥ (3¥/mm), which is twice the price of juvenile abalone produced at prefecture laboratories. ORI is one of the largest producers of *H. discus hannai* juveniles and produced about 2 million juveniles in 1981.

The ORI is comprised of two facilities. One laboratory is located on the property of the Tohoku Electric Power Station in Shiogama, Miyagi Prefecture. The other laboratory is located on Mohne Bay, near Kensennuma, Miyagi Prefecture. This chapter describes the Mohne Bay laboratory because it is the principal laboratory producing juvenile abalone. The laboratory building sits on a very narrow edge of flat land next to Mohne Bay (Figure 1). The limited surface area available for laboratory space is a disadvantage of the site, but the water quality is excellent since the bay is very deep. The unique features of the Mohne Bay site required the development of special tanks, buildings, and techniques.

Most of the information in this chapter has previously been unpublished and was obtained from several sources. I had the opportunity to visit the ORI on several occasions to discuss the operations with Tetsuo Seki.[1] In addition, Liz Kumabe wrote an excellent unpublished in-house report on the techniques used at ORI.[2] Unless otherwise cited, the information presented in this chapter comes from the unpublished report, personal conversations with Tetsuo Seki, and personal observations of the ORI facility.

BROOD STOCK

Selection

Fewer than 100 adult abalone are kept are ORI due to the prohibitive amount of food and space needed to maintain these animals. Half of the brood stock comes from animals with known life histories that were raised at ORI. Selection of animals for brood stock is subjective; there is no guarantee that individuals with certain external physical characteristics will produce better juveniles. However, brood stock animals are selected for shell shape, growth rate, previous spawning success, and fertilization rate.

The preferred shell shape is cupped with few convolutions, although this criterion is based more on aesthetics than a scientific basis. Individuals are also selected based on fast shell growth and production of large quantities of gametes with a high fertilization rate. These criteria assume that selection for these physical factors may eventually produce favorable adults. Also, "albino" individuals, although very rare, have been produced from some crosses. The "albino" is a light reddish brown and does not produce the characteristic green shell color when fed *Laminaria* spp. These individuals have been successfully grown to

FIGURE 1.   ORI facility at Mohne Bay, Kensennuma, Miyagi Prefecture. Left — main building for conditioning, larval rearing and settlement. Center — multilayer structure. Right — building housing pumps, filters and water heater. Foreground — raft system.

juvenile size and attempts are underway to commercially produce "albino" individuals as a speciality item. Crosses between "albino" and normal green-shelled abalone produce three types of offspring; a normal green shell, shallow thin red shell with a slower growth rate, and a purple shell with deep shell shape and good growth rate.

The other half of the brood stock is obtained from commercially caught abalone to help maintain genetic variability in the culture population. These animals are selected for preferred shell shape and color, length of 9 to 10 cm, and a healthy appearance. Only those individuals still healthy after a week are tagged and retained. The tag is used for identification and as a marker for shell growth, and causes no harm to the animal. The animal is protected from injury during drilling by pushing the body and mantle away from the anterior edge of the shell. A hole is drilled through the shell, the tag is placed on the outer surface of the shell, and a plastic bolt is inserted through the shell and tag from the inside out. The tag is secured by tightening a nut on the bolt, cutting off the excess bolt length, and melting the end. This is the best tagging method because the tag can remain on the animal indefinitely.

Food

Natural Kelp

Adult abalone are fed fresh *Laminaria japonica* and *Undaria pinnatifida* when they are available and frozen kelp at other times. The primary growing season of *L. japonica* ends in late summer. The blades begin degenerating, with only a small portion of the plant left by September. A small secondary growth begins at this time in natural populations but is absent in cultured populations because the whole plant is harvested. This natural secondary growth is used to feed abalone through the fall and occasionally *Eisenia bicyclis* is obtained from southern Japan. Oyster fishermen collect these secondary growth fronds from their rafts and sell them to ORI. The supply soon disappears and *L. japonica* becomes scarce from early November to February. The feeding rate of the abalone conveniently decreases with lower ambient sea water temperature in winter, thus helping reduce the demand for kelp. The abalone are fed *Ulva* spp. or frozen *L. japonica* until March when *U. pinnatifida* has grown large enough for use as food.

Freezing or drying has no apparent effect on the food value of *Laminaria* spp. The advantages of using frozen vs. dried *Laminaria* spp. concerns the cost of storage and labor. The labor required for drying *Laminaria* spp. is greater than the cost for frozen storage. Also, dried *Laminaria* spp. needs to be rehydrated before it is fed to abalone since the dry fronds float. Frozen *Laminaria* spp. can be used without any preparation.

Laminaria Culture

ORI maintains a long line culture of *L. japonica* at the tip of Karakuwa Peninsula. ORI began culturing *L. japonica* to ensure a reliable supply of food for the adult abalone throughout the year. *L. japonica* is no longer collected from the wild, which would remove kelp from natural populations of abalone and defeat the purpose of abalone seed production. The culture techniques used at ORI were developed by Hokkaido *Laminaria* fishermen.

Laminaria spp. culture beings in late September when the thalli produce ripe sporangia. Mature *L. japonica*, which are ripe and ready to release spores, are selected. Portions of blades are air dried on newspaper for 2 to 3 hr at ambient temperature. The blades are then placed in a tray filled with filtered, UV-sterilized sea water for about 30 min to collect spores. It is important not to allow broken ends to contact the water because mucus from the thalli can entrap spores and make them nonviable. Spores are filtered out of the water with a 40-μm screen and counted. The optimal spore density is about 30 to 50 spores visible in a 10 $\times$ field of vision. Too few spores provide sparse growth and too many spores quickly use up the nutrients in the culture water.

Initially, the sea water in the culture tank is sterilized by heating the water to 80°C for 30 min, and then allowed to cool to 17 to 19°C. Thirty milliliters (30 mℓ) of *L. japonica* spore stock solution is added for every liter of water (3% final concentration) in the tank and the water is thoroughly stirred. Two kilograms of string is placed into the tank, which will produce about 40,000 kg of fully grown *L. japonica*. Spores are allowed to settle on the string for about 15 hr. The development and settlement of the spores is monitored by hanging a glass slide in the spore solution. The glass slide is checked after 15 hr and if the spores are still active, the settlement step is extended. The water temperature is initially 17 to 19°C but is decreased to 15°C after the spores have set on the sting.

After 24 hr the string is transferred to a new culture tank containing nutrients for algal growth. The main ingredient in the nutrient solution is potassium iodide (KI). Two liters of nutrient solution is added for each 100 ℓ in the culture tank (final concentration — 2%). It is important to have enough nutrients in the water at this early stage so later growth will be good. No additional nutrient solution is added during the first 20 days. After 20 days, the water is replaced with clean sea water containing a 2% nutrient concentration. Each subsequent week, 50% of the sea water is changed and nutrient solution (final concentration — 2%) is added.

Aeration is gently started on the second day of culture and is gradually increased to a more vigorous level. The light intensity, initially 2,000 lx, is increased to 4000 lx after 1 week and 6000 lx after one more week. The photoperiod is usually 12 hr light/12 hr dark but can be changed to 18 hr light/6 hr dark to accelerate growth.

When thalli on the string reach 1 cm in length (about January) the string is cut into short sections. These sections are tied at regular intervals on a larger rope used for long-line culture. Transferring the thalli at the proper time is very important because they will not survive if too young. The young blades require a short conditioning period in a calm bay before the ropes are placed in the open ocean. The blades reach a harvestable length of 1 m by March or April. The growth during the spring is approximately 5 cm a day.

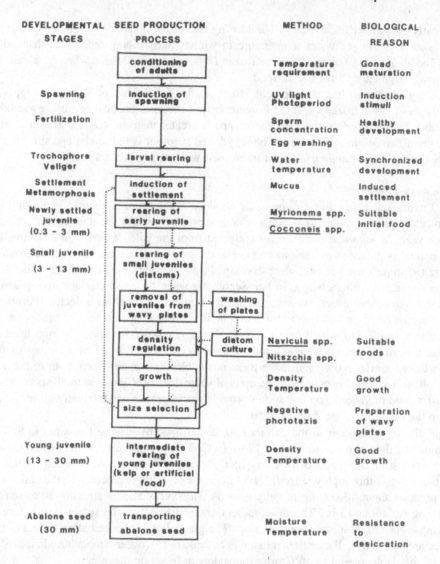

DEVELOPMENTAL STAGES	SEED PRODUCTION PROCESS		METHOD	BIOLOGICAL REASON
	conditioning of adults		Temperature requirement	Gonad maturation
Spawning	induction of spawning		UV light Photoperiod	Induction stimuli
Fertilization			Sperm concentration Egg washing	Healthy development
Trochophore Veliger	larval rearing		Water temperature	Synchronized development
Settlement Metamorphosis	induction of settlement		Mucus	Induced settlement
Newly settled juvenile (0.3 – 3 mm)	rearing of early juvenile		*Myrionema* spp. *Cocconeis* spp.	Suitable initial food
Small juvenile (3 – 13 mm)	rearing of small juveniles (diatoms)			
	removal of juveniles from wavy plates	washing of plates		
	density regulation	diatom culture	*Navicula* spp. *Nitszchia* spp.	Suitable foods
	growth		Density Temperature	Good growth
	size selection		Negative phototaxis	Preparation of wavy plates
Young juvenile (13 – 30 mm)	intermediate rearing of young juveniles (kelp or artificial food)		Density Temperature	Good growth
Abalone seed (30 mm)	transporting abalone seed		Moisture Temperature	Resistance to desiccation

FIGURE 2. Flow chart of abalone seed production at ORI. (—) Flow of abalone; (- - -) flow of wavy plates. (From Seki, T., *International Symposium on Coastal Pacific Marine Life*, Office of Sea Grant, Bellingham, Washington, 1980, 45. With permission.)

Conditioning

Hatchery production is conducted from late spring to late summer, the period when the ambient air and water temperatures are most favorable for larval and juvenile rearing (Figure 2). Hatchery and seed production must occur during a short time span to ensure that juveniles are more uniform in size and to simplify culture maintenance. Brood stock are removed from outdoor tanks and placed into conditioning tanks in February to obtain ripe individuals for spawning by May. Sexes are separated and 50 individuals are placed into one of three conditioning tanks, one for each sex and a reserve tank.

The conditioning tanks are constructed of synthetic wood frames (1.5 m × 0.9 m × 0.8 m, water capacity of 1000 ℓ) lined with polyvinyl chloride (PVC) coated canvas (Figure 3). Four PVC wedge-shaped shelters for the animals are placed widthwise across the bottom of the tank. Water enters the tank from a perforated pipe alone one side of the bottom and air is supplied along the opposite side. This arrangement generates water circulation through the shelters. The water current in the tanks carries smaller fecal material and detritus to the

FIGURE 3. Conditioning tanks.

overflow. Water flow is dependent upon the size and number of individuals in the tanks. At ORI, 600 ℓ/hr will support 50 individuals with an average length of 9 cm and a total weight of less than 5 kg. An artificial photoperiod is produced by a centrally located 40-W light bulb, which generates about 150 lx at the bottom of the tank and about 8 lx inside the shelters (simulating the light found at 7 to 8 m depth, the preferred depth of adult abalone). The photoperiod is maintained at 12 hr light/12 hr dark to match the photoperiod during the natural spawning season. The photoperiod is also phase shifted so that the light period begins at 1:00 a.m. The photoperiod does not affect gonad maturation but helps synchronize the spawning time of adults.

The phase shift in the photoperiod required a change in the design of conditioning tanks. Tanks were orginally lined with a light-colored canvas but this material reflected light into the shelters. The canvas color was changed to black, which reduced the reflection of light. This change was very important to ensure the shift in the photoperiod and synchronization of spawning.

One week after the brood stock are placed in the conditioning tanks, sea water temperature is raised from ambient (~5°C) to 20°C. This temperature is maintained for 3 months to guarantee complete gonad maturation. A reserve supply of 50 animals is kept in a third conditioning tank with ambient sea water and only changing the photoperiod. A reserve supply is needed in case problems arise (e.g., all the abalone in the other tanks prematurely spawn). Ambient sea water temperature increases above 7.6°C (the biological zero point of this species) by April, and it is possible to calculate the level of gonad maturation in the reserve tank by using the effective accumulative temperature (EAT) formula. Abalone conditioned at 20°C have an EAT of 1500°C-days and are completely mature by the end of May. The ambient water temperature at Mohne Bay will generate an EAT of 1500°C-days during September, which is the natural spawning period.

Abalone in the conditioning tanks are fed fresh *L. japonica* or *U. pinnatifida* daily, and any uneaten kelp from the previous day is removed. The kelp blades are torn into large

FIGURE 4.   Spawning tanks set up for beginning of spawn induction. Adult abalone are placed into the clear tanks on middle shelf. Empty tanks on the bottom are used during fertilization and larval rearing.

pieces, about 20 cm in length, before being fed to the abalone. The exact size is not too important with large individuals. Although the actual consumption of kelp is usually around 8 to 10% of the animal's body weight per day, 30% of the animal's body weight is provided each day to ensure sufficient food for all individuals. The feeding rate is calculated each day. A feeding rate below 4% of the body weight per day, indicates that rearing conditions are unsatisfactory or there is some adverse factor affecting the abalone. Finding individuals out of the shelters during daylight hours is another indication of poor conditions.

HATCHERY TECHNIQUES

Spawning

The induction of spawning occurs on a regular schedule at the beginning of each week during the summer (June, July, and August). The tanks (usually six) used for spawning are set up the preceding night and water flow is adjusted to 50 ℓ/hr (total for all tanks) (Figure 4). The water flow should be the same into each tank. The tanks are 21 cm × 37 cm (top), 19 cm × 35 cm (bottom), side 27.5 cm, outlet hole 6 cm from top and 1.5 cm in diameter. The water overflows through a small tube outlet at 15 ℓ.

It is critical to begin induction of spawning at the proper time, allowing the correct time lag to synchronize the UV technique with the artificial photoperiod. Just before 10:00 a.m., six gravid individuals (three of each sex) are selected. The gonad maturation is rated on a 0 to 3 scale with maximum maturity rated 3. Only individuals rated 2 or 3 are selected for spawning. Each animal is measured with a caliper, weighed, and tag number recorded. This information is kept on all spawnings for future reference. At 10:00 a.m. individuals are placed between layers of damp cotton gauze at room temperature (about 20°C) for 1 hr. This treatment alone has sometimes caused spawning of very ripe adults. Water level in the spawning tanks is reduced to 2 ℓ just before the adults are placed in the tanks. This smaller

FIGURE 5. Close-up view of spawning tank with thermostat probe.

amount of water allows better control over water temperature during spawning induction. At 11:00 a.m. each individual is placed in a separate tank. This isolates the released gametes and prevents uncontrolled fertilization. Some laboratories spawn males and females in the same container but this does not allow control over the timing and uniformity of fertilization. Controlled fertilization is required to obtain synchronized development of the larval stages and a predictable time schedule for the necessary water changes. The water flow is maintained at less than 50 ℓ/hr for all six tanks combined or 4.3 ℓ/hr/tank. The water flow is low enough not to disturb the spawned eggs which are slightly higher density than sea water and settle on the bottom of the tank. The water level in each tank slowly rises to 15 ℓ.

The water used for spawning induction passes through both a 108-W and 13-W UV unit. This water then enters either of two holding tanks, one refrigerated to 15°C and the other heated to 25°C. A thermostat probe, inserted into one of the spawning tanks, measures the water temperature (Figure 5). The oscillation of water temperature during the induction cycle is controlled by a timer that rotates a plastic wheel. The wheel is cut (the radius of the wheel determines the water temperature) to set the thermostat to the desired temperature. Depending on the water temperature in the spawning tank, the thermostat selects either hot or cold water to obtain the desired temperature. This method gives more precise control over the water temperature than would be possible if the water was preheated to the desired temperature before entering the tank. The water temperature is maintained at the exact level without a time lag. The initial water temperature in the spawning tanks (20°C) is raised to 23°C within 30 min, maintained at 23°C for 2 hr and then lowered to 20°C within 2 hr.

The three treatments (desiccation, UV-irradiated sea water treatment, and thermal shock) contribute to the induction of spawning in *H. discus hannai*. This method greatly increases the control and success of spawning induction by using a combination of factors, and assures spawning within a few hours. The conditioned adults usually spawn with 3 hr after being placed in the UV-irradiated sea water. Completely spawned adults are returned to the raft system. Partially spawned individuals are sometimes returned to the conditioning tank and become ripe again in about a month.

Fertilization

Eggs must be fertilized within 1 hr after release since viability decreases with time. This time limitation stresses the importance of the synchronized spawning of both sexes. Sperm also declines in viability with time but may be continuously released over a long period. This causes the viability of the sperm in the spawning tank to be variable. Therefore, using more sperm than necessary guarantees a complete fertilization of the eggs. A synchronous fertilization assures uniformly developing larvae and allows maximum production with minimal work. Also, uniform development is required for any type of research conducted on the larval stages.

The eggs (270 μm in diameter) are siphoned from the bottom of the spawning tank with a glass tube and rubber tubing, and are filtered through a 276-μm nitex screen to remove large detritus and fecal matter. The eggs flow directly into an identical 20-ℓ spawning/larval tank. The clear tanks allow viewing of the eggs and later the larvae. Also, the light weight of the tanks make the frequent water changes easier. The quantity of eggs in each container is adjusted to obtain a monolayer of eggs on the bottom, i.e., 400,000 eggs can lie on the bottom of the 20-ℓ container. Each tank is marked at 2 and 15 ℓ. The monolayer ensures each egg has maximum exposure to sperm for fertilization. The eggs from a single female are usually split into two containers.

The eggs are fertilized with a mixture of sperm from all spawning males. One or two liters of "sperm" water is removed from each male spawning tank and combined in a large container, decreasing the possibility of nonviable sperm being used for fertilization. The sperm density is also important for successful fertilization. A concentration of less than 100,000 sperm per milliliter produces less than 100% fertilization, and a concentration of over 1×10^6 sperm per milliliter produces polyspermy and dissolves the egg jelly layer. The ideal sperm concentration is 400,000 sperm per milliliter. (See Chapter 4, Spawning and Fertilization.)

The sperm density of the combined "sperm" water is measured with a hemacytometer. The required volume of this solution is added to each tank of eggs to obtain a final concentration of 400,000 sperm per milliliter in 2 ℓ (i.e., 8×10^9 sperm). Water level in the tank is immediately increased to 2 ℓ. After 2 min., the tanks are vigorously stirred, stopping fertilization, and water level is increased to 15 ℓ with 20°C sea water. A difference between the air temperature and water temperature is not critical unless it causes circulation of the water, which will disturb the settling of the eggs. The eggs are allowed to settle (approximately 10 min), the water decanted, and the tank refilled to 15 ℓ. The tank is decanted ten times to remove all sperm in the tank. If sperm remains in the rearing water after fertilization, it may cause abnormal development, possibly by enzymes from the sperm or bacterial decomposition of the sperm.

Larval Development

All water used in larval rearing is 1-μm filtered and UV sterilized. Maintaining very clean conditions is very important during the larval development period. All instruments and equipment are thoroughly cleaned and washed with UV-sterilized water before being used. Cleanliness cannot be stressed enough.

Tanks are placed in a dark, temperature-controlled room after the eggs have been thoroughly washed and excess sperm removed from the water. The time of hatch out, as well as all larval stages, is controlled by the water temperature. A water temperature of 20°C is ideal for quick development and timing of the necessary procedures during larval culture. Timing all required procedures to occur during normal business working hours, saves money otherwise spent for overtime to employees working nights or weekends. At 20°C, eggs fertilized on Monday will be ready for settlement by Friday afternoon.

The morning following fertilization (Tuesday), the trochophore larvae hatch out within 1 or 2 hr of each other. Healthy trochophores are separated as soon as possible from the egg cases, unfertilized eggs, and other debris on the bottom of the tank. The water quality quickly deteriorates from bacterial contamination and possibly from the enzymes secreted to break down the egg case for hatch out.

The healthy larvae begin swimming near the surface after hatch out. The top 75% of the water is decanted into a clean tank with a single smooth motion. Shining a light across the mouth of the container makes the trochophores visible during the decanting process. This fraction, called the "first trochophore", should be absolutely free of all egg cases. The remaining 1/3 to 1/4 of water left in the container is allowed to settle and decanted again. This fraction, called the "second trochophore" will probably be contaminated with some egg cases but care should be taken to avoid it. The "second trochophore" fraction may allow a little larger production of larvae if they do not die from contamination.

The new tank is slowly filled to 15 ℓ with isothermal, UV-sterilized water. The added water must be within ± 1°C of the temperature in the tank, so the trochophores are not killed by temperature shock. Larval density can range from 150,000 to 700,000 larvae per 15-ℓ tank. The trochophore larvae are returned to the dark, temperature-controlled room, and remain in a static (no aeration, no water exchange) culture until the larval shell is completed. The larval body and forming shell edge are both very delicate before the completion of the larval shell and can easily be damaged by excessive water movement. The time of occurrence of each larval stage can be predicted for a given temperature by using the EAT required for each larval stage (Table 1).

After completion of the larval shell, the water in the rearing tank is changed. The water is poured through a 40-µm filter sitting inside a low rectangular tray, guaranteeing that larvae are always in water, and the larvae are retained on the screen (Figure 6). The larval shell might be damaged if water is not present, cushioning the larvae from the surface of the filter. The larvae are thoroughly washed with flowing isothermal UV-sterilized sea water, taking care not to force the larvae against the screen. The larvae are gently swished by the flowing water and are then poured into a clean rearing tank. The water in the rearing tank is replaced, removing waste products that would retard larval development. There is usually a higher density of larvae on the bottom of the container than in the water column immediately before the water is changed. However, almost all larvae begin swimming within a few minutes after the water is changed. Healthy larvae have a characteristic swimming behavior. Groups of larvae repeatedly form vertical spiral columns in the water with individuals swimming to the top and then slowly sinking to the tank bottom. During the period from completion of larval shell to settlement, the water is changed three times a day and the tanks are changed every day. The tanks are sterilized with 100-ppm solution of $HClO_2$ for 12 hr.

The number of water changes during larval development depends on the water temperature. At 20 larvae per milliliter, water is changed three times a day (9 a.m., 12 p.m., and 3 p.m.) when the water temperature is 20°C or higher, once a day when the water temperature is between 18°C and 20°C, and changed only when necessary (veligers remain on the bottom of the tank) when the water temperature is below 18°C.

SETTLEMENT

Preparation of Settlement Plates

The substrata used for settlement are translucent, corrugated, hard plastic sheets (33 cm in height by 40 cm in width, with corrugations 1 cm in depth). The substrata are called "wavy plates". These plates are very durable and reusable. Fouling organisms and dirt are removed from used plates by a machine with double brushes and strong water pressure.

Clean plates are placed into plastic-coated metal racks (top width — 42 cm, bottom width — 28 cm, sides — 32 cm, and vertical height — 33 cm) capable of holding 15 plates (Figure

Table 1
TIMING OF CRITICAL LARVAL STAGES
OF HALIOTIS DISCUS HANNAI WITH
EAT

Water temperature (°C)	Hatch out (hr)	Completion larval shell (hr)	Settlement (hr)
10.0	66.7	162.5	429.2
10.2	61.5	150.0	396.2
10.4	57.1	139.3	367.9
10.6	53.3	130.0	343.3
10.8	50.0	121.9	321.9
11.0	47.1	114.7	302.9
11.2	44.4	108.3	286.1
11.4	42.1	102.6	271.1
11.6	40.0	97.5	257.5
11.8	38.1	92.9	245.2
12.0	36.4	88.6	234.1
12.2	34.8	84.8	223.9
12.4	33.3	81.2	214.6
12.6	32.0	78.0	206.0
12.8	30.8	75.0	198.1
13.0	29.6	72.2	190.7
13.2	28.6	69.6	183.9
13.4	27.6	67.2	177.6
13.6	26.7	65.0	171.7
13.8	25.8	62.9	166.1
14.0	25.0	60.9	160.9
14.2	24.2	59.1	156.1
14.4	23.5	57.4	151.5
14.6	22.9	55.7	147.1
14.8	22.2	54.2	143.1
15.0	21.6	52.7	139.2
15.2	21.1	51.3	135.5
15.4	20.5	50.0	132.1
15.6	20.0	48.8	128.8
15.8	19.5	47.6	125.6
16.0	19.0	46.4	122.6
16.2	18.6	45.3	119.8
16.4	18.2	44.3	117.0
16.6	17.8	43.3	114.4
16.8	17.4	42.4	112.0
17.0	17.0	41.5	109.6
17.2	16.7	40.6	107.3
17.4	16.3	39.8	105.1
17.6	16.0	39.0	103.0
18.8	15.7	38.2	101.0
18.0	15.4	37.5	99.0
18.2	15.1	36.8	97.2
18.4	14.8	36.1	95.4
18.6	14.5	35.5	93.6
18.8	14.3	34.8	92.0
19.0	14.0	34.2	90.4
19.2	13.8	33.6	88.8
19.4	13.6	33.1	87.3
19.6	13.3	32.5	85.8
19.8	13.1	32.0	84.4
20.0	12.9	31.5	83.1

Table 1 (continued)
TIMING OF CRITICAL LARVAL STAGES
OF HALIOTIS DISCUS HANNAI WITH
EAT

Water temperature (°C)	Hatch out (hr)	Completion larval shell (hr)	Settlement (hr)
20.2	12.7	31.0	81.8
20.4	12.5	30.5	80.5
20.6	12.3	30.0	79.2
20.8	12.1	29.5	78.0
21.0	11.9	29.1	76.9
21.2	11.8	28.7	75.7
21.4	11.6	28.3	74.6
21.6	11.4	27.9	73.6
21.8	11.3	27.5	72.5
22.0	11.1	27.1	71.5
22.2	11.0	26.7	70.5
22.4	10.8	26.4	69.6
22.6	10.7	26.0	68.7
22.8	10.5	25.7	67.8
23.0	10.4	25.3	66.9

7). The racks are placed in the raft tanks to accumulate a film of diatoms. The length of time needed to obtain sufficient diatom growth is dependent upon the time of the year. High water temperatures and increased light during the summer produces a thick diatom growth within 2 weeks. The diatom growth rate is greatly reduced during other seasons, requiring the plates to be left in the tanks for over a month.

Colonizing diatom communities are classified into two basic groups. The first diatom community consists of long filamentous diatoms, mainly *Navicula* spp. and *Nitzschia* spp. Juvenile abalone (>10 mm) are allowed to graze on the plates covered with the primary diatom community, thus encouraging the development of a secondary community of encrusting diatoms, *Cocconeis* spp. and *Myrionema* spp. Juveniles graze on the primary diatom community for 2 weeks (20°C, 15-mm juveniles) and the plates become covered with a greenish yellow film of encrusting diatoms.

The grazing juveniles also cover the plates with mucus during this period. The mucus secreted by grazing juveniles is a factor necessary for inducing settlement and metamorphosis of the larvae of this species. Settling larvae are believed to favor a surface with a mucus coating and encrusting diatoms. The secondary diatom community is also an ideal food source for newly settled juveniles. These small encrusting diatoms are more easily handled by the developing mouth parts of the juveniles. Because newly metamorphosed larvae have not developed mouth parts or a mouth opening capable of feeding on diatoms larger than 10 μm, smaller particles, such as bacteria, are thought to be their principal food.

Preparation of settlement plates influences the time schedule that must be maintained during the seed production season. The juvenile-grazed plates from an entire raft tank (approximately one third of the plates are discarded) are needed for each weekly larval settlement. Timing is critical and requires an accurate prediction of diatom growth rate since the plates go directly from the raft to the settlement tanks. Therefore, each raft tank is labeled and assigned a specific date for use in settlement.

The plates used for settlement are brought into the lab and the grazing juveniles are removed. Juveniles are removed from the plates with a combination of thermal shock and the anesthetic, ethyl *p*-aminobenzoate ($NH_2C_2H_4COOC_2H_5$, mol wt = 165.19). (A stock

FIGURE 6.   Hatchery worker changing the water in the larval rearing tank. Water hose supplies continuous flow of UV-sterilized sea water during water change.

solution is prepared by dissolving 100 g of 95% ethyl alcohol. The working solution is obtained by mixing 0.5 to 1 mℓ of stock solution with 1 ℓ of sea water, or 1 ℓ of stock solution is added to a tank containing 1.8×10^3 ℓ of water.)

The racks of plates are first submerged in a tub lined with a net (Figure 8). The water temperature is usually maintained at approximately 12°C above ambient water temperature. This increase reaches the lethal thermal level during the warm summer months when the ambient water temperature is above 20°C. The temperature difference is slightly reduced to only 6 to 8°C above ambient during this period. The juveniles do not fall off the plates at these temperatures but relax their attachment and can be easily shaken or brushed off.

The plates are left in the water for less than 1 min. This time limit is dependent upon the size of the juveniles. The plates are vigorously shaken and quickly checked for any remaining juveniles. Each plate is rinsed in clean sea water, scrubbed to remove any large debris or filamentous algae attached to the surface, and inspected to determine if it is suitable for settlement. The racks are also scrubbed and then plates and racks are put back together. The racks of plates will be used for inducing settlement of the larvae.

The anesthetized juveniles are immediately removed from the tub with hand nets and

FIGURE 7.   Wavy plates in rack.

FIGURE 8.   Hatchery workers removing juvenile abalone from wavy plates in preparation for settlement induction. Woman on right is shaking the racks to dislodge the juvenile abalone. Woman on left is removing the anesthesized juvenile abalone from the tank.

FIGURE 9.   Settlement tanks prepared for settlement induction. Racks of diatom and mucus covered wavy plates are placed horizontal in the settlement tanks.

placed in a water table to recover. The juveniles are allowed to reattach to new wavy plates. Each wavy plate is placed in the center slot of a rack of diatom-covered plates and returned to the raft tanks for further growth. The juveniles will evenly distribute themselves naturally while grazing on the diatoms.

Induction of Settlement

Larval settlement is induced weekly during the hatchery season. Temporary settlement tanks are set up only during the hatchery season. The settlement tanks are constructed of a synthetic wood frame (2.4 m × 1.5 m × 0.5 m) lined with PVC-impregnated canvas that is held in place by plastic grips at the top. Twenty racks, each containing 15 wavy plates (300 plates total), are placed in each tank. The plates are placed horizontally in the water (1-μm filtered, UV sterilized) and the water level in the tank is adjusted above the surface of the wavy plates (Figure 9). Wavy plates are not put on the bottom or sides of the settlement tanks since larvae are very selective and only settle on suitable substrata.

Each plate supports 2000 larvae, so approximately 600,000 larvae are poured into each tank. The larvae are distributed evenly throughout the water by manually lifting each plate rack above the surface of the water. After distributing the larvae, each rack is tipped slightly underneath the water to remove any trapped air bubbles. Four 1-m long, 40-W lamps are hung 20 to 30 cm above the surface of the water immediately after introducing the larvae for settlement. The lights are left on continuously, ensuring good diatom growth and inhibiting diatom respiration, which would produce anoxic conditions on the surface of the wavy plate and cause asphyxiation of the larvae.

The settlement process is monitored by cutting off a small piece of a plate and examining it under a microscope. The rate of settlement is directly correlated with the water temperature. During the first part of the hatchery season (late June) when the water temperature is about 17°C, at least 17 hr are required for completion of settlement and initiation of water circulation. As the water temperature increases during July and August, settlement occurs more rapidly and water circulation is started after about 10 hr.

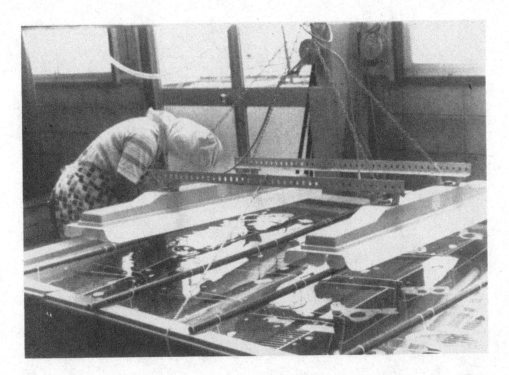

FIGURE 10. Settlement tank after completion of larval metamorphosis. Racks of wavy plates are hung vertically from plastic pipes, and lamps are placed above the tanks to ensure good diatom growth.

The next day the settled larvae on the plates are checked for completion of metamorphosis. Complete metamorphosis is indicated by the loss of the velar cilia cells, and the beginning of the peristomal shell fringe. Metamorphosis is assumed to be complete when the velar cilia cells are gone, even though there may not be peristomal shell growth visible. The plates are placed vertically in the water column after completion of normal metamorphosis, so that the parallel corrugations on the plate surface are aligned perpendicular to the tank bottom (Figure 10). If the corrugations are aligned parallel to the bottom, the curved surface provides a trap for oxygen produced by diatom photosynthesis and a shelf for fecal material produced by the developing larvae, thus reducing the available living space for the larvae.

Following metamorphosis, water flow is begun (500 ℓ/hr) through a perforated pipe along one length of the tank bottom. Aeration is supplied through a similar perforated pipe along the opposite side. The combination of water flow and aeration allows complete water circulation in the entire tank. Water exits into an outlet pipe at the tank top and any remaining planktonic larvae are collected on a 40-μm filter.

After 1 week in the settlement tanks, the number of plates in each rack is reduced to eight (i.e., a plate in every second slot) and the racks are transferred to a small outdoor tank for one additional week. The indoor settlement tanks are drained, scrubbed with salt (NaCl), rinsed, and prepared for the next batch of larvae to be settled. The additional time larvae spend in the outdoor tank allows development of a shell shape and positioning for better water-flow dynamics over the body. The juveniles are about 500 to 600 μm in length at this time. Two weeks after settlement, the racks are moved to the raft tanks.

SEED PRODUCTION

Raft System

The raft tank system is unique to the ORI and was built in 1978 for a total cost of 6 million ¥ ($24,000 to $30,000). This design was necessary since there was very little flat

FIGURE 11.　Foreground — raft system floating in Mohne Bay. Background — oyster racks, typical of those used by fishermen for intermediate culture of juvenile abalone.

land and no room for tanks unless placed on the bay. The raft system (24 m × 70 m) is built with recycled plastic-coated (increases the resistance to sea water) iron tubing and large plastic-wrapped styrofoam for flotation (Figure 11). The raft holds 19 tanks and two rows of 20 cages, although the number of tanks and cages can easily be changed. The front of the raft floats in a few meters of water while the depth at the far end of the raft exceeds 15 m due to the steeply sloped bay floor. Mohne Bay is very sheltered with little wave action, other than from passing boats, to disturb the raft.

The construction of the tanks is also unique. The tanks are made of PVC-coated canvas, and can easily be removed for cleaning and maintenance. They are also inexpensive to replace if damaged. One disadvantage to having the tanks floating in the water is that fouling organisms settling on the under surfaces cause the tanks to become very heavy. Each tank is 22 m × 2 m × 1 m, and holds 160 racks (2400 plates total) (Figure 12).

Sea water sent to the raft is only sand filtered. Water enters each tank at the surface near the shore end and exists through four tube openings at the far end. Ninety thousand liters of sea water per hour goes to the raft, which means each tank receives over 4500 ℓ/hr. Through experience it has been determined that 1000 ℓ/hr can support 1000 juveniles (2 cm). This recommended amount is a standard value which should be followed until the sea water quality and efficiency of the facility (i.e., aeration) can be established. The water flow used at ORI is 500 ℓ/hr per 1000 juveniles [i.e., ~8000 juveniles (2 cm) per tank].

Diatom growth is stimulated by adding inorganic nutrients to the water in each tank. The most important nutrient, ammonium sulfate $[(NH_4)_2SO_4]$, is a source of nitrogen. Two plastic buckets, filled with a mixture of dry inorganic nutrients, hang partially submersed in each tank. The buckets have a 1-mm hole in the bottom and slowly leak out 5 kg of nutrient mixture every 1.5 months.

Aeration is continuously supplied through a perforated plastic pipe structure that also supports the wavy plastic racks. The racks sit on perforated hard plastic sheets laid on top

FIGURE 12.   Tank used for juvenile rearing on raft system. Right — tank with racks of wavy plates. Left — tank after cleaning and drying.

of the pipe structure framework (Figure 13). The plates are placed vertically in the water with the bottom of the plates at a maximum depth of 40 cm (Figure 14). The light intensity at 40 cm is the minimum amount required by the diatoms for growth. Plate depths greater than 40 cm stunt diatom growth. The opaque, bottom sheets also serve as shelters for the negatively phototactic juveniles during the daylight hours.

While the raft system has many advantages, it does have one major disadvantage. The maintenance of the raft tanks is the single most labor-intensive job at ORI. The tanks are cleaned of fecal material which can produce toxic chemicals, especially oxidized sulfur compounds that are very harmful, and excessive diatom growth on the bottom and sides of the tanks which consume supplemental nutrients supplied for the diatom growth on the wavy plates. The frequency of cleaning is dependent upon water temperature, season, and size of juveniles in the tank.

The tanks are cleaned approximately weekly to remove most of the debris in the tanks because the water flow and aeration do not sufficiently remove this material. Aeration and water flow is stopped, the sides and bottom of the tank are brushed free of any adhering material, and the hanging plastic frameworks with the wavy plate racks are pulled to the side. A small water pump removes the detritus and a net retains any abalone in the water that might have been vacuumed up during the cleaning (Figure 15). This procedure requires very little labor but does not completely clean the tank.

Periodically, the raft tanks are completely cleaned to remove diatoms from the walls and fouling organisms from the underside. This procedure is very labor intensive. All aeration framework and wavy plate racks are removed and transferred to a previously cleaned tank. The water is removed from the dirty tank and a wooden platform is placed beneath one end. The wood frame allows for removal of juveniles that have fallen to the tank bottom, and provides support while scraping off the fouling diatoms. The platform is moved down the length of the tank as necessary. The tank is scrubbed with salt and dried before refilling

FIGURE 13.   Empty tank showing the perforated plastic pipes that supply
aeration and support the racks of wavy plates. The racks sit on a hard sheet
of plastic on top of the plastic pipes.

(Figure 12). The high cost of labor indicates this job should be simplified with specially
designed equipment.

The cages on the raft system are used for raising juveniles larger than 13 mm. The cages
are 2 m × 1 m × 2 m, made of a flexible reinforced nylon net with a 5-mm mesh, a base
of hard plastic with a larger mesh of 6 mm, and framed with plastic pipe. These cages are
collapsible, and can easily be stacked and stored on land. Two thousand juveniles are raised
in each cage until they reach an average shell length of 3 cm. About 2 kg of *Laminaria* sp.
and other suitable macroalgae are placed into the cages every 2 days. Only kelp blades are
given as food since the holdfasts may contain unwanted organisms and material. Blades are
torn into small pieces to increase the amount of blade edge available since juveniles begin
eating at the edge of the kelp. The tank depth (2 m) ensures that light intensity at the bottom
of the cage is insufficient for algal photosynthesis to produce gas. This prevents the seaweed
from floating and thus unavailable to feed the abalone. Four shelters on the bottom of each
cage house the juveniles during the daylight hours. Each shelter is collapsible and is made
of four rectangular vinyl plates held together in an elongated ''M'' formation. They can be
easily straightened out by unnotching the restraining strings, and stacked and stored flat.

FIGURE 14.   Racks of wavy plates sitting in a tank on the raft system.

Cage maintenance is easy due to its construction, and the fact that it is collapsible and portable. Fouling organisms are periodically cleaned from the cages since water circulation can easily be inhibited by any slight interference. Good water circulation is critical for prevention of anoxic conditions due to the metabolic activity of the abalone and the build up of fecal material. Calcium carbonate ($CaCO_3$) building sponges are the greatest problem in cleaning the cages. These fouling organisms are removed by drying the cages and then pounding the net mesh to crush the $CaCO_3$ coating. The greatest amount of fouling from these organisms occurs during July and August. During the cool months of the year, there is no fouling from these organisms and no need to change the cages.

Juveniles Rearing on Diatom Plates

Controlling the density of juveniles on the diatom-covered wavy plates is critical, and must correspond with the balance between the rate of diatom production and grazing. The diatom growth on one plate sufficiently feeds the newly settled juveniles until they reach a length of 2 to 3 mm. An inadequate amount of food is indicated by finding individuals on the sides of the tanks or off the plates during daylight hours.

The continuously increasing rate of consumption by the growing juveniles requires that within the first 2 months, the settlement plates containing juveniles be alternated with clean wavy plates. Diatoms start growing on the clean plates and juveniles naturally spread out. After a few months, the juvenile density on the plates is thinned again by half. The plates from one rack are divided to fill two racks, alternating the plates with clean wavy plates. The juveniles are not removed from the original plates since there is sufficient food provided by the balance between diatom growth and juvenile grazing. The juvenile density is approximately 200 per plate at this stage.

The juveniles are sized and separated into three groups (small, medium, and large) when they reach an average length of 5 mm. Separating the animals makes handling easier and ensures the proper amount of food is provided. The large size group needs more food than

FIGURE 15.   Worker vacuuming the tanks.

the other size groups and their density must be reduced. A small handful of large individuals are allowed to attach to a clean wavy plate, which is then inserted into a rack containing only diatom-covered plates. The juveniles quickly spread out among the available plates. The plates will not need to be changed until the amount of grazing exceeds the diatom growth (approximately 2 months). Changing the plates is simplified by the fact that juveniles (~2 to 3 mm) exhibit negative phototaxis. This behavior develops after formation of the first respiratory pore. The juveniles move to the under size of the bottom plate just before dawn which leaves the plates relatively absent of any juveniles.

The juvenile density on each plate is decreased to between 50 and 70 individuals and this level is maintained until the juveniles reach 10 mm in length. When juveniles reached an average length of 10 mm they are separated into size classes for further growth. From 10 to 15 mm the density is maintained at 20 to 30 individuals per plate. Juveniles are grown in the raft tanks until they are 13 to 15 mm in length. It takes approximately 5 months with the warm water temperatures during summer to reach this size. At 13 to 15 mm, juveniles are transferred from the plates to cages on the raft system. Juvenile mouth parts have sufficiently developed by this time, allowing the food to be changed from a strictly diatom diet to a diet of macrolagae (usually *L. japonica*) or artificial food for abalone.

FIGURE 16.   An empty tank in the multilayer structure. Plastic pipes on the bottom supply aeration. Cages hang from the edge of the tank.

Multilayer Structure

The multilayer structure was built to create additional space needed for rearing juvenile abalone, but in a limited surface area. The multilayer structure consists of three floors of tanks. There are eight tanks on each floor with four rows of tanks on two layers. The tanks on the top floor receive fresh sea water input directly from the sand filter. Water flows out of each tank on the top floor into the corresponding tank of the same row and same layer position in the floor directly beneath it. Thus, there are actually eight series of sea water flowing through the multilayer structure. The multilayer structure receives 40,000 ℓ/hr with each tank (or actually series of tanks) receiving 5000 ℓ/hr. The capacity of each tank is 1200 ℓ of sea water, thus the water in each tank is completely exchanged within 1 hr. During the summer, the water in the tanks must be completely exchanged within 1 hr, or the quantity of fecal material and other metabolic wastes reach levels harmful to the juveniles.

Each tank (6 m × 1 m × 0.25 m) is made of industrial plastic-coated wood planking. Nine cages (1 m × 0.6 m × 0.2 m, 0.15 m water depth) can be placed into each tank. Two types of cages are used for raising the juveniles. One is made of flexible, reinforced synthetic netting with a molded hard plastic net-like bottom. The other is completely made of the molded, hard plastic simulated netting. Both cage types are framed along the top edge with supporting industrial plastic tubing. Two bent vinyl shelters (60 cm × 50 cm) with eight parallel folds and regular spaced perforations of 2.5 cm diameter are placed in the cages. The multilayer structure enables use of cages with a variety of mesh sizes, which allows raising 7-mm long juveniles in the system. Continuous aeration is provided by a series of perforated plastic pipes along the bottom of the tanks that provide several streams of air bubbles across the center of each cage (Figure 16).

Maintenance of the multilayer structure is simpler, and the removal of excess diatom growth and fecal material is not as labor intensive as in the raft tanks. The most obvious advantage is that it is located on land. The cages and aeration pipe frames can be easily

FIGURE 17.   Artificial abalone food. The cups are used to measure the food given to each cage.

removed, and the shallow tanks facilitate any needed work. A mechanized system to automatically distribute the artificial food pellets to the cages has been discussed to further reduce the labor.

The multilayer structure is a supplementary facility to the raft system and, additionally, has many advantages over the raft system. The rearing capacity of the multilayer system far exceeds that of the raft system since the food source is not dependent on natural diatom growth on wavy plates (the limiting factor in the raft system). Juveniles can be maintained in high densities with an abundance of artificial abalone food in the multilayer system, greatly increasing the number of juveniles raised per land surface area. The artificial food can be eaten by juveniles as small as 7 mm in length and produces an excellent growth rate (Figure 17). The artificial food pellet is broken into small pieces before feeding it to 7-mm individuals. An intact pellet does not provide enough edge area and would cause variation in growth rate in each cage. A greater percentage of food per whole animal weight is given to juveniles since the broken pieces of artificial food dissolve faster than whole pellets. The cool water temperatures during the winter inhibits this disintegration to some extent. The juveniles grow large enough to eat the whole food pellet by the time the water temperature rises again. The ORI pays 470 ¥/kg (~$2.50/kg) for artificial food.

The quantity of juveniles is additionally increased in the multilayer structure by stacking tanks on three floors. Juvenile abalone are transferred from diatom plates in the raft system to the multilayer system when their average shell length is 7 mm ± 1 mm. They remain in the multilayer system until they reach the shell length needed for intermediate culture. It takes approximately 2 to 3 months in the summer and 6 months in the winter for juveniles to grow from 7 to 15 mm. The size and quantity of juveniles held in the multilayer system varies throughout the year depending on the season and demand for abalone seed. The multilayer structure is also used as a convenient temporary storage area for 30-mm juveniles that will be sold in the near future. Repeated handling and removal of juveniles from the wavy plates can be harmful, so they are temporarily placed in the multilayer tanks rather than being returned to the raft system. Also, the juveniles with the slowest growth rate from the last seed production season are put in the multilayer system until they reach 30 mm. The demand for abalone seed makes it profitable to retain the slow-growing juveniles until they are sold. The juveniles are sold for 3 ¥/mm when 30 mm in length. The time from fertilization to 30 mm is about 1.5 years. The survival rate to 30 mm depends on when the eggs were fertilized; July — 20 to 25% (highest survival); August — 18%; and September — 15%.

The capacity of the multilayer system has been calculated for each juvenile size class. Juveniles entering the system have a median length of 7 mm with a minimum length of 5 mm. This system is capable of supporting a theoretical pure size class of 1,200,000 juveniles 5 mm in length, 700,000 juveniles 7 mm in length, and 420,000 juveniles 10 to 15 mm in length.

Intermediate Culture

The raft system holds appoximately 400,000 juveniles of 15 to 30 mm size, which is only a small fraction of the total annual production. The ORI expanded production by developing a "feed lot" system with the local oyster fishermen. This method is called "intermediate culture" and represents growing the juveniles to a larger size before release. The "feed lot" system of intermediate culture began in 1975 with 15 fishermen, each culturing about 1000 juveniles. The initial success of this method encouraged expansion by the oyster fishermen. The quantity of juveniles cultured by each fisherman now ranges from 40,000 to 70,000. The abalone are not actually counted since this would be very time consuming and labor intensive. The quantity of juveniles is estimated by weight. The juveniles are first sorted with large sieves in 1-mm increments. The average weight of each size of juvenile is known and the total number can be estimated with an accuracy of ±2%. In 1981, 1 million juveniles (13 mm) were raised in approximately 400 net bags. The juveniles grew approximately 15 mm during the 5 months of intermediate culture and the survival rate was about 98%.

Fishermen place cages given by ORI on their oyster racks for intermediate culture of juveniles. Each cage can hold from 2000 (>20 mm) to 3,000 (15 to 20 mm) juveniles. An oyster raft holds 28 cages, so each fisherman reserves about two rafts for the few months during the summer required for intermediate culture. The fishermen receive 15-mm juveniles for intermediate culture at the end of June and by November, the juveniles have grown to an average shell length of approximatley 25 mm. ORI pays the fisherman 1 ¥/mm of growth.

The fishermen are urged to long-line culture *Laminaria* sp. and are discouraged from harvesting kelp from the wild. Harvesting naturally occurring kelp not only reduces the food supply for abalone in nature, it is also counterproductive to the efforts of seed production. Two baskets of kelp (∼2 kg) are given to each cage every other day to feed the juveniles. This method allows the fishermen to easily supply the correct amout of kelp to each cage. Oyster fishermen like the intermediate culture because it is very profitable for a relatively short duration of labor. Also, the intermediate culture is very successful and profitable for ORI.

Transportation of Juveniles

The best method for transporting juveniles is in aerated sea water. The cost of transporting water and aeration equipment, however, makes the method economically infeasible for large quantities of juveniles. An economical transportation method has been developed for large quantities of either juveniles or adults. The abalone are packed between damp pieces of foam rubber and placed in styrofoam-insulated boxes. The foam bits are first soaked overnight in sea water and excess water is squeezed out before being used. Layers of juveniles are alternated with layers of damp foam pieces and each layer is separated by plastic netting. One styrofoam box can hold 10,000 juveniles with an average length of 10 mm. If the temperature in the box is kept below 10°C, there will be 100% survival after 24 hr. Survival for 24 hr allows transportation of juveniles anywhere in Japan. When the air temperature is higher than 10°C, "blue ice" blocks must be placed in the styrofoam box.

When the juveniles are released into the open ocean or when they are transported long distances, abalone are attached onto oyster shells before they are placed in styrofoam boxes. Oyster shells provide a substratum for the juveniles to adhere, which reduces the rate of desiccation during transportation, and are economical and ecologically neutral. The oyster shells with the attached juveniles can be released directly into the open ocean. The oyster shells also help protect the young abalone from predation after release. Juveniles released without attachment to a substratum probably have low survival rates.

PHYSICAL PLANT

Aeration and Water Filtration

Aeration for the whole facility is provided by two root blowers (compressed air is not needed). Each blower produces a pressure of 0.1 kg/cm^2 at 1350 rpm and supplies 4.75 m^3 of air per minute.

Sea water is pumped directly from Mohne Bay rather than from the open ocean. The water quality of Mohne Bay is very good and the bay has very little fresh water input. Water temperature and salinity are taken daily at the surface, 3, 5, 7, and 10 m. There are three sea-water intake pipes (an intake pipe for each pump/filter system) located on the raft in deep water and the depth of each pipe can be adjusted to take advantage of seasonal conditions, and obtain the optimum temperature and salinity. The intake pipes are placed at 3 m during the winter but are raised to 1 m when the weather becomes warmer and surface water temperature increases due to the thermocline in the bay. However, spring is the spawning period of copepods, mussels, and other fouling organisms. The water depth of the intake pipe is selected to balance the favorable warm water layer and the increasing quantities of fouling organisms in the water column. The surface water temperature reaches 22°C by the beginning of summer but dense populations of fouling organisms requires that the intake pipe be positioned at 3 m where the water temperature is around 19°C.

Wide slits along the sides of each intake pipe opening prevents the entrance of large debris and other material that may clog the pipe. The pipes are cleaned frequently during the summer and fall to remove large bivalves, such as, mussels and scallops that entered at spat. The amount of fouling is indicated by pressure gauges on the intake pipes. The three intake pipes allow stopping individual pumps for removal of fouling material or for maintenance work without stopping the water flow to the culture tanks. A valve located inside the mouth of each pipe prevents sea water from automatically siphoning out of the pump, which is located above sea level, when the pump is stopped. If this valve was not present, the pump would require refilling with water before being started again.

All sea water at ORI is filtered before it is used. Each pump/filter system is independent and has a capacity of 50,000 ℓ/hr. The filtered sea water from each pump/filter system can be independently channeled for particular uses.

The minimum size of filtration is approximately 100 μm, which is the size of gastropod eggs and copepod nauplius larvae. Sand grains of 0.6 mm in diameter are used in the filter because they remove 100% of 100-μm particles and require less frequent flushing to maintain a constant filtering rate than smaller sand grains. However, the irregular-shaped sand grains eventually become polished by the constant flow of water, and cause the sand grains to coalesce and act like larger particles. ORI replaces the sand each year prior to the beginning of the seed production season.

Five values on each filter control the water flow; two valves for normal flow, two valves for backwashing the filter, and one valve for bypassing the first 10 min of regular filtered sea water after backwashing the filter. (This water should not go to the culture tanks since backwashing fills the filter with unfiltered water.)

Backwashing can be done either manually or electronically with a timer. Backwashing is required when the pressure gauges on the input and outlet pipes reach a predetermined level or can be done automatically every day with a timer that operates the appropriate valves to backwash the filter for 40 min. The turbidity of Mohne Bay is extremely low (a Secchi disk is visible at 15 m) and backwashing is not necessary every day.

The majority of the filtered sea water goes directly to the raft system. The high volume of water ensures good diatom growth on wavy plates and removes fecal material and metabolic wastes produced by the juveniles. One pump/filter system sends 50,000 ℓ/hr directly to the raft system while another pump/filter system sends an additional 40,000 ℓ/hr indirectly to the raft system. The water from the second filter system first goes to a storage tank. About 10,000 ℓ/hr is used for hatchery and conditioning operations before the remaining 40,000 ℓ/hr is sent to the raft system. The third pump/filter system sends 40,000 ℓ/hr to the multilayer system. The reduced water flow compensates for the additional power needed for pumping the water up to the third level of the structure.

The hatchery and conditioning operations usually require higher quality sea water than the culture tanks, so the water in the storage tank can be processed (e.g., UV irradiated, heated) before it is used. An additional gravity sand filter in the conditioning room helps remove an unknown substance that causes the larvae to develop a condition called "the abnormality". It causes many different effects in the larvae and is related to the amount of offshore water entering the bay.

The water heating system produces water temperature up to 40°C. A small flow of sea water from the large storage tank enters the boiler's insulated vacuum drum first chamber. An oil fuel heating element along the base of the drum heats the sea water to boiling, which is below 100°C due to the high pressure. The steam passes into an upper second chamber containing a coiled titanium condensing tube. The boiler is capable of heating 5000 ℓ/hr at a maximum of 100,000 kcal/hr. A valve next to the water outlet of the boiler can provide 40 to 50°C water for removing juveniles from settlement plates.

Heated water from the boiler is sent to two small storage tanks that independently control the water temperature. The water temperature is adjusted to the desired temperature by the addition of cool water. The water from one of the small storage tanks is used for the conditioning tanks and the other small storage tank is used for research. The cool water used for adjusting the temperature comes directly from the main storage tank.

The water temperature of the storage tank used for conditioning is kept at 20 to 20.5°C, depending on the ambient air temperature. The small difference in water temperature in the storage tank and the conditioning tank (20°C) compensates for the heat loss during water transport. About 60 ℓ/hr is UV irradiated for spawning induction and larval rearing. This water requires an additional temperature-control system to obtain the precise temperatures needed during these procedures. The temperature control system also allows preprograming the temperature fluctuation necessary during spawning induction.

UV Light System

There are two UV-light (2537 Å) units, 13 and 108 W, giving a total of 121 W. However, the important value for induction of spawning is the net amount of UV light which reaches the water. The UV system used at ORI efficiently transmits UV light energy into the water column and can sterilize 500 ℓ/hr. The 108-W unit has four UV lamps submerged in the flowing filtered sea water. Water enters through the base of the cylindrical container with the UV lamps, flows upward, and exits out the top. Each lamp is sealed in a quartz sheath to protect against changes in temperature and humidity. Periodic cleaning of the submersed quartz tubes eliminates interference of the UV light. The net amount of light entering the water column is 27 W (~30% efficiency) from each 90-W lamp. Each lamp has an estimated life of 8000 hr at optimum strength (>90% of the original amount of light from the lamp). This is approximately 3 years of use at ORI.

The present system is a large improvement over older models. Condensation is eliminated from the quartz tubing by having the lamps sealed and the efficiency is increased by submersion of the lamps into the filtered sea water.

The amount of UV irradiation needed for spawn induction is based on the strength of the UV lamp, water flow, and quality of the water. The optimum amount of UV irradiation at ORI (1 μm filtered sea water) is 2.42 W/ℓ. UV irradiation = strength of UV light (W/hr)/ water flow into spawning tank (ℓ/hr) = 121 W/hr/50 ℓ/hr = 2.42 W/ℓ.

FUTURE PLANS

In the future, the abalone might be completely raised to adult size by the fishermen. This will require the hatchery to initially produce 45-mm juveniles 1 to 1.5 years. Fishermen would then raise the juveniles for 2 years rather than the 4 years required when juveniles are placed in the open ocean. This type of intermediate culture would probably require large cement cages which would rest on the bottom of the ocean, have *Laminaria* sp. fed by a pump, and could hold about 100,000 adults.

Animals reach 3 cm in 1.5 years with the present culture methods and 9 cm (the fishing size limit) after 4 years in nature. However, the growth rate of *H. discus hannai* can be greatly increasing by raising them in warm water. In 1963, Takeo Imai began growing juvenile abalone in warm water effluent at the Tohoku Electric Power Station. The abalone were found to grow 4 to 5 times faster than those raised at the ORI facility at Mohne Bay.[3] Under ideal conditions (constant 20°C all year) *H. discus hannai* will grow to 6 cm in the first year and 12 cm in the second year. Juveniles transferred from the Mohne Bay facility to the Tohoku Electric Power Station facility during the winter and returned to the Mohne Bay facility during the next summer, grew to 4.5 cm in 1 year. Transferring juveniles between facilities cannot be done on a commercial scale. Although the Tohoku Electric Power Station facility uses hot water effluent to maintain a constant 20°C water temperature during the winter, the facility is unusable during the summer because the ambient water to the tanks reaches 26 to 28°C. The pipe from the sand filter is very long and the sun heats the water before it gets to the tanks.

The ideal location for an abalone seed production facility would be next to the solid-refuse burning facility of a city with a population around 50,000 to 60,000 people. This would be just large enough for a continuous flow of burnable matter but not large enough for another means of disposing of the refuse.

REFERENCES

1. **Seki, T.,** personal communication, 1981.
2. **Kumabe, L.,** Methods used at the Oyster Research Institute, unpublished report, 1981.
3. **Shaw, W. N.,** The culture of molluscs in Japan, *Aquaculture Magazine,* 9(1), 43, 1982.
4. **Seki, T.,** An advanced biological engineering system for abalone seed production, in *International Symposium on Coastal Pacific Marine Life,* Office of Sea Grant, Bellingham, Washington, 1980, 45.

ABALONE AQUACULTURE IN CALIFORNIA

Kirk O. Hahn

INTRODUCTION

To understand the importance of abalone to Californians and the development of abalone aquaculture in California, it is necessary to be familiar with the role abalone has played with the culture and traditions of California.

H. lomaënsis, the ancestor of the modern black abalone (*H. cracherodii*), lived during the Upper Cretaceous period (over 80 million years ago) and is the oldest known abalone species in the world. Modern abalone species have changed very little since ancient times. Sea otters began living along the Pacific coast of North America in recent geological time, preying upon abalone and other animals. Also, native Indians within the last several thousand years began living along the coast, and harvested both the sea otter and abalone. This food web was stable as long as no part was seriously altered.[1]

The first major factor upsetting the ecological balance of California abalone was the harvesting of sea otters for their fur. Harvesting began in the 1700s and continued until the sea otter was almost extinct by the beginning of this century. In addition, the native Indian populations were reduced in numbers by disease and cultural changes caused by Spanish missionaries. Due to the lack of major predators, the abalone populations exploded and soon dominated the intertidal zone with animals two to three deep in many areas.[1]

Chinese immigrants in California for the gold rush and building the railroads were the first group of settlers who realized the great value of abalone and began harvesting the animal. The Chinese air dried the meat and shipped it back to China, since American settlers did not eat abalone. By 1879, the annual harvest was in excess of 2 million kg.[1]

Laws were passed in 1900 to stop abalone harvest in water less than 20 ft deep. This shift from the intertidal zone to the sublittoral zone caused a transition from Chinese shore fishermen to Japanese-American hard-hat divers. The Japanese-Americans had a long history of diving for abalone in Japan. The abalone industry was dominated by Japanese-American divers until the beginning of World War II, when all Japanese-Americans in California were placed in "relocation camps". By the end of the war, the Japanese-Americans had lost everything and the abalone industry was being run by other people.[1]

Although fishing methods changed with each new group of fishermen, there were few regulations to limit the harvest. The abalone fishery steadily declined from its inception. The abalone fishery was maintained at a high level only by moving south into unexploited areas along the coast. The abalone fishery center in 1929 was Monterey, by 1940 it had moved to Morro Bay, and by 1960 the center had moved to Santa Barbara where it remains today. Santa Barbara has remained the principal fishery center because of its proximity to the Channel Islands, a major commercial fishing area.[2] In addition to the commercial fishery, there is an important and large sport fishery for abalone in California. The harvest of the sport fishery is equal to or exceeds the commercial fishery.

Annual commercial abalone harvests declined by 80% from 1960 to 1975.[3] The causes for this rapid decline were overfishing by commercial and sport fishermen: pollution from agricultural, commercial, and urban discharge; and an increase in the sea urchin population.[4] At the same time as the decline in the abalone harvest, the sea otter, once thought to be extinct along the California coast, was becoming more numerous. Many authors blame the decline in the abalone populations solely on the sea otter, ignoring the fact that the sea otter and abalone lived in ecological balance for thousands of years.[5-7] Only after the introduction of an abalone fishery in 1850 did abalone populations decline to low levels. The abalone fishery was based on the unnaturally high numbers of abalone due to the almost complete elimination of sea otters by hunters. By the time sea otters reestablished themselves along their traditional territory, abalone populations had already been reduced to severe low levels.[2]

Although current abalone populations are very low and meat retails above $60/kg, there exists a high demand for abalone meat in California. U.S. abalone consumption (almost all is in California) is valued at $30 million (retail).[3] The fishery is unable to meet this demand without further decline to the abalone populations along the coast of California. To alleviate this situation, abalone aquaculture is being developed in California to supply abalone to the public. There are two different paths by which these cultured abalone will ultimately reach the public. The first path is using hatchery-reared juveniles for reseeding depleted areas to restore the natural abalone populations to levels capable of maintaining a fishery. The second path is commercially raising abalone to market size and selling it directly to the public. These two paths are not mutually exclusive, since most of the abalone aquaculture companies in California sell juvenile abalone to the Department of Fish and Game for reseeding, while also selling marketable abalone directly to the public.

AQUACULTURE

The first laboratory to culture red abalone in California was built in 1964 at Morro Bay. The success of this laboratory created an interest in abalone aquaculture by private business, and university and government scientists. In 1970 the California Department of Fish and Game began research on red abalone aquaculture at the Granite Canyon Laboratory, near Monterey, and in 1974 the Ab Lab began operation near Ventura. Southern California Edison, a major California electricity producer, has also sponsored abalone culture projects which use warm-water effluent.[8]

Most of the effort in abalone aquaculture in California has concentrated on red abalone, *H. rufescens*. This species was selected because it is the largest abalone species in the world, the most popular native species, has high quality meat with a high market value, good growth rate, and easily controlled reproduction. Green abalone (*H. fulgens*) is also being cultured in California. This species is found in the warmer waters along the coast of California and Baja California, and grows fast with special culture conditions.

Development of culture techniques for the warm water species (*H. corrugata, H. fulgens,* and *H. cracherodii*), which ensure good survival and growth, are necessary for successful abalone aquaculture in the warmer regions of California. There is also interest in crossing these species with red abalone to produce a hybrid with better qualities (e.g., meat quality, growth, etc.).[2]

The culture techniques used for culturing the red abalone at two of the facilities in California, Granite Canyon and the Ab Lab, are discussed in separate chapters in this book. The culture of green abalone, although more recent and at smaller levels than red abalone, has a potential for great success. Green abalone culture techniques are similar to red abalone but differ in many important aspects. The techniques used for culturing green abalone using thermal effluent at the Redondo Beach, Southern California Edison electric power facility, will be briefly discussed.

Haliotis fulgens Culture Techniques

Adults are naturally ripe from late spring to midsummer and again from early to midfall.[9] The brood stock is conditioned during the winter by maintaining the animals in 20 to 28°C water.[10] The brood stock is fed *Egregia laevigata* and *Macrocytis pyrifera*, which are preferred by green abalone and produce the best growth.[9,11] Spawning induction is possible throughout the year with either UV-irradiated sea water or hydrogen peroxide.[10] Ambient water is used from fertilization through embryonic development to hatch out. After hatch

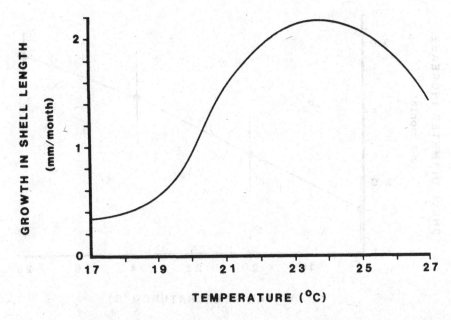

FIGURE 1. Mean size of juvenile (initial size, 1.0 ± 0.2 mm) *H. fulgens* reared for 1 month. (From Leighton, D. L., Byhower, M. J., Kelly, J. C., Hooker, G. N. and Morse, D. E., *J. World Maricult. Soc.*, 12(1), 170, 1981. With permission.)

out, larvae are raised at 18 to 20°C in 12-ℓ polyethylene containers throughout the 5 to 6 day larval development period. Normal larval development of green abalone occurs from 10 to 24°C, although development is faster at the higher temperatures. The optimum water temperatures are 20 to 24°C.[10,12] Water temperatures above 25°C cause abnormal development and death.[12] The time from fertilization to settlement at temperatures within the physiological range are from 3.5 days at 24°C to 12 days at 14°C. Larvae raised at 12°C require more than 14 days before settlement.[10]

Sea water in the rearing container is changed daily and antibiotic prophylaxis is applied (Penicillin G potassium, 50 ppm, and Streptomycin sulfate, 50 ppm) up to the 4th day of culture. Larval settlement is induced in 100-ℓ polyethylene containers and the newly settled juveniles are fed cultured diatoms (*Nitzschia* sp., *Amphora* sp., and other pennate diatoms). Juveniles (3 to 5 mm) are transferred from the settlement tank to a 1300-ℓ cylindrical fiberglass tank for continued growth. Rearing densities are approximately 5000/tank. Juveniles are fed cultured diatoms, macroalgae (*Egregia, Macrocystis, Ulva,* and *Gigartina*), and adventitiously colonizing microflora. Water flow is about 10 ℓ/min with the flow directed tangentially to create a rotatory current of 5 to 10 cm/sec at the periphery.[10]

Juveniles are more tolerant to extremes in water temperature and show the best growth rate at 20 to 28°C (Figure 1). The growth rate of juvenile (1 to 2 cm) abalone raised at 28°C is 2.9 mm/month which is approximately 1.6 times the rate at 16°C. Rearing juveniles at higher water temperatures increases the growth rate 1.5 to 2.0 times over individuals reared at ambient temperatures (14 to 20°C). The increase in shell length is linear over the temperature range of 16 to 28°C (Figure 2). The relationship between shell length and body weight is log W = 3.04 (log L) − 3.94, where W is the total body weight (grams) and L is the shell length (millimeters). The weight gain at 28°C is 2.02 times greater than animals raised at 16°C. The Q_{10} values for length increase and weight gain over the range 16 to 26°C are 1.55 and 1.94, respectively. Warm water reduces the time for production of 1 to 2-cm individuals from 1 year to 6 months, and individuals become reproductively mature at 1.5 years (4 to 5 cm.).[10]

Larvae, juveniles, and adults are not adversely affected by power plant effluent. The use

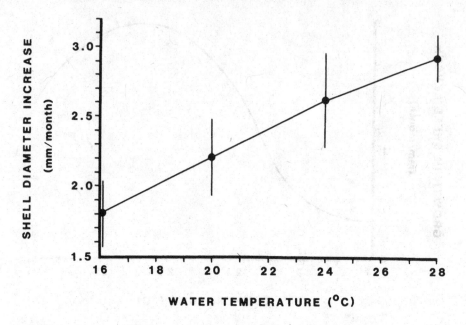

FIGURE 2.　Mean shell diameter increase of juvenile (initial size, 13.9 to 14.1 mm; S.D., 1.8 to 2.3 mm) *H. fulgens* reared for 4 months. Error bars represent ± 1 SD. (From Leighton, D. L., Byhower, M. J., Kelly, J. C., Hooker, G. N., and Morse, D. E., *J. World Maricult. Soc.*, 12(1), 170, 1981. With permission.)

of thermal effluent reduces culture costs of warm water species and increases growth to allow more crops per year. Although the growth rate is increased with elevated water temperatures, the rate is still below natural growth rates. Natural growth rates average around 3.5 mm/month (summer — 5.5 mm/month and winter — 2.5 mm/month). The reduced growth rate in the hatchery suggests there are density-dependent factors and/or nutritional physiological conditions that are slightly suboptimal for rapid growth.[10]

GROW-OUT METHODS

There are three different grow-out methods being used in California, each having advantages and disadvantages. The method of choice should be selected after consideration of production, labor, and equipment costs, and percent survival needed for a profit.

One method uses on-land tanks, raceways, or ponds with intensive feeding. This method requires a large initial investment in land, facilities, and equipment, and has high pumping, operation, and employment costs. This method allows greater control over the culture process and simplifies the animal maintenance in return for the additional effort and expense.[2]

A second method uses cages that are suspended from buoys or piers in protected areas (harbors or bays) or anchored to the ocean floor. This method facilitates raising abalone in high densities with good water quality and reduces operation costs. The culture facility can be reduced in size and less water pumping is required than in land-based systems. The disadvantages of this method are the difficulties associated with feeding animals (which are kept in containers suspended in the ocean) with macroalgae, maintenance of these containers, and loss of equipment and animals from storms, boats, and poaching.[2]

The third method, ocean ranching, has become very popular in recent years. Hatchery-reared juvenile abalone seed are released into subtidal areas leased from the state. This method requires minimal initial investment, and has low maintenance and grow-out costs. Although the final harvest rate from open ocean ranching is lower than in other culture systems (as a result of predation and other losses), the potential economics of scale and the reported success of similar efforts in Japan are stimulating further investigation of this strategy, particularly in Southern California.[3] To date this method has not been successful in California[13] (see chapter — Abalone Seeding.)

REFERENCES

1. **Howorth, P. C.**, *The Abalone Book*, Naturegraph, Happy Camp, Calif., 1978, 1.
2. **Hooker, N. and Morse, D. E.**, Abalone: the emerging development of commercial cultivation in the United States, in *Crustacean and Mollusk Aquaculture in the United States*, Huner, J. V. and Brown, E. E., Eds., AVI Publishing, 1985, 365.
3. **Morse, D. E.**, Prospects for the California abalone resource: recent development of new technologies for aquaculture and cost-effective seeding for restoration and enhancement of commercial and recreational fisheries, in *Ocean Studies*, Hansch, S., Ed., Calif. Coastal Comm., San Francisco, 1984, 165.
4. **Cicin-Sain, B., Moore, J. E., and Wyner, A. J.**, Management approaches for marine fisheries: The case of California abalone, *Calif. Sea Grant Pub.*, 54, 1, 1977.
5. **Burge, R., Schultz, S., and Odemar, M.**, Draft report on recent abalone research in California with recommendations for management, State of California, The Resources Agency, Dept. Fish Game, 1975.
6. **Cicin-Sain, B.**, Sea otters and shellfish fisheries in California: the management framework, in *Social Science Perspectives on Managing Conflicts Between Marine Animals and Fisheries*, Cicin-Sain, B., Grifman, P. M., and Richards, J. B., Eds., Univ. Calif. Coop. Ext. Serv., San Luis Obispo, 1982, 195.
7. **Hardy, R., Wendell, F., and DeMartini, J.**, A status report on California shellfish fisheries, in *Social Science Perspectives on Managing Conflicts Between Marine Animals and Fisheries*, Cicin-Sain, B., Grifman, P. M., and Richards, J. B., Eds., Univ. of Calif. Coop. Ext. Serv., San Luis Obispo, 1982, 328.
8. **Ebert, E. E. and Houk, J. L.**, Elements and innovations in the cultivation of Red abalone *Haliotis rufescens*, *Aquaculture*, 39, 375, 1984.
9. **Leighton, D. L.**, A floating laboratory applied to culture of abalone and rock scallops in Mission Bay, California, *Proc. World Maricult. Soc.*, 10, 349, 1979.
10. **Leighton, D. L., Byhower, M. J., Kelly, J. C., Hooker, G. N. and Morse, D. E.**, Acceleration of development and growth in young Green abalone *(Haliotis fulgens)* using warmed effluent sea water, *J. World Maricult. Soc.*, 12(1), 1970, 1981.
11. **Leighton, D. L.**, Studies on food preference in algivorous invertebrates of southern California kelp beds, *Pac. Sci.*, 20, 104, 1966.
12. **Leighton, D. L.**, The influence of temperature on larval and juvenile growth in three species of southern California abalones, *Fish. Bull.*, 72(4), 1137, 1974.
13. **Tegner, M. J. and Butler, R. A.**, The survival and mortality of seeded and native Red abalones, *Haliotis rufescens*, on the Palos Verdes peninsula, *Calif. Fish Game*, 71(3), 150, 1985.

ABALONE CULTURE IN AN URBAN ENVIRONMENT

John McMullen and Tim Thompson

INTRODUCTION

Ab Lab was founded in 1972, principally as a hobby with its initial interest directed toward replenishing dwindling wild stocks of abalone with seed. Early efforts focused on a closed-system approach conducted in a backyard. This system was abandoned after 1 year when success became no more than a measure of how long the wild adult animals could live within the system.

In 1974, Ab Lab obtained a lease at the mouth of a commercial harbor in Port Hueneme, California. Efforts were then geared toward developing a system of culture for the red abalone, *Haliotis rufescens*; from spawn to table. In 1981, Ab Lab became the first to market an aquaculturally produced abalone (to the best of our knowledge) and has become the largest commercial hatchery in the world. Today, we market not only products for restaurants, but also provides "seed" (6 to 8 mm) abalone to other growers in California.

This chapter will cover aspects of our culture system in an urban environment. Ab Lab utilizes a two-phase system of culturing abalone: an on-land hatchery which raises abalone from egg to 8 mm, and a near-shore containment system from seed to market size. It should be noted here that throughout our development, we have been a privately funded business operation. The techniques we have developed have been with economic constraints in mind. To a scientist or academic, perhaps better ways of performing the same functions could be achieved; we wish to emphasize that the economic "bottom line" dictates how we conduct our biology.

SITE SELECTION

The Ab Lab hatchery is located at the mouth of a deep-water commercial harbor in Port Hueneme (Figure 1); approximately 50 mi. (80 km) north of Los Angeles. The harbor sits at the head of a submarine canyon, Hueneme Canyon. Commercial activities in the harbor include commercial fishing, unloading of South American produce, disembarkment of Japanese automobiles, staging vessels for offshore oil drilling, as well as use by the U.S. Navy.

Despite the apparent heavy industrial use in the harbor, several natural indicators suggested that location of a hatchery there could be successful. First, the water is generally clean; underwater visibility averages 10 ft. Second, a natural population of black abalone, *H. cracherodii*, existed. Third, there is a large stand of "giant kelp", *Macrocystis pyrifera*. Finally, the harbor mouth supports a diversity of marine life. All these factors combined indicated that the water was of good quality.

After 12 years of culture activity, we have experienced no pollution-related problems. However, we are gravely concerned about increasing levels of TBT found in mussel tissue grown in the harbor. Routine analysis of cultured abalone tissue for heavy metals and organic pollutants have either been nondetectable, or well within FDA safety standards.

WATER DELIVERY AND FILTRATION

In the hatchery portion of our operation, it is imperative that the young abalone be given a clean, competitor- and predator-free environment. To that end, we copied a filtration unit designed by U.S. Naval engineers for ultra-pure water systems. Although this seems excessive, the benefits of increased growth and survival outweighed the cost of system installation.

The primary intake pump is located 50 ft from the waters edge, having a maximum lift of 18 ft. The intake line consists of a single, 6-in. steel pipe. Water is drawn by a 25-hp

FIGURE 1.   Aerial view of Port Hueneme, California. Industrial activities in the harbor include staging for offshore oil platforms, housing of a commercial fishing fleet, unloading of South American produce, disembarkment of Japanese-built automobiles, as well as use by the U.S. Navy. Ab Lab is located at the mouth of the harbor.

Fybroc pump equipped with a fiberglass impeller. To screen out larger foreign material (e.g., rocks, fish, urchins, etc.) the intake pipe is fitted with a screen cap, and straining basket is installed post pump/presand filter.

The filtration system is two phase: water passes through two roughing filters filled solely with anthracite, and two polishing filters filled with plastic rods, anthracite, and sand. This system removes particles greater than 100 μm.

After filtration, the seawater is spilled into two 2100-gal fiberglass tanks. From these tanks, a second 7.5-hp 1700-rpm stainless-steel pump is used to drive the water to the culture tanks. Because of the fine filtration delivered, a regular program of maintenance is required to maintain optimum performance. Pressure gauges to all filters are monitered daily, and when required, the filters are backflushed.

BROOD STOCK MANAGEMENT, SPAWNING, AND LARVAL DEVELOPMENT

To ensure a successful hatchery operation, it is of foremost importance that a line of reliable breeding abalone is established and managed. At the Ab Lab, we maintain both wild and F_1 hatchery grown animals as brood stock. To faciliate record keeping, each animal is given a number by either affixing the tag with Brolite epoxy, or by use of a stainless-steel machine screw.

For maturation of red abalone, our experience has been that temperature, vigorous water motion, and adequate food supply are more important than photoperiodicity. The system of brood stock management, as practiced in Japan, is not necessary for red abalone, but has been used more successfully on the green (*H. fulgens*) and pink (*H. corrugata*) abalone. We induce maturation solely on a diet of *Macrocystis*, and do find it necessary to vary the diet to the animals.

For spawning, we have used both the UV method of Kikuchi and Uki[1], and the peroxide technique developed by Morse and his colleagues.[2] Interestingly, prior to the publication of the landmark publication of Kikuchi and Uki, Ab Lab had been consistently achieving red abalone spawns, when other California researchers were unable to do so. The reason was that a UV unit had been utilized simply to provide sterile water for the larvae. It was not until 1976 that the work of Kikuchi and Uki became known to us, at which time we realized the reason for our earlier success.

Males and females are spawned in separate 5- or 20-ℓ buckets, depending on the size of the animal. Both UV and hydrogen peroxide have worked consistantly within certain parameters. When using UV, we have found a temperature inhibition for male abalone above 19°C, although females will spawn at that temperature. When using peroxide, it is important that non-UV water is utilized as the combined affect will inhibit spawning. With either stimuli, the animals spawn within 2.5 to 3 hr after induction begins.

For fertilization, eggs are siphoned from the spawn buckets, passed through a 200-μm filter screen to eliminate fecal material, and collected on a 70-μm screen. A sperm bath is made up; the sperm density is adjusted "by eye" based on long-term experience. The screen containing the eggs is immersed in the sperm bath, after 15 sec an aliquot of eggs/sperm is removed and examined under the microscope. The eggs are monitored, and when an average of ten sperm cells per vitalline layer is observed, the eggs are removed from the sperm solution.

After fertilization, the eggs are gently washed with UV-sterilized water, and then placed in 100-ℓ fiberglass tanks for incubation. In our experience, a density of ten eggs per milliliter is optimal. Generally, a 75 to 80% hatch out, yielding 750,000 larvae/tank is achieved.

Hatch out occurs within 14 to 17 hr after fertilization. To separate the viable larvae from unfertilized eggs and egg cases, the larvae are siphoned from the upper three fourths of the incubation tank onto a 70-μm screen. Usually we discard the last one fourth of the tank, but when additional larvae are necessary, we will combine these with the lower one fourth from the other incubation tanks, refill that tank, and resiphon the swimming larvae.

Larval development of red abalone takes 6 to 7 days, depending on the culture temperature. During this development time, culture water is changed twice daily using UV-sterilized sea water. The larvae are siphoned onto the 70-μm screen, and then gently rinsed. Bacteria are rarely a problem; we do not use antibiotics in our larval culture.

During development, we daily observe the larvae with the microscope. It is our experience that some larval batches may develop differently than others, and the appearance of the larvae is an indicator of subsequent settling success. Larvae with well-developed mantles and full visceral masses have been correlated by our staff as yielding optimal success. Conversely, emaciated larvae, as recognized by distended viscera and poor mantle definition, yield poor results. If the later condition exists, larvae are discarded and spawning is reinitiated.

We note the end of the larval period when the greater proportion of the larvae begin to show strong creeping behavior. To prepare for settlement, larvae are again gathered in the 70-μm screen, rinsed, and dispensed into 4-ℓ buckets for density counts.

SETTLEMENT AND GROWTH MANAGEMENT TO 8 MM

For the first 6 months of postlarval development, until approximately 6 to 8 mm, all growth occurs in on-land circular tanks (Figure 2). During this period, the growing abalone feed on adventitious diatoms and benthic algae growing on the tank wall surfaces. Despite our best efforts, our survival averages only 5%, with a range of 3 to 18%. With a 6-month hatchery growth cycle, we push two sets per year through the tanks, yielding a current production of 620,000 animals per year.

The hatchery is arranged in a series of 50-gal circular tanks. For efficient light transmittance, we have found that stacking the tanks four high is best. Water is applied to the top

FIGURE 2.   Hatchery tanks for growth of abalone from settlement to 6 to 8 mm. Staff is engaged in cleaning operations.

tanks at a rate 1.5 to 2 gal per minute per series. Flow is directed circularly; the resultant current creates a vortex which picks up loose detrital material and sends it to waste.

Algal culture on the tank wall is accomplished by allowing a natural set of adventitious diatoms and benthic algae, and then managing those algal communities by manipulating light, grazing densities, and fertilizer applications. Creation of the correct algal communities is accomplished principally by experience, i.e., an "old farmer's" sense of which parameter to change. We do not practice external culture of diatoms as described by Ebert and Houk.[4]

In practice, new tanks are initiated by allowing an initial diatom film to form. Depending on prevailing conditions in the ocean, this is accomplished in 4 to 10 days. At this point, 1-in. grazing abalone are introduced to the tanks. The number of abalone per tank depends upon the sunlight levels reaching the tanks. Generally, top tanks require 20 to 25 abalone, with lower tanks requiring less. Grazing abalone accomplish two functions: (1) simply by the action of their grazing the algal community is restricted to single-cell diatoms and monostromatic benthic algae, and (2) they provide settlement-inducing mucus trials. In our system of culture, it is important the larvae settle on the tank walls. As most of the grazing occurs on the side walls, this ensures settlement there.

Controlling the level of light to the tank has been the most important and most effective means of determining the algal community. We accomplish this by utilizing greenhouse shading material, or black polyethylene plastic, stretched over plastic drip-irrigation tubing formed into hoops (Figure 3). By using various screening materials, we can very precisely regulate the light from total darkness to ambient light. These shade hoops are simply made, inexpensive, and stack neatly when not in use.

Utilization of fertilizer is a rare practice; only in late winter when light, nutrient, and

FIGURE 3. Shading "hoops" used to regulate light intensity to hatchery tanks.

temperatures levels are low, or in tanks where the grazing density is exceeding the capacity of the tank to provide food. When application is necessary, we use only nitrate and silicate applied at a rate of 10 ppm each. Flow is shut off to the tanks, fertilizer is applied and allowed to sit for 1 hr before resumption of flow.

Once the correct algal community has been established, spawning is initiated. The day of settlement the tanks are prepared by removing the grazing abalone, vigorously rinsing the tanks to remove copepods, and refilled. After the tanks are cleaned and filled, flow is shut off, and larvae are introduced. Under best-care conditions, settlement occurs very rapidly. Often, however, red abalone larvae will take 10 to 12 hr to settle. As such, normal operating procedure is to introduce the larvae in the late afternoon, and not start water flow again to the tanks until the following morning. In some instances, where the tanks have not been correctly prepared, we can have settlement failure. In these cases, grazers are reintroduced and tanks are correctly reconditioned for a later settlement.

For the next 4 to 6 months, algal food for the abalone is managed as described above. In addition, once per week the tanks are rinsed to remove detritus and copepods; very gently at first, but more vigorous as the abalone gets larger. As necessary, tank positions in the vertical column will be changed. Bottom tanks generally have the highest number of settled abalone, but the weakest algal growth. In these cases, the bottom and top tanks are switched.

By the fourth month after settlement, 10 to 20% of the settled larvae have reached 6 mm, and are effectively outcompeting the other abalone for the limited food resource. Once this begins to occur, the first culling harvest is performed. In our experience, 6 mm is the minimum size of red abalone that can be induced to feed on kelp. Hence, the term "kelp weanable" is sometimes used to describe those animals.

The abalone are anesthetized, removed from the tank, sorted, and then counted. Determining the number of abalone is done by weight. One thousand abalone are hand counted, shaken as free of water as possible, and weighed. This gram weight then serves as the basis for determining the total number. Abalone smaller than 6 mm are returned to the hatchery

FIGURE 4.   Abalone seed bags. Six- to eight-millimeter abalone are distributed 1000/bag for shipment to other growers. Kelp is added to the bag, and
then sealed shut.

tanks at a rate of 2000/tank. Larger abalone are placed in our grow-out containers, or are
shipped out to the other growers who contracted for the seed.

This culling procedure is then repeated in the fifth and sixth months; this accounts for
90% of the settled abalone. The remaining 10% ''runt'' abalone are either discarded, or
placed in our grow-out system; these are not sent out as contract seed abalone.

In preparation for shipment of contracted seed abalone to other growers, the seed abalone
are packed 1000/net bag (1-mm mesh). A single blade of kelp is enclosed, and the bag is
sealed shut (Figure 4). The bags are then placed in a shallow water table, horizontally, and
supplied with vigorous aeration. To ensure that the correct number of contracted abalone
have been placed into the bag, after the seed abalone are separated to the shipment bags,
two bags are randomly selected, and all abalone are again hand counted. Abalone are shipped
within the week that they are harvested from the tanks.

Seed abalone are shipped by sandwiching the bags between moist paper towels in a plastic
bag, filling the bag with oxygen, and then packing them in an ice chest. In this manner,
we have shipped seed abalone with up to 24-hr transport time with no mortality.

After the last cull of the hatchery tank has occurred, a judgment call is made on the algal condition. If a high number of crustose corraline algae, *Ralfsia* sp., or other undesirable community has been established, those tanks are bleached, and scrubbed clean. If conditions in the tanks are suitable for settlement, the tanks are scraped free of serpulids, and larvae are reintroduced to the tanks.

GROW OUT: FROM SEED TO MARKET

Grow-out techniques for rearing the seed abalone to a marketable size include ''farming'' in onshore culture sites, open ocean reseeding, and containment rearing in offshore growing areas (i.e., raft or long-line culture). Here at Ab Lab, we practice containment rearing. The three methods are contrasted below, followed by a discussion of our system.

On-Land Rearing
The first abalone farms in California have utilized an onshore culture method to produce marketable abalone.[5] In this form of culture, the abalone are maintained in raceway systems similar to those developed for freshwater trout farming. While this system allows for isolation of the abalone from natural-environment predators and competitors, the animals become dependent upon mechanical water- and air-delivery systems to maintain the environment. Breakdowns or electric failures can result in the loss of inventory. Furthermore, this system requires high start-up capitalization costs, and the large volumes of air and water pumped through the system create expensive utility bills.

Ocean-Floor Rearing
Out planting to the ocean floor entails placing juvenile abalone on a suitable substrate and allowing them to grow in the wild to an adult size. This technique has the advantage of avoiding the high capitalization and energy costs associated with on-land rearing. However, this method is less than ideal because: (1) stocking the underwater habitat with abalone seed is difficult and time consuming; (2) it is impossible to isolate the stock from animal and human predation; (3) there are difficulties in monitoring the feeding regime and health of the crop, and even more difficulties in altering that environment; (4) harvest of marketable stock requires costly underwater diving labor.

Containment Rearing
The third alternative involves an offshore farming method which utilizes containment pens suspended from piers, docks, or long lines. Some of the advantages of this method include: proper natural growing conditions (e.g., salinity, pH, temperature, and dissolved oxygen); low capitalization costs and no energy utilization; protection in the containment pens from both marine and human predators; controlled food supply (kelp) for optimal feeding; ability to polyculture abalone with other aquacultural systems (e.g., salmon pens, oyster or mussel long lines); easy access for feeding, managing, and harvesting; moderate labor costs; and absolute ownership and owner identification.

Ab Lab's container system of culture utilizes modified plastic drums that allow a free flow of water through the vessel (Figure 5). These containers are then suspended from fixed dock rafts, or may be long lined like oysters. The seed abalone are stocked at a specific density, and once per week are fed sufficient amount of kelp to last an additional week. Doing this ensures both optimal feeding and optimal flow rates through the containers. At the time of feeding, the containers are also checked for crabs or octopuses that may have inadvertantly entered the unit. One barrel takes only 10 to 15 min per week to service.

In southern California, it requires only 15 to 18 months from the time the containers are stocked until the time the abalone are harvested for sale to restaurants. Ab Lab's current market size is 2 in (50 mm), but we are planning to increase that size limit to 2-1/4 in. (to yield an increase of 40% meat weight).

FIGURE 5.   Containment pens used for culture from 6 to 50 mm. Pens are modified plastic barrels outfitted with plastic mesh at the top and bottom. These pens are suspended from a pier in Port Hueneme.

PARASITES, PREDATORS, AND ABIOTIC DISORDERS

Perhaps due to the fact that abalone aquaculture is relatively young (less than 30 years old), there are no reported diseases of biological origin in commercial hatcheries. A variety of symbionts will occupy the shell, and occasionally tissue, of wild-caught abalone. In our 13 years of production experience, we have never had a biologically induced disorder of epidemic proportions. More commonly, problems we encounter are a result of poor husbandry. When we encounter biological problems, they have been either predators which inadvertently enter the culture system, or as shell-inhabiting symbionts.

In the grow-out system, the important parameters to manage are predator control and to be sure that water flow through the barrels is unrestricted. When the seed abalone are first introduced into the barrel, we use a 1/8-in. screen mesh on the bottom of the barrel. As the animals grow in size, the screen is changed out to 1/4-in. to allow better flow. When adding kelp to the barrels, it is important that only a sufficient amount of kelp be introduced for feeding. Excess kelp results in impedance of the water movement; the kelp dies and rots creating a stagnant condition in the barrel. High mortalities have occurred in our barrels when these conditions occur. On occasion, small cancer crabs and octopuses are inadvertantly introduced with the kelp. The only means of managing this problem is to examine barrels weekly, and remove those predators at that time.

Polychaetes, particularly species of *Polydora,* are found as shell-inhabiting symbionts of many commercially important marine molluscs,[6] including abalone.[7] Although the infesting worm does not directly necrotize tissues, its presence does cause sufficient irritation as to inhibit growth. Ab Lab has identified *Polydora websterii* inhabiting the shell of red abalone. In rare cases, the infestation was severe enough to cause the animal to form misshapen shells; forming dome-like shells as opposed to the normal ear-shaped one. Applications of the drug Trichlorofon has yielded promising results in eradication of the worm, and a return to normal growth (work in progress). Further testing is required, however.

MARKETS, PREPARATION, AND RECIPES

Ab Lab pioneered the concept of "cocktail" abalone and introduced the product to southern California restaurants in 1981. The rarest of all seafoods, these 2-in. succulent abalone are in increasing demand in the U.S. and abroad. Chefs and consumers alike extoll the tenderness of the product. The mild, delicate flavor appeals to the discerning palate of the seafood connoisseur.

Following a year of nuturing, the abalone are removed from the ocean and placed live in holding tanks to await shipment. There are size limits determined for the harvesting of wild abalone (minimum 5 in.; see Section 8304 of the California Fish and Game Department). Being a cultured product, cocktail abalone are exempt from any size restrictions and are unavailable from the natural fishery. This makes Ab Lab the sole producer of cocktail abalone in the world.

The abalone arrive at the restaurant live in plastic containers of 24 each. A fresh product may be assured in refrigeration for up to 7 days. Ab Lab assures live delivery, ensuring the utmost in freshness. The cocktail abalone have been delivered throughout the nation, including New York, Michigan, Hawaii, Colorado, Texas, and Washington. Ab Lab has also delivered live abalone to England, Canada, Chile, China, and Japan.

The majority of our restauranteur clientele offer the cocktail abalone as an appetizer with three abalone being the amount of a typical serving. Several of our customers have experienced the pleasure of an ongoing request from their diners for the abalone appetizer. Some restaurants have had great success in using Ab Lab abalone as an entreé, six or seven per serving (Figure 6).

Abalone are shucked by running a knife between the meat and the shell. The entrails are separated from the meat and the head is removed. The meat, known as the "foot", is then placed between two moist towels and lightly struck once or twice with a culinary tenderizer in order to relax the muscle. The tenderized foot is then dredged through flour and egg prior to being sauteed for 10 sec to a side in very hot butter or oil. The shells can be used to make a very attractive presentation. We present two recipes in the Appendix as developed by two of our prestigious clients.

As cocktail abalone have not had a long culinary history nor wide restaurant exposure, innovative chefs have found them to be ideal for the creation of new and exciting presentations. For the health conscious, fresh abalone are virtually fat and cholesterol free.

RESEARCH IN PROGRESS, FUTURE PLANS

As in any small business, research continues at the Ab Lab to define better, more efficient means of growing abalone. In conjunction with researchers at the University of Southern California, we are examining larval nutrition and how it relates to overall survival and growth in the hatchery. With researchers from the University of California, Los Angeles, we are

FIGURE 6.   Cocktail red abalone as prepared at the Waterside Inn, Santa Barbara, Calif. The abalone are served on a buerre blanc sauce (see Appendix 1), and served with golden caviar in the shell. (Photo courtesy of D. B. Pleschner).

conducting experiments on the effects of exogeneously applied biochemicals on larval and juvenile abalone. Recently, we have developed new techniques for abalone anesthesia that greatly facilitates our handling of the abalone, reducing time and saving money.

Ab lab is currently in negotiations for property to expand its hatchery operations. Furthermore, we have engaged in experimental projects to grow red abalone in the Pacific northwest and South America. By 1988, we hope to be the largest producer of red abalone in the world.

DISCUSSION

Culturing of abalone in an urban environment has both disadvantages and advantages. Many would-be aquaculturalists feel that only remote, pristine environs can provide the best chance for success in a venture; the most obvious reason being the desire to locate away from possible sources of pollution. Here at our facility, we are constantly concerned with harbor activities that could have adverse effects on operations. Routinely, water quality analyses are obtained.

Contrasted with this, however, are the problems that arise when locating a facility far from urban areas. To begin with, even under the best of circumstances, maintaining an adequate, skilled labor force is difficult. Locating a facility in isolation limits the labor pool from which you can draw. A second problem is electrical power. Where does one get electrical power if it does not exist initially? If there are no electrical lines leading to your proposed site, will it be economically feasible to bring in new power lines, or should you generate your own power? One needs to consider emergency problems — a blown electrical motor, a broken pump impeller, a disabled air compressor, etc. Speed of repair being critical, how fast can repair parts be obtained? Consider logistics, i.e., receiving supplies and shipping product. These are only a few examples of the negative aspects associated with locating a facility in remote areas and should be addressed when considering an aquaculture site.

In our 13 years of experience in Hueneme Harbor, we have successfully managed the problems and reaped the benefits of working in the urban envrionment. We believe we will continue to expand and prosper here.

ACKNOWLEDGMENTS

We wish to thank Mssrs. Michael Hutchings of the Waterside Inn and Dave Skaggs of the Ranch House Restaurant for allowing us to use their original recipes here. Thanks go to Ms. Diane Pleschner for use of her photo. Specials thanks to our wives, Michele McMullen and Sue Thompson, for their continued support and help in all our endeavors.

Also, special thanks to Mr. Earl Ebert whose help and encouragement meant so much in the first several years of Ab Lab's development.

APPENDIX

BABY ABALONE & ALASSIO SAUCE

Make the following sauce:

Cook until just done but not mushy

$^{1}/_{2}$ cup olive oil
2 cups celery, minced
1 carrot, minced
1 $^{1}/_{2}$ Bell Pepper, minced
5 green onions, minced
$^{1}/_{2}$ tsp. fish herb blend
$^{1}/_{2}$ tsp. herb salt

When done add and heat to serving temperature:

2 tbs. Lemon juice
$^{1}/_{2}$ cup Bechamel sauce
1 $^{1}/_{2}$ cubs crab meat, chopped fine

To serve, saute Abalone in butter very quickly on each side.

Serve immediately on the above sauce which has been spread over a bed of rice. Garnish lightly with finely chopped parsely and lime wedge.

From the Ranch House Restaurant, P.O. Box 458, S. Lomita, Ojai, California 93023. With permission.

Preparing Abalone

Prepare the abalone for cooking by running a small knife between the meat and the shell. Remove the intestines and cut off the head, located at the front area. Place the abalone foot between two moist towels and lightly pound to even them and to tenderize the meat.

Buerre Blanc Sauce

6 shallots, peeled and chopped

> 2 cups dry white wine
> 1 cup whipping cream
> 1 ¹/₂ lbs. unsalted butter
> 1 tablespoon lemon juice
> salt to taste
> freshly ground white pepper to taste

Reduce wine and shallots until almost dry in a heavy-bottomed sauce pot (stainless steel or anodized aluminum is preferred.) Add cream and reduce by one-half. Cut butter into small pieces and whisk into the reduction over medium heat. DO NOT BOIL. Adjust seasonings, add lemon juice and strain. Keep warm.

Additions to Buerre Blanc Sauce

> 1 cup enoki daki mushrooms (Japanese snow drop mushrooms)
>   or regular mushrooms, cut julienne
> ¹/₂ cup fresh tomatoes, peeled, seeded, and diced
> 2 tablespoons fresh, chopped dill, tarragon, or parsley (in that
>   order of preference)

Add the above to the buerre blanc sauce, mixing well. Reserve.

Cooking the Abalone

Season the abalone, dredge in flour and dip into beaten eggs. Saute the abalone quickly, 5 to 10 seconds, per side in very hot, clarified butter.

Serving

Spoon sauce onto plate, top with abalone. Garnish with a tomato rose and dill sprig. The abalone shells may be filled with golden caviar and used as additional garnish.

From Michael's Waterside Inn, 50 Los Patos Way, Santa Barbara, California 93108. Michael Hutchings Proprietor/Chef. With permission.

REFERENCES

1. **Kikuchi, S. and Uki, N.,** Technical study on artificial spawning of abalone, genus *Haliotis* II. Effect of irradiated sea water with ultraviolet rays on inducing to spawn, *Bull. Tohoku Reg. Fish. Res. Lab.*, 33, 79, 1974.
2. **Morse, D. E., Duncan, H., Hooker, N., Morse, A.,** Hydrogen peroxide induces spawning in mollusks, with activation of prostaglandin endoperoxide synthetase, *Science*, 196, 298, 1977.
3. **Morse, D. E., Hooker, N., Duncan, H., Jensen, L.,** γ-Aminobutyric acid, a neurotransmitter, induces planktonic abalone larvae to settle and begin metamorphosis, *Science*, 204, 407, 1979.
4. **Ebert, E. E. and Houk, J. L.,** Elements and innovations in the cultivation of Red abalone *Haliotis rufescens*, *Aquaculture*, 39, 375, 1984.
5. **Oakes, F.,** The abalone farm, inc., *Aquacult. Dig.*, 10, 6, 1985.
6. **Kinne, O.,** Introduction, in *Diseases of Marine Animals*, Vol. II, Biologische Anstalt Helgoland, Hamburg, 1983, 467.
7. **Kojima, H. and Imajima, M.,** Burrowing polychaete in the shells of abalone *Haliotis diversicolor aquatilis*, chiefly on the species of *Polydora*, *Bull. Jpn. Soc. Sci. Fish.*, 48(1), 31, 1982.

ABALONE CULTIVATION METHODS USED AT THE CALIFORNIA DEPARTMENT OF FISH AND GAME'S MARINE RESOURCES LABORATORY

Earl E. Ebert and James L. Houk

INTRODUCTION

Seven abalone species and one subspecies occur along the west coast of the North American continent. The range of distribution of the various species, both latitudinally and bathymetrically, is temperature dependent. One species, the pinto, *Haliotis kamtchatkana*, occurs as far north as Sitka, Alaska and is the object of an important, although limited, commercial fishery in British Colombia. The pinto abalone is very similar, morphologically, to the northern Japanese abalone, *H. discus hannai*. The pinto abalone ranges southward along the Oregon, Washington, and California coasts to near Pt. Conception; however it is only sparsely distributed and not fished commercially in these areas. The threaded abalone (*H. kamtchatkana assimilis*), a subspecies or geographic form of the pinto, ranges from the Monterey, California region southward to Baja California. But, it is relatively small in size, nowhere numerous, and not sought commercially. One other abalone species, the flat *H. walallensis* occurs from British Columbia southward into Baja California waters; but it also is sparsely distributed throughout its range and not important commercially. Another abalone species, the white (*H. sorenseni*) is of limited commercial importance due to the greater depth that they inhabit (to about 60 m), and because they are sparsely distributed except at a few of the islands offshore southern California and Baja California. The white abalone ranges from Pt. Conception, California southward into central Baja California. The remaining four abalone species (red, pink, green, and black) are important commercially and recreationally, to California and Baja California. Pink and green abalone, *H. corrugata* and *H. fulgens* predominate in the Baja California fishery. Geographically, red abalone range from southern Oregon to central Baja California. Bathymetrically they range from the low intertidal zone to 40-m depths, following a north-southward direction. Pink abalone range from about Santa Maria Bay, Baja California, northward to about Pt. Conception, California and the adjacent offshore islands, and to depths of about 60 m. The green abalone has a similar latitudinal range (Magdelena Bay, Baja California to Santa Barbara, California), but bathymetrically they are a shallow-water species, ranging from the intertidal to 10-m depths. Black abalone, *H. cracherodii*, occur from central Baja California northward to northern California, however, they are uncommon north of San Francisco. Black abalone occupy the shallowest depth distribution of all eastern Pacific Coast haliotids. They seldom occur below 7 m and are most dense in the low tidal region.

In California the commercial catch composition has experienced a dramatic shift. Red abalone, *H. rufescens*, historically comprised most of the commercial catch. The fishery for pink abalone developed during the 1940s.[1] Up to 1970, pink and red abalones comprised close to 90% or more of the annual commercial catch. However, these two species experienced a sharp decline in commercial landings during the early 1970s. The decline of the pink abalone catch resulted from over exploitation, poor fishing practices, and habitat degradation; while the decline in the red abalone catch was principally related to the sea otter. The sea otter, a serious shellfish predator, has experienced a dramatic comeback from near extinction in California, and has impacted the entire central California commercial red abalone fishery. Concomitant with the sharp decline of the fishery for pink and red abalone, a fishery for black abalone, *H. cracherodii* developed. This species had been used principally as bait in the lobster fishery prior to 1969. Black abalone landings peaked in 1973 at about 2 million pounds. The commercial abalone harvest in California averaged over 4 million pounds from 1950 to 1970, and peaked at 5.4 million pounds in 1957. However, the fishery began a sharp decline thereafter and recently (1983) the annual harvest fell below 1 million pounds. This decline has served as an impetus for fishery enhancement through aquaculture.

Pioneering efforts to culture North American haliotids began in 1940.[2] However, it was not until the mid 1960s that abalone hatchery development ensued.[3] The State of California Department of Fish and Game, initiated abalone culture studies in 1970.[4] Presently, there are three private abalone hatcheries cultivating abalone in California, and one private hatchery in Canada (Victoria, Vancouver Island). Also, the Mexican government has recently shown a renewed interest in cultivating abalone in order to replenish the badly overexploited abalone stocks that occur along the western shore of Baja California and adjacent islands.[11]

Abalone species cultivated include the red, green, pink, and pinto. The pinto is cultured exclusively at the hatchery on Vancouver Island (Pacific Trident Mariculture, Ltd.). The pink abalone is cultured, to a limited extent, at the California Department of Fish and Game's pilot-production hatchery, and more recently at Southern California Edison Company, at their steam generation power plant at Oxnard, south of Santa Barbara. The green abalone species is cultured in the private hatchery at Port Hueneme (the Ab Lab), and at Southern California Edison. The pinks and greens are relatively warm water species. The red abalone is the principal species cultured (exclusively by one private hatchery), and the most valuable. It commands the best market price and attains the largest size of all haliotid species (about 30 cm in length).

The California Department of Fish and Game's Marine Culture Laboratory (MCL) at Granite Canyon (central California, near Monterey) began to investigate the feasibility for culturing the red abalone in 1970. The broad objectives of the project have been to develop technological and biological methods, determine economic feasibility, and information dissemination. Because the MRLs broad mission is multispecies oriented, with limited staff, abalone culture studies have waxed and waned as priorities shifted. A pilot-production hatchery for abalone was finally constructed in 1983.

Initially, the MRL abalone culture project drew very liberally from the extensive Japanese literature on abalone culture.[5-10] The technology that has evolved is a synthesis of pioneering Japanese culture methods, our research findings, and the accretion of knowledge from "trial and error". One broad difference that has been apparent between Japanese and North American abalone cultivation practices is the method by which abalone larvae are "caught" on substrates at metamorphosis. The Japanese method typically employs "wavy plate" (corrugated) panels that are preconditioned (have a diatom film). These corrugated panels are arranged in bundles and placed in holding tanks to receive the larvae. California hatchery methods generally use larvae settling tanks whereby the larvae settle directly on the tank surfaces. This difference in hatchery methods may have evolved because early (1970 to 1972) experiments in California with corrugated panels were largely unsuccessful. We concluded that abalone larvae were sensitive to the fiberglass particles contained in the corrugated panels. The Japanese use PVC-corrugated panels. As far as is known, corrugated PVC is not manufactured in the U.S. The abalone hatchery on Vancouver Island did adopt the Japanese "wavy plate" panel method.

Herein we present the abalone cultivation methods that are being practiced at the California Department of Fish and Game's Granite Canyon Laboratory. These methods, in part, are continually being subjected to experimentation to seek refinements, or new methods are being sought. We wish to emphasize this because while many advances in abalone culture have been made in recent years, some major gaps remain in the technology overall. These are pointed out and discussed in the following sections. After the cultivation methods section we present a short section on abalone diseases followed by a discussion on the "state of the art".

CULTIVATION METHODS

Cultivation methods are divided into seven elements. These are presented concurrently for two species, the red and pink abalone. Although the pink abalone is a warmer water species, where culture differences exist, we point these out within the specific element.

Brood Stock

Brood stock management methods provide a year-round source of sexually mature abalone. We generally use F_1 or later generation stocks, of 8- to 12-cm shell lengths. Fecundity estimates for fully mature red abalone at these sizes may range from 5×10^5 to 4×10^6, respectively. Brood stock are maintained in 15-ℓ plastic containers according to geographic origin and parentage. The red abalone receive ambient temperature seawater (9 to 15°C), at a 3 to 4 ℓ/min flow rate, that is sand-filtered to about 20 µm, UV treated, and with a natural photoperiod. Pink abalone receive a similar treatment except the seawater temperature is maintained at 18°C ± 2°C. Brood stock containers are cleaned weekly, and fresh kelp, *Macrocystis* spp., is supplied to excess at the same time. The provision of filtered seawater to brood stock, and possibly in combination with the UV treatment, generally protects the brood stock from unsolicited spawnings that occur elsewhere in the laboratory hatchery where these treatments are not used. Unsolicited abalone spawnings typically occur during unsettled weather conditions associated with a fall in barometric pressure, a shift in wind direction to the south, and hence, a warmer water regime.

Brood stock selection for ripeness and spawning is based on gonadal bulk. We use a subjective indexing system that ranges from 0 to 3. Index #3 abalone are fully mature and spawnable. The day prior to a planned spawning the abalones are selected, their shells scrubbed, and they are isolated without food (kelp).

Spawning Induction, Fertilization, and Early Development

We use the UV spawning-induction method.[8] Male and female abalone are held in separate containers. The 1-µm-filtered seawater flow rate is adjusted to 150 mℓ/min. The UV unit is activated by a timer set to come on at 0600 hr on a "spawning day". A 3- to 4-hr exposure to the UV-treated seawater is generally sufficient to trigger spawning.

After spawning the ova are collected by siphon and distributed into 15-ℓ plastic containers. They are generally fertilized within 1 hr postspawning. Sperm viability is lost well in advance of the ova.[11] The ova are slightly more dense than seawater and settle to the bottom of the container. Two or three decantations are made to remove excess sperm. Following the last decantation the culture container is refilled. One-µm-filtered, UV-treated seawater at 15°C is used for all culture-water exchanges. It is desirable to end up with a single layer of ova on the culture container bottom. Left undisturbed, at 15°C the ova develop to the two-cell stage in 2 hr and cleave at hourly intervals thereafter. At 16 hr postfertilization the first larval stage (trochophore) ensues.

Larval Cultivation

Optimum larval culture temperatures for red abalone range from about 14 to 18°C. Poor culture success is obtained above 20°C. The lower temperature limit for larval cultivation has not been determined; however, relatively good culture success may be obtained at 10°C, although the larval development rate is retarded, and it takes about 14 days postfertilization to attain the benthonic stage. We routinely culture red abalone larvae at 15°C.

Optimum culture temperatures for pink abalone larvae range from 18 to 21°C.[12] We culture these larvae at 18°C.

Larvae remain in the hatching container, undisturbed, for about 30 hr postfertilization. At about 16-hr postfertilization they attain the trochophore stage and ciliary movement becomes evident. Two hours later the trochophore is rotating within the egg membrane, the membrane becomes highly elastic, ruptures, and the trochophore rises to the culture surface layer. The shell gland becomes evident at the trochophore stage, and development of the larval shell commences once the trochophore erupts from the egg membrane. This signals initiation of the veliger stage. However, until the veliger-stage larval-shell development is completed, they should not be transferred as they are more vulnerable to damage and mortality.

After 30-hr postfertilization the veligers are siphoned into a screened container (90 μm), that is partially immersed in a water bath. Care must be taken not to siphon or transfer "contaminants" from the bottom of the larval-hatching container. Veligers are then washed into a 1-ℓ beaker and their numbers estimated. Following this, they are transferred to an 8-ℓ larval culture apparatus,[3] 40,000 per culture (5/mℓ). The larvae receive 3-μm-filtered, UV treated seawater, 200 to 300 mℓ/min flow rate, for the approximate 6-day rearing period. Following this the veligers are nearing their benthonic, postlarval period. This stage can be determined by microscopic examination or by gross observations in the culture. Microscopic examination should reveal that the cephalic tentacles have four branches; and the larval foot is sufficiently developed such that the veliger can pull itself upright and glide about. The cilia are used for propulsion. Macroscopically, the veligers are observed to concentrate around the perimeter of the culture, at the culture water surface. They are reconcentrated in a 1-ℓ beaker, their density estimated, and they are distributed into postlarval rearing tanks.

Nutrition

Probably the most critical element to any culture endeavor is that of nutrition. Abalone larval culture is relatively uncomplicated due to the short rearing period and the absence of a dietary requirement. However, upon metamorphosis the postlarvae subsist on diatoms. These diatoms should be small, less than 10 μm in greatest dimension, so that early postlarval stages can ingest them. Single diatom species may be cultured and provided to the postlarvae, or multiple diatom species can be size selected, cultured, and offered. We prefer the latter method.

Details of the diatom collection system are provided in Ebert and Houk.[3] In general, the system entails stage filtration of seawater from the main intake pumps, proceeding from sand filters (20 μm) to cartridge filters (1-μm nominal rating). Those diatoms passing through the 1-μm cartridge filters are "caught" on a plastic substrate, resuspended, and fed to newly settled postlarvae. Larger diatoms, those not passing through a 1-μm bag filtration, are retained for juvenile abalone.

Abalone-rearing tanks for postlarvae and early juvenile stages may be inoculated with additional diatom slurries or covered to inhibit diatom growth, in response to the abalone grazing rate. Typically the grazing rate of postlarvae is relatively low and diatom cell production exceeds the grazing rate, and the tanks are covered. However, following the onset of respiratory pore development (about 50-day-old juveniles, and 2.1-mm-long shell lengths) the grazing rate rapidly accelerates. The tanks are uncovered, and shortly thereafter diatom slurry inoculums are required. It is critically important to maintain sufficient diatoms in the rearing tanks during the early juvenile stages (about 2 to 5-mm shell lengths). Otherwise, abalone of these sizes are easily stressed, their growth will be stunted, they may crawl above the culture water level and desiccate or succumb to prevailing bacterial populations, including the pathogenic *Vibrio* spp., that may not otherwise affect them. Abalone, at least up to 12 mm or greater shell lengths, prefer diatoms; but it becomes increasingly more difficult to provide adequate quantities, with our culture system, much beyond 5-mm shell lengths. Therefore, the diatom diet is supplemented with macroalgae (kelp) for abalone in the 5- to 12-mm size range; and only kelp is provided thereafter.

Two major canopy-forming laminarian kelps occur along the central California coast; an annual species, the bull kelp, *Nereocystis leutkeana* and a perennial, the giant kelp, *Macrocystis* spp.

FIGURE 1.   Abalone postlarval and juvenile-stage rearing tanks. Each tank incorporates a
separate cartridge filter that has a water distribution spray bar threaded to the filter outlet.

We conducted a 1-year study to determine whether giant kelp or bull kelp was more nutritious for red abalone. One-year-old abalone, averaging about 21 mm in shell length were used; and ambient temperature seawater (monthly − temperature range = 10.0 to 12.3°C). After 1 year the abalone on a bull kelp diet averaged nearly 40 mm in shell length while those provided with giant kelp averaged 44 mm in shell length Moreover, abalone on the giant kelp diet exhibited a significantly better food conversion efficiency. Their feeding rate, expressed as the consumption per mean abalone weight, averaged about 5.9%; compared to abalone offered bull kelp where the feeding rate averaged 16.8% of their mean body weight. As a result of this study we feed giant kelp to juvenile and later growth stages of red and pink abalones.

Postlarval and Early-Juvenile Stage Cultivation

Cultivation of these abalone stages is accomplished in circular fiberglass tanks having a dished bottom and a white gel-coat finish. The tanks are about 107 cm wide by 25 cm deep at the sidewall. The bottom slopes downward to about a 33 cm depth at the center bottom drain opening. The drain opening accommodates about a 3.8-cm diameter standpipe. Tank volume is about 260 ℓ and offers about 17,000 cm^2 of wetted surface area. Each tank has a separate cartridge filter with a water distribution spray bar threaded to the outlet. We incorporate 30 of these tanks, in two tiers, in our pilot-production operation. These tanks are located in a greenhouse that incorporates translucent plastic siding in order to maximize solar illumination (Figure 1) and hence diatom production.

Preparation of the postlarval tanks for abalone generally commences 24 hr before seeding.

Tanks are scrubbed, rinsed, and filled with 1-μm-filtered, UV-treated seawater at 15 or 18°C (pink abalone). A diatom slurry inoculum is added to each tank (see Nutrition section), and the tanks receive aeration. The following day, when the abalone larvae are introduced, the diatom inoculum has formed a faint tannish-colored film. Abalone are seeded at densities ranging from 1.5 to 2.0/cm² (25,000 to 35,000 per tank). We do not use an abalone settlement inducer.[13,14] A 90-μm screen is placed over the tank drain and 1-μm filtered seawater, 15 or 18°C, is provided at a reduced flow rate (about 0.2 to 0.3 ℓ/min) for about 48 hr or until most of the larvae have settled and metamorphosed. Following this, the drain outlet screen is removed and the water flow rate is increased to 5 to 6 ℓ/min. About 2 weeks after seeding the abalone, the tanks are drained and rinsed, and at weekly intervals thereafter. The water filtration level is reduced to 3 μm, 3 to 4 weeks postseeding, and to 20 μm after about 8 to 10 weeks in these tanks.

When the red abalone are 2 weeks old they average about 0.5 mm in length, by 6 weeks about 0.8 mm long, and the first respiratory pore notch forms at about 8 weeks, at 2.1-mm shell lengths. When 4 months old they average about 5.0 mm long, and at 6 months about 8.0 mm. Generally, the abalone are transferred to weaning tanks (see Weaning section) when they attain 5- to 8-mm lengths.

Pink abalone postlarvae are cultured in a manner that is identical to red abalone; but a higher culture temperature (18°C) may be more feasible. Comparable postlarval culture tests run over a 6-month period at 15°C and 18°C revealed that pink abalone grew slightly faster at 18°C (Figure 2). However, after 4 months the growth rate of both test groups slowed down. This reduced growth rate corresponded to the time at which abalone had denuded the tank surfaces of diatoms, and diatom additions (slurries) could not adequately meet the nutritional demand.

Abalone mortality in the settlement tanks is rather extensive, particularly during the initial few weeks, and may exceed 70 or 80%. Most of this mortality consists of veligers that fail to metamorphose. The mortality rate declines following respiratory pore development. Abalone survivorship in these tanks generally average 8 to 12% of those introduced during the 4- to 6-month rearing period.

Weaning

This cultivation stage evolved in our system largely because of an inability to provide sufficient quantities of diatoms for juvenile abalone (>5 mm shell lengths) in the settlement tanks. Therefore, these size abalone are frequently transferred to weaning tanks. These tanks are 2.4 m × 0.57 m × 0.28 m and offer just over 30,000 cm² of wetted surface area. Abalone habitats are added. They consist of PVC pipe sections of varying lengths, generally about 8 to 10 cm in diameter, that are cut in half lengthwise. These habitats provide shelter for the increasingly light-sensitive abalone, and secure kelp fronds. Abalones in the weaning tanks receive raw seawater at ambient temperature at a 6 to 8 ℓ/min flow rate. These tanks are drained and rinsed weekly, and supplied with fresh giant kelp fronds. Diatom inoculums are frequently added, and the weaning tank surfaces are not scrubbed so that other algae (e.g., corallines) become established and provide additional abalone forage. Abalone mortality in the weaning tanks is minimal. Upon reaching about 12-mm shell lengths (2 to 3 months in these tanks) the abalone are transferred to growout tanks.

Grow Out

We use round and trough-type tanks constructed from fiberglass in the grow-out operations. Both tank configurations work equally well, are stocked at similar densities, and have abalone habitats that are designed to fit the tank configurations (Figure 3). The round tanks are 1.05 m in diameter by 0.58 m deep and hold about 400 ℓ. They offer about 4.0 m² of surface area including the habitat. The trough tanks are 2.07 m × 0.56 m × 0.53 m deep and also

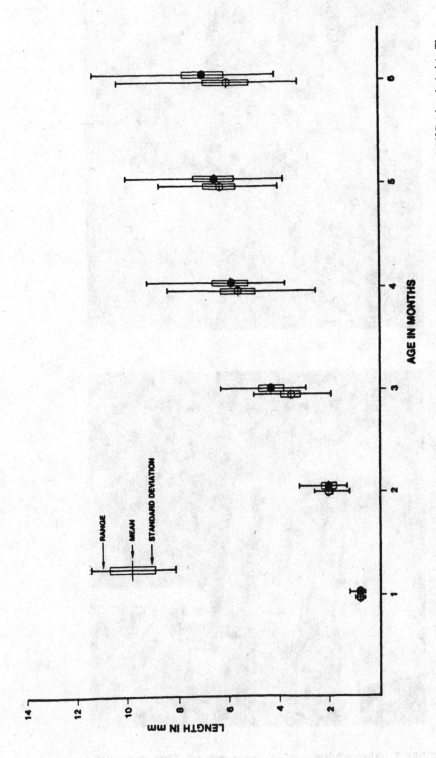

FIGURE 2. Comparative growth rates of postlarval pink abalone cultured in 260-ℓ settlement tanks at 15°C (open circle) and 18°C (closed circle). The slowed growth rate after 4 months, for both test groups, corresponds to the time that diatoms became limiting.

A

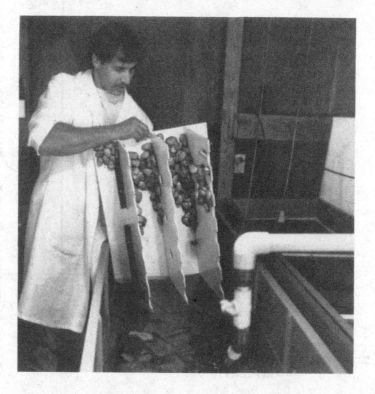

B

FIGURE 3.   Abalone habitats used for abalone grow out. These are designed to increase vertical (preferred) surface area for abalone and to provide shelter.

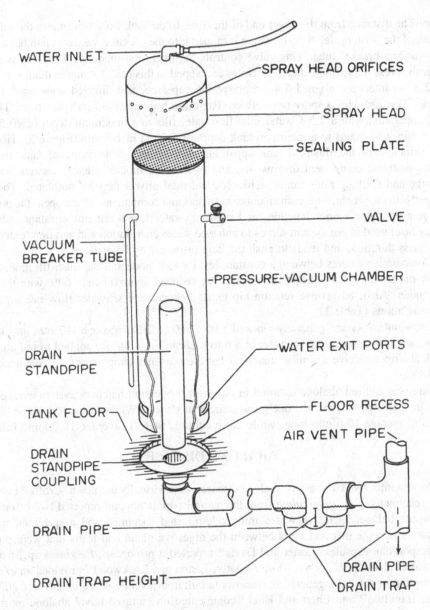

FIGURE 4. Flush-fill tank apparatus components. The spray head is used in the round grow-out tanks; but is also an optional component used for water distribution.

hold about 400 ℓ (at standpipe depth). They offer about 6.9 m² of surface area including the two habitats. Also, the round tanks incorporate a unique flush-fill tank-cleaning apparatus.

The flush-fill tank apparatus, was designed to enhance water circulation, help maintain food (kelp) freshness, keep the food proximal to the abalone, and to facilitate sediment and waste removal. The apparatus is constructed from plastic pipe (Figure 4). For convenience, pipe diameters are presented in nominal pipe size (in inches); otherwise the metric system is used. The pressure-vacuum chamber consists of 4-in. pipe with a PVC sealing plate at the top, and four water exit ports (1.0 cm by 3.5 cm) near the base, level with the tank floor. The pressure-vacuum chamber fits into a recess in the tank's floor that also accommodates a 2-inch diameter drain standpipe, 19 cm long. A hole is tapped in the pressure-vacuum chamber 30 cm about the base to accommodate a ¹/₄-in. street-ell. A ¹/₄-in. pipe, 25 cm long, is threaded into the street-ell, forms an airway, and serves as a vacuum-breaker tube. The distance from the lower end of this tube to the tank floor determines the minimum depth of the flush cycle. A $1/4$-in. valve is tapped into the pressure-vacuum chamber opposite the vacuum-breaker tube. This valve controls the operation mode. A 4-in. cap is sleeved over the pressure-vacuum chamber. Holes are tapped in this cap, 2.4 mm in diameter, spaced at 2.3-cm intervals, aligned 5.4 cm above the cap face, and directed downward at a 10° angle. This provides a spray-head effect. The drain trap consists of $11/2$-in. pipe. The tank and apparatus, with a 7.5-ℓ water inlet flow rate, fills to a maximum depth of 20.5 cm in 18.5 min, and drains to a minimum tank depth of 5.2 cm in 1.5 min (Figure 5). However, the principle of the flush-fill tank apparatus allows for a wide range of tank sizes and configurations, component dimensions, and operational modes. Hence, various water exchange and flushing rates can be achieved, and tidal prisms may be simulated. The valve tapped into the pressure-vaccum chamber is an optional component. When open, the pressure-vaccum chamber cannot function, and the tank water levels remains constant. Also, the spray head used in our system serves to enhance water distribution and aeration, extend kelp freshness duration, and tends to push the kelp proximal to the abalone.

Comparative studies between a constant-level water mode and the flush-fill mode, in the grow-out tanks, disclosed that tank wastes are reduced approximately 60% with the flush-fill mode. Also, an inverse relationship exists between the seawater flow rate and waste accumulations (Table 1).

Grow-out tanks are generally stocked with 2500 to 3000 abalone (12 mm and larger). Raw, ambient temperature seawater at a 6 to 8 ℓ/min flow rate is supplied to red abalones. Pink abalones receive a similar treatment but the seawater temperature is maintained at 15 or 18°C.

One-year-old red abalone cultured in our pilot-production hatchery system average 20.5 mm in shell length. Pink abalone grow somewhat slower. When 1 year old, those cultured at 15°C average 16.6 mm long, while those cultured at 18°C average 18.5 mm (Table 2).

ABALONE DISEASES

Prior to intensive cultivation, abalone diseases were virtually unknown. Crofts[15] examined 400 specimens of *H. tuberculata* from the natural population and reported two instances of diseased abalone. Inclusive was a trematode-infested specimen and a specimen with an unknown parasite that had cysts between the digestive gland and testis that were possibly haplosporidian capsules. Lester and Davis[16] reported a protozoan, *Perkinsus* sp., from the Australian abalone, *H. ruber*. More recently, Elston and Lockwood[17] provided an excellent description of the pathogenesis of vibriosis in cultured juvenile *H. rufescens* in California. Also, Hamilton[18] and Ebert and Houk[3] commented on cultured larval abalone mortalities associated with the presence of the pathogenic bacteria, *Vibrio* spp., and possible treatments. We do not view the bacteria (*Vibrio* spp.) problem associated with abalone larvae as serious principally because stringent sanitary practices, treatments, or phased larval-rearing periods have been relatively effective in combating the problem. Also, the larval-rearing duration is relatively short. However, the disease problem (vibriosis) reported by Elston and Lockwood[17] in juvenile abalone is serious and costly because of the expense invested in cultivating abalone to the juvenile stage.

In the past 14 years we experienced three epizootics among juvenile or older red abalone (>8 mm shell lengths). The first disease outbreak (April to June, 1979) was confined to one grow-out tank. Abalone in the tank ranged from 2 to 3 cm long. Mortality was 84%. Grossly characterized, infected abalone typically reposed hunched up on the substrate, with cephalic and epipodial tentacles retracted, and the sole of their foot lacked adhesiveness. Fresh mortalities were examined by a department biologist (Ron Warner) who has a background in shellfish pathology. He found[21] from gross observations, the presence of lesions

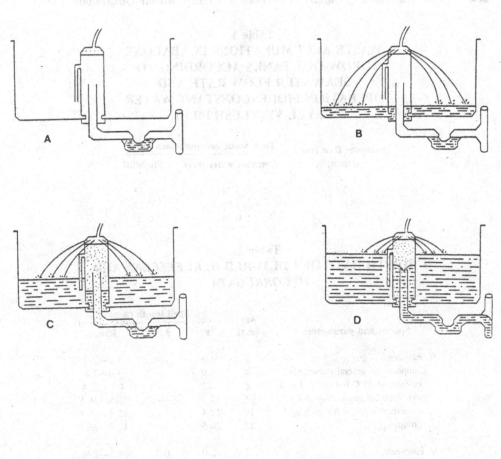

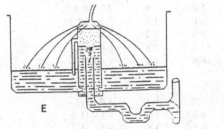

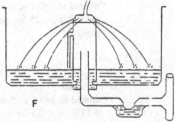

FIGURE 5. A sequential presentation of events of the automatic flush-fill tank apparatus. (A) Drain trap primed for water to be turned on, (B) The water level rises both in the tank and the pressure-vacuum chamber at the same rate. The spray head water coverage strikes the water surface near the tank periphery. (C) Water in the tank has risen above the orifice of the vacuum breaker tube, trapping and compressing air in the pressure-vacuum chamber that, in turn, presents a resistance that suppresses water rise in the pressure-vacuum chamber. Spray head water now contacts the surface closer to the tank center due to the rising water level. (D) The water level in the pressure-vacuum chamber has risen above the lip of the drain standpipe has filled the drain line, and the drain cycle commences. The air compressed in the chamber can now escape, pressure is released, and a vacuum forms in the chamber. The water spray head has swept the water surface on the fill cycle and will reverse the pattern on the drain cycle. (E) Vacuum inside the chamber maintains the draining level while the water level in the tank continues to fall. (F) The falling water level has exposed the orifice of the vacuum breaker tube, allowing air to enter the chamber, breaking the siphon, and the water level will commence to rise.

on the foot surface of all specimens. Histological examination showed multiple lesions that extended into the foot musculature, and a loss of columnar epithelium with resultant necrosis of the underlying musculature. Also, Mr. Warner noted a marked hemocytic response in several of the lesions. One specimen had bacterial colonies interspersed with necrotic epithelial and muscular debris.

Table 1
WASTE ACCUMULATIONS IN ABALONE GROW-OUT TANKS ACCORDING TO SEAWATER FLOW RATE AND OPERATION MODE (CONSTANT WATER LEVEL VS. FLUSH/FILL)

Seawater flow rate (ℓ/min)	Tank waste accumulations (mℓ/ℓ)	
	Constant water level	Flush/fill
3	3.74	1.62
6	1.05	0.33
9	0.74	0.29

Table 2
GROWTH RATE OF CULTURED *H. RUFESCENS* AND *H. CORRUGATA*

Species and parameters	Age (mo)	Shell length (mm)		
		x̄	±SD	Range
H. rufescens	2	1.8	—	1.2—2.3
Composite of several spawnings;	4	5.0	—	3.0—7.8
cultured at 15°C for initial 3 to 4	6	8.1	—	4.5—12.4
mo; Ambient temp. (9—15°C)	8	13.0	—	10.5—16.5
thereafter; n = 50 for each age	10	17.4	—	15.3—18.7
group.	12	20.5	—	12.2—26.1
H. corrugata	2	2.0	0.3	1.2—2.6
Cultured in 260-ℓ settlement tanks	4	5.6	1.4	2.5—8.4
at 15°C; n = 25 for each age	6	6.0	1.8	3.2—10.3
group.	8	8.8	2.2	5.4—15.3
	10	12.4	2.8	6.9—18.1
	12	16.6	3.1	8.4—22.1
H. corrugata	2	2.0	0.5	1.3—3.2
Cultured in 260-ℓ settlement tanks	4	5.9	1.4	3.7—9.2
at 18°C; n = 25 for each age	6	6.9	1.7	4.1—11.3
group	8	9.2	2.8	5.3—16.4
	10	13.3	3.4	7.1—20.4
	12	18.5	3.7	9.0—24.9

The second abalone epizootic occurred during March and April of 1984. This epizootic spread from one to five growing tanks, and over 5000 abalone died before the disease abated. Infected abalone were grossly characterized similar to the 1979 disease outbreak.

The third juvenile abalone epizootic initiated in March 1985 and did not abate until July. This was the most serious disease outbreak. Ultimately, red abalone in all weaning and grow-out tanks became infected. Abalone in the weaning tanks (8- to 12-mm shell length) were most susceptible and mortalities exceeded 90%. This epizootic was studied more intensively than the previous two, particularly from an etiological standpoint; and was the most difficult to treat. A vibriosis was implicated in all growing tanks. In the weaning tanks the vibriosis was characterized similar to that reported by a department biologist (Ron Warner) during the 1979 disease outbreak, and by Elston and Lockwood.[17] Larger-sized diseased abalone (>2-cm shell lengths) also showed an obvious vibriosis on the sole of the foot in some cases; but others showed evidence of having been cannibalized — the sole of their foot was gnawed away. Still others did not exhibit any external lesions or necrotized epidermal tissue. These abalone displayed a distended buccal region and an extended, semirigid foot muscle. Internal microscopic examination revealed a severe vibrosis throughout the digestive tract. It was theorized that these abalone succumbed from ingesting infected tissue of their cohorts.

During each epizootic, stepped-up sanitary practices were instituted when diseased abalone first appeared. Tank drain and rinse frequencies were, on occasion, done daily; often bidaily. Frequently, the abalone received a brief tap-water spray treatment (this is occasionally done with healthy abalone to rid the tanks of copepods, amphipods, or isopods). Abalone mortalities were removed daily. Tissue samples were examined microscopically and/or plated on TCBS agar to confirm the presence of vibrios. Therapeutic treatments were also tested. These included nitrofurazone, prefuran, and neomycin sulfate. Treaments were administered at irregular intervals. There was no clear evidence that any of the treatments were effective; although the mortality rate slowed during the treatment period, it flared up when applications ceased.

Elston and Lockwood[17] suggested that juvenile red abalone were predisposed to the vibriosis. Evidence suggests that the three epizootics that occurred in our pilot-production hatchery may have been mediated by stress. In two of the epizootics the abalone were being maintained at high densities with inadequate or insufficient habitats in the tanks. These abalone often cut one another with their anterior (sharp) shell edge. In the most recent epizootic (March to July, 1985) a recirculating seawater system was tested just prior to the appearance of diseased abalone.

It is of interest that all three epizootics occurred during a similar time annually when seawater temperatures are lowest. Moreover, two other California-based facilities are known to have experienced juvenile abalone die offs in 1985 at the same time we did. Also, Kraeuter and Castagna[19] reported that the majority of disease outbreaks in juvenile hard clams, Mercenaria mercenaria occurred in early spring when seasonally low water temperatures approached 10°C. But this was incongruous with known levels of vibrios in the seawater during the warmest months. They speculated that the clams were stressed during the spring when food reserves were minimal, and they may have been more susceptible to a bacterial infection. In essence, at least four facilities including one on the East coast where molluscs are being cultivated (either pelecypod or a gastropod) are reporting juvenile-stage shellfish mortalities during the spring when water temperatures are lowest. Kraeuter and Castagna[19] also noted that clam larval mortality was highest during peak temperature periods. This is also the case with abalone larvae in California hatcheries. The association of high water temperature with vibriosis is widespread and well recognized among shellfish cultivators. Some hatcheries avoid rearing larvae during this period.

From the foregoing we suggest that the springtime juvenile mollusk mortality phenomenon may not only be widespread, but the causative mechanism may be similar. If this be the case, then the low food-reserve condition noted by Kraeuter and Castagna[19] would not be implicated because this condition is not applicable to abalone. However, the disease problem may still be predisposed to a physiological condition related to low seawater temperature, corresponding low metabolic rate, and high nutrient levels in the incoming seawater supply. Temperate-region oceanographic regimes characteristically have peak nutrient levels in early spring, concomitant with upwelling. Moreover, a second peak in nutrients occurs in late summer off California that corresponds with the high larval-mortality period.

Elston[20] reported on oxygen intoxication in juvenile abalone when exposed to supersaturated conditions. Although he demonstrated this experimentally, and noted that a vibriosis followed the oxygen supersaturation exposure, he pointed out that abalone have been affected by gas-bubble disease (supersaturated air) due to mechanical failures in the seawater pumping system. We have had similar occurrences in our laboratory. We did not measure oxygen levels in our seawater during any of the aforementioned epizootics. We presumed oxygen levels were sufficient because of the inverse relationship between water temperature and oxygen-carrying capacity. Kraeuter and Castagna[19] suggest that the exact cause and mechanism of the spring mortality should be investigated. We concur — the magnitude of the problems and implications are enormous.

DISCUSSION AND STATE OF THE ART

The decline of the commercial abalone fishery in California, due to multiple factors, has lent impetus for the development of an abalone cultivation industry. However, the growth of this fledgling industry has been slow. Reasons for this slow growth range from the difficulties in securing prime oceanfront sites for hatchery operations at reasonable costs, to regulatory constraints, and to the lack of research and development funds. Considerable progress has been made in abalone cultivation practices, particularly in the areas of brood-stock management, spawning induction, and larval rearing. Further refinements are needed, principally in the area of nutrition, genetics, and disease problems.

Two salient aspects of our brood-stock management system are (1) when possible only laboratory-raised abalone (F_1 or later generation) are used and (2) the seawater treatment (sand filtration and UV) generally protects brood stock from unsolicited spawnings. Wild-population abalone, particularly those species that have a somewhat constricted natural spawning period annually (e.g., pink abalone), frequently do not respond to a spawning-induction stimulus, either when they are collected ripe (bulky gonads), or after they have undergone a conditioning period in the laboratory. By constrast, all F_1 and later generation laboratory-raised red and pink abalone may become reproductively mature during their second or third year in the laboratory, and readily respond to a spawning-induction stimulus.

Maintenance of sexually mature brood stock year round, is a valuable asset to any cultivation endeavor. Occasionally, about every 5 to 10 years, we experience unusually heavy plankton blooms in our laboratory receiving seawater. The sand filters and UV seawater treatments become ineffective in plankton removal, and an unsoliciated brood-stock spawning occurs. This last occurred in January 1985 during an unusually warm winter period. The plankton bloom, consisting principally of the dinoflagellate, *Gymnodinium* sp., left a noxious odor throughout the laboratory-hatchery complex. All ripe abalones spawned, and it took 3 to 4 months of conditioning before they returned to a spawnable state.

Procedures for abalone spawning induction, ova fertilization, development to the larval stage, and larval rearing may vary amongst various cultivators. Often these differences are predictated by personal preference or the laboratory hatchery seawater system and equipment specific to a site. Regardless, the technology and biology is sufficiently advanced such that most cultivators are experiencing a relatively high degree of success with these elements. Occasionally there are bacterial problems, principally *Vibrio* spp. Careful sanitary practices and the use of filtered (3 μm or less), UV-treated seawater are generally adequate preventative measures. We seldom use antibiotics, although neomycin sulfate (50 mg/ℓ), may effectively control vibriosis. More than one treatment during the larval-rearing period should be avoided because it may arrest development, although the larvae appear healthy, outwardly. We have maintained such larvae up to 35 days at 15°C. Also, vibrios are more prevalent in laboratory hatcheries along the California coast during the warmer water (September to December) period. It may be feasible to avoid larval rearing during this time.

Abalone nutrition deserves the most attention (and needs) of all abalone cultivation elements. These needs for improvement extend from the diatom dependency period (postlarvae to juvenile), to adult sizes. For postlarval and juvenile stages the evolution in our laboratory has proceeded from using single species diatom cultures to the present diatom-slurry method. This method is adequate for part of the year, but it is dependent upon available nutrients in the seawater that vary seasonally. Therefore, a consistent diatom production is not possible. Nutrient additions increase diatom production, but their cost-benefit ratio may be prohibitive. Our goal is to develop a nutritious, single-species diatom culture capability as a backup to the diatom-slurry method. Giant kelp provides adequate nutrition for juvenile and later growth stages of abalones. Although extensive kelp beds occur along much of the central and southern California coastal shelf and this kelp is generally available, it is unwieldly (bulky) to handle and requires relatively large amounts to sustain a modest-size hatchery. Research is needed on the development of a compounded diet that contains essential nutrients and can be presented in a technologically and biologically feasible form.

At least three methods are used to settle larval abalone and rear the early growth stages: the method outlined in this paper and Ebert and Houk[3]; and two methods that use settlement inducers. One inducer, γ-aminobutyric acid (GABA) has been used principally for experimental purposes.[13] The other inducer consists of mucoproteins from the sole of the abalone's foot. These mucoproteins are deposited when larger-size abalone pregraze the diatom film in a settling tank, followed by the introduction of larvae that are attracted to these surfaces.[14] The three abalone larvae settlement methods are currently undergoing comparative testing in our laboratory, both from a technological and biological viewpoint, for application to commercial hatchery methods.

Although abalone mortality rates, from larval settlement through the postlarval stages to the development of the first respiratory pore, continue to be extensive, we do not view this as a serious problem. Sexually mature adult abalone are generally available year round, they are fecund and an abundant supply of larvae are readily available. Therefore, we anticipate the high larval and early post-larval mortality rates, and compensate for this accordingly with an initially high stocking density. We strive for consistency in yield per tank, per time period, and have steadily improved in this regard. Our main limitation in the settlement tanks relate to available substrate and diatom densities.

As pointed out earlier, Japanese culture methods typically use preconditioned, corrugated plastic panels for larval settlement. An advantage of this method is that panels can be manipulated according to settlement success and diatom density. Our abalone settling tanks offer only the tank surfaces, and it is often difficult to adjust food bases (diatom densities) to postlarval and juvenile abalone densities. We have been testing various methods to increase substrate surface area. In concept, what is evolving is a system midway between our existing one and that of the Japanese corrugated-panel approach.

Paradoxically, juvenile and adult abalones prefer very subdued light or total darkness, but light is essential to diatom growth. For juvenile diatom-dependent abalone stages various shelters (e.g., pipe sections) can provide a dark haven, yet not inhibit diatom growth on light-exposed exterior surfaces. This method is used in our weaning tanks. However, once abalone pass from the weaning phase to grow out it is desirable to maintain them in darkness, similar to the practice of mushroom farmers. In darkened conditions, the adult abalone feeding period is extended and their growth rate is accelerated.[3]

REFERENCES

1. **Cox, K. W.**, California abalone, family Haliotidae, *Calif. Dept. Fish Game, Fish Bull.*, 118, 1, 1962.
2. **Carlisle, J. G., Jr.**, Spawning and early life history of *Haliotis rufescens* Swainson, *Nautilus*, 76(2), 44, 1962.
3. **Ebert, E. E. and Houk, J. L.**, Elements and innovations in the cultivation of red abalone, *Haliotis rufescens*, *Aquaculture*, 39, 375, 1984.
5. **Murayama, S.**, On the development of the Japanese abalone, *Haliotis gigantea*, *J. Coll. Agricult.*, Tokyo Imp. Univ. 13(3), 227, 1935.

6. **Ino, T.,** Biological studies in the propagation of the Japanese abalone (genus *Haliotis*), *Bull. Tokai Reg. Fish. Res. Lab.*, 5, 1, 1952. (In Japanese with English summary.)

7. **Shibui, T.,** On the normal development of the eggs of Japanese abalone, *Haliotis discus hannai* Ino, and ecological and physiological studies of its larvae and young, *Bull. Iwate Pref. Fish. Exp. St.*, 2, 1, 1972.

8. **Kikuchi, S. and Uki, N.,** Technical study on artificial spawning of abalone, genus *Haliotis*. II. Effect of irradiated sea water with ultraviolet rays on inducing to spawn, *Bull. Tohoku Reg. Fish. Res. Lab.*, 33, 79, 1974.

9. **Kikuchi, S. and Uki, N.,** Technical study on artificial spawning of abalone, genus *Haliotis*. III. Reasonable sperm density for fertilization, *Bull. Tohoku Reg. Fish. Res. Lab.*, 34, 67, 1974.

10. **Kikuchi, S. and Uki, N.,** Technical study on artificial spawning of abalone, genus *Haliotis*, V. Relation between water temperature and advancing sexual maturity of *Haliotis discus* Reeve, *Bull. Tohoku Reg. Fish. Res. Lab.*, 34, 77, 1974.

11. **Ebert, E. E. and Hamilton, R. M.,** Ova fertility relative to temperature and to the time of gamete mixing in the red abalone, *Haliotis rufescens*, *Calif. Fish Game*, 69(2), 115, 1983.

12. **Leighton, D. L.,** The influence of temperature on larval and juvenile growth in three species of southern California abalones, *Fish. Bull.*, 72(4), 1137, 1974.

13. **Morse, D. E., Hooker, N., Jensen, L., and Duncan, H.,** Induction of larval abalone settling and metamorphosis by γ-aminobutyric acid and its congeners from crustose red algae. II. Applications to cultivation, seed-production and bioassays; principal causes of mortality and interference, *Proc. World Maricult. Soc.*, 10, 81, 1979.

14. **Seki, T. and Kan-no, H.,** Induced settlement of the Japanese abalone, *Haliotis discus hannai*, veliger by the mucous trails of the juvenile and adult abalones, *Bull. Tohoku Reg. Fish. Res. Lab.*, 43, 29, 1981.

15. **Crofts, D. R.,** Haliotis, *Liverpool Mar. Biol. Comm. Mem. Typ. Br. Mar. Plants Anim.*, 29, 16, 1929.

16. **Lester, R. J. G. and Davis, G. H. G.,** A new *Perkinsus* species (apicomplexa, perkinsea) from the abalone *Haliotis ruber*, *J. Invertebr. Pathol.*, 37, 181, 1981.

17. **Elston, R. and Lockwood, G. S.,** Pathogenesis of vibriosis in cultured juvenile Red abalone, *Haliotis rufescens* Swainson, *J. Fish Dis.*, 6, 111, 1983.

18. **Hamilton, R. M.,** Relationships Between Harpacticoid Copepod *Tigriopus californicus* Swainson, Master's thesis, Humboldt State Univ., Arcata, Calif., 1984, 34.

19. **Kraeuter, J. N. and Castagna, M.,** Disease treatment in hard clams, *Mercenaria mercenaria, J. World Maricult. Soc.*, 15, 310, 1984.

20. **Elston, R.,** Histopathology of oxygen intoxication in the juvenile red abalone, *Haliotis rufescens* Swainson, *J. Fish Dis.*, 6, 101, 1983.

21. **Warner, R.,** personal communication.

ABALONE FARMING IN KOREA

Sung Kyoo Yoo

INTRODUCTION

Whenever harvesting of marine animals is mentioned, Koreans promptly think of abalones prior to any other species. Abalones have been a very popular and delicious food since early days. "Jeon-bok-juk", abalone soup prepared by boiling rice and abalone flesh together in water, has been highly esteemed not only for the ill or infants but as a delicacy or epicurean food for gourmets. Abalones are also consumed raw, especially by the coastal inhabitants.

Since even the supply of about 500 mt annually cannot satisfy the enormous demand, the price of abalone in the local markets is astonishingly high. The abalone fishery in Korea still employs rather primitive harvest methods which are mostly dependent on women skin divers who do not use special equipment. Rapidly increasing industrial pollution and frequent reclamation of the coastal areas may lead to a steady decline in abalone production and even a loss of suitable farming beds (Table 1).

The continuous high market price, however, has caused the coastal inhabitants and fisheries scientists to become interested in developing abalone culture. Before 1974, the abalone farming in Korea was a incomplete type of culture in which larval development and the provision of food for juveniles and adults were left entirely to nature. They released seeds collected in the near shore and stocked them in the waters surrounded by many reefs to protect against escape until harvest.

In 1974, abalone seed production was initiated at the Fisheries Seedling Production Station of Jeju using a controlled system. Four additional stations (Figure 1) were established to begin an extensive national project for seed production to increase recruitment of abalone stocks throughout the coastal area (Table 2).

The number of seed abalone produced in these stations increased every year and reached a total of 1,155,300 individuals in 1983. Seedling production stations also provide valuable extension services to disseminate the modern cultivation techniques. Seeds (2 cm) are distributed to the coastal fishermen without charge or are released into the sea.

Only a few farmers still carry out the entire rearing process themselves because juvenile seed produced by the government is available free of charge for release to restock the sea.

INDIGENOUS ABALONE SPECIES

Five species of abalones are found in Korea and four of these species are commercially valuable — *H. discus hannai*, *H. discus*, *H. gigantea sieboldii*, and *H. gigantea*. The northern (cold water) and southern (warm water) species are separated by the isothermal line of 12°C at 25 m depth in February.

H. discus hannai is found north of the isothermal line and inhabits the shallowest water depth, about 4 to 5 m, and reaches 130 to 140 mm in length. This species normally spawns at temperatures of 15 to 20°C during the period August to October. Although the smallest in size, this species has had the greatest success in culture.

H. discus, *H. gigantea sieboldii*, and *H. gigantea* are all southern species. *H. discus* inhabits a water depth of 4 to 10 m and grows to 170 to 250 mm in length. *H. gigantea sieboldii* and *H. gigantea* are the largest species, 200 to 250 mm in length and inhabit the deepest depth, 15 to 30 m. These three species prefer the same temperature range but, being southern species, spawn later than *H. discus hannai*, from mid September to November.

Table 1
THE YEARLY CULTURED
AND HARVESTED
ABALONE PRODUCTION
IN KOREA

	Production (mt)	
Year	Culture	Harvest
1962	—	369
1963	—	391
1964	—	1089
1965	—	449
1966	—	544
1967	—	414
1968	—	450
1969	—	318
1970	8.8	378
1971	18.6	553
1972	13.6	959
1973	9.8	2346
1974	1	517
1975	0.2	568
1976	7	615
1977	24	582
1978	—	419
1979	203	458
1980	98	540
1981	—	558
1982	82	453

FIGURE 1.   Abalone seed production station at Geoje.

Table 2
GOVERNMENT FISHERIES SEEDLING
PRODUCTION STATIONS AND THEIR
PRODUCTION CAPACITY

Station	Number of tanks	Water capacity (m³)	Year founded
Jeju	101	588	1972
Jumunjin	68	392	1978
Yeosu	57	249	1979
Pohang	38	153	1980
Geoje	162	644	1982

CONDITIONING

H. discus hannai grows to 7.5 cm in 5 to 7 years in northern Korea. However, Yoo et al.[1] reported that, when transplanted into southern warm waters, it grows faster and reaches 11 cm in 4 to 6 years. The degree of fatness, the ratio of gonad weight to total body weight, is highest from February to July and lowest in September to October.

Feeding activity is the highest in the spring and decreases during summer to reach the lowest in winter. When the kelp is given as food, feeding rate (ingested food amount/body weight of individual abalone) is 17.6% at 20.3°C in the spring and 7.5% at 10.9°C in the winter. Below 7°C, motility for feeding is hardly found. The adult abalones used for spawning are stocked in tanks and given abundant kelp. If they are reared under favorable conditions, the gonad will rapidly develop and spawning will occur about 2 weeks earlier than normal.

Adults which do not show active feeding during the spawning season will not have ripe gonads. Healthy individuals larger than 120 mm are preferred for spawning.

The average number of eggs released by an adult abalone is about 5.2 million in *H. gigantea* and 0.3 million in *H. discus hannai*. Effective accumulative water temperature (EAT), calculated from the first day of stocking in the tanks, is important for determining when the gonads will be mature. *H. discus hannai* usually spawns after artificial induction when it has experienced an EAT above 1500°C days and *H. gigantea* requires an EAT above 3500°C days.

Adult abalones are maintained at low temperatures to suppress spawning until it is required. Both sexes will mature fairly rapidly if the water temperature is raised to 20°C. Maturity can be easily determined by examination of the gonads, which are green in mature females and milky white in males.

The most active specimens are usually the best spawners. For spawning and fertilization, males and females (ratio of 1:4) are placed in 1.5 m × 2 m × 0.8 m indoor concrete tanks and induced by UV-irradiated seawater together with thermal shock and dessication. Raising the water temperature by 5 to 10°C for 50 to 100 min, followed by a return to the original water temperature is usually effective, but the procedure may need to be repeated.

FERTILIZATION

In still water, eggs and sperm tend to sink to the bottom of the tank. Excessive sperm may cause a poor fertilization rate by causing the loss of the egg membrane. The concentration of the sperm is maintained at less than 17.3×10^6 per mℓ. Quick fertilization after spawning is desirable since fertilization rate is dependent on the time elapsed after spawning and ambient water temperature. If fertilization is delayed until 3 hr after spawning, the fertilization rate is about 80% at 17°C but only 25% at 23°C.

The spawned adults are removed from the tank and their wastes are siphoned out.

LARVAL BIOLOGY

Fertilized eggs of *H. discus* discharge the first and second polar bodies 7 to 8 min after fertilization at 17°C. Cleavage begins 100 min after fertilization, the morula stage, a globular mass of blastomeres formed by cleavage of the egg, after 6 to 7 hr and the trochophore stage, cilia-rotating larvae within the egg membrane, after 13 hr. Hatch out occurs after 20 hr.

Hatched trochophore larvae are 170 μm (diameter) × 230 μm (length) and swim along the bottom of the tank. When a light is illuminated on the water, the trochophore larvae are concentrated on the surface since they are positive phototactic. It is relatively simple to collect larvae by filtering the surface water column with fine mesh when the water is changed and to remove dead or unfertilized eggs by siphoning out the bottom.

They develop into veligers with a shell 27 to 28 hr after fertilization. The larvae continue to swim around by ciliary movement until they begin to creep along the bottom after 6 days. During the creeping period cilia are shed and new shell is deposited around on the originally formed shell. The first respiratory pore appears about 130 days after fertilization in the young abalone 2.3 to 2.5 mm in size.

In *H. discus hannai* at 17°C, cilia form 8 to 10 hr after hatching, the veliger stage occurs at 24 hr, and the creeping stage after 2 or 3 days. When the young abalone reaches 1.8 to 2.0 mm in size, the first respiratory pore appears in 42 days.

The shell grows rapidly after settlement with individuals 3-mm long having two respiratory pores and individuals 3.7 mm having three respiratory pores.

During the larval period the larvae do not feed. After drifting stage of the larvae, sessile diatoms, such as *Cocconeis*, *Amphora*, and *Platymonas* spp., are eaten. As the larvae grow larger, they begin eating diatoms such as *Navicula* and *Licmophora*. When they are 3 to 5 mm, soft and small algae *Endarachne binghamiae* and *Ulva pertusa* are eaten.

Below the water temperature of 7°C, feeding activity of *H. discus hannai* is very sluggish. Young abalone will eat 7.5% of their body weight of kelp per day at 11°C, 14.7% at 16°C, and 17.6% at 20°C. The increments of body weight for a month at these temperature regimes are 12.8%, 27%, and 32.1%, respectively. The optimal temperature for growth of *H. discus hannai* is 15 to 20°C.

The amount of oxygen consumed at 23°C in *H. discus hannai* ranges from 52.2 to 85.3 mℓ per individual, but at 14°C decreases to 16.8 to 46.6 mℓ. Oxygen consumption is always higher during the night which is probably related to their nocturnal behavior. Below 24°C, oxygen consumption is usually proportional to temperature.

No difference in oxygen consumption is found with salinities above 25‰. However, below 23‰ the oxygen consumption decreases sharply with decreasing salinity.

Ino[2] stated temperature resistance of abalone larvae. Thirty-five-hour-old larvae of *H. gigantea sieboldii* show some resistance against temperature changes ranging from 8 to 28°C, with the tendency of strong resistance below 13°C. At 5°C and 32.4°C all the larvae tested died in 24 and 8 hr, respectively. He also reported that the normal larval development occurs in the salinity range of 24 to 36‰.

Yoo[3] stated the relationship between salinity and larval development. The optimal salinity for the normal larval development of *H. discus hannai* was 35‰. Most of the tested larvae showed abnormal development below 29‰. In addition, retarded development and intestinal corrosion of the larvae occurred at the salinity of 42‰.

GROWTH OF ABALONE

Growth is quite different depending on the species and the environmental conditions for seed production. The growth of *H. discus* and *H. discus hannai* produced in different locations

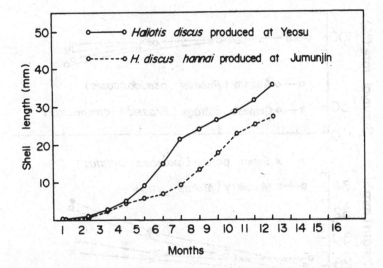

FIGURE 2. Growth of two species of abalones produced at Yeosu and Jumunjin Seedling Production Stations (From C. K. and Chung, S. C., *Bull. Mar. Resour. Res. Inst. Cheju Natl. Univ.*, 8, 41, 1984. With permission.)

is shown in Figure 2. *H. discus* seed produced at Yeosu Seedling Production Station at the end of May, grew to 20 mm by September, and 35 mm after a year. *H. discus hannai* seed produced at Jumunjin station in the middle of September reached more than 20 mm by the following June and about 30 mm after a year.

Marine algae, *Ulva* spp. and *Undaria* spp., are used as the principal food for the young abalones. Rho[4] tested ten species of terrestrial plants as potential food for young abalones of 20 to 50 mm in length. Chinese cabbage, mulberry, acasia, and sweet potato leaves were preferentially eaten (Figure 3). When the young abalones reach about 5 mm and begin the transition from eating diatoms to macroalgae, Chinese cabbage, and mulberry leaves can be used as an alternative food source to alleviate the seasonal scarcity of macroalgae. When feed efficiency is compared with growth rate of abalones fed on five species of terrestrial plants, the best food is chinese cabbage leaves followed by mulberry, bean, acasia, and ivy leaves (Table 3).

The survival rate from larvae to juveniles of about 3 cm in length, suitable size for stocking and release into the sea, is less than 1%. However, 10-mm abalone reared in tanks show a much higher survival rates of 50 to 100%. No known disease or parasite problems exist at this point and the major cause of mortality during settlement is believed to be a lack of suitable food. Seed production in Korean hatcheries is increasing every year.

Induction of spawning has always been a big problem to successful seed production of abalones. The annual seed production went up and down due to unskilled spawn induction. Experimental spawnings were attempted in spring 1974. By repeated trial and error, successful spawning was possible a month earlier than normal (March to May) by controlling adults in tanks.

Seed production was always unsatisfactory due to a lack of an optimal and suitable food supply. However, the survival rate of small abalones, 2 to 3 mm, has been consistently increasing since the discovery in 1982 of various suitable food sources, such as *Amphora*, *Ulva*, and terrestrial plants.

Abalones are raised to 35 to 40 mm in tanks using the methods described above and then transferred to plastic culture baskets covered with cloth netting which excludes predators. The baskets are hung in the open ocean using a long-line method (Figure 4). The growth rates are shown in Figure 5. The abalones have a higher survival rate in the culture baskets (85 to 90%) than the abalones released to the sea bottom (35 to 40%).

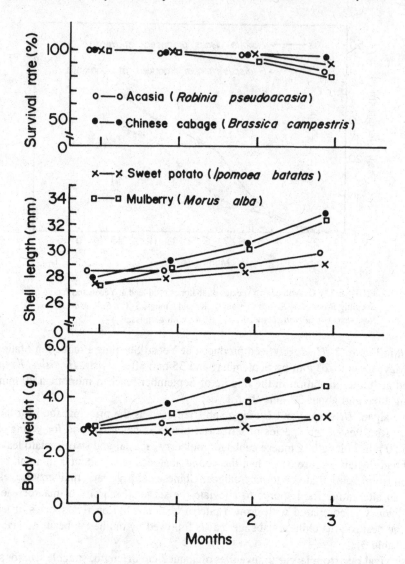

FIGURE 3. The effectiveness of terrestrial plants as food source for young abalone rearing. During the culture period, water temperature in the culture tanks ranged from 20 to 25.07°C. (From Rho, S., Master's thesis, University of Pusan, 1982, 1. With permission.)

The four commercial species have to be grown for 4 years in order to reach a marketable size.

SUMMARY

The abalone culture industry in Korea has come a very long way within a decade, from a dwindling overfished resource to a fishing and culture items. With the desire for this luxury seafood production, the government fishery bodies, cooperatives, and scientists reacted quickly and sensibly to the situation of decreasing natural stocks and catches. As a measure to increase abalone production, the Korean Fishery Administration is planning to establish six more seedling production stations in the major coastal areas.

Much more seed abalone produced in these stations will be distributed to the coastal farmers or released into the sea for recapture. To secure these released seeds, government prohibited catching individuals less than 7 cm in length.

Table 3
FOOD EFFICIENCIES OF TERRESTRIAL PLANTS FED YOUNG ABALONES ABOUT 28 mm IN SHELL LENGTH

Species	Daily increment Length (μm)	Daily increment Weight (mg)	Daily feeding rate (body wt%)	Feed coefficient[a]	Survival rate (%)
Bean	17.47	5.07	5.70	34.26	90.00
Chinese cabbage	58.79	28.25	10.79	16.56	100.00
Acasia	27.80	8.96	7.40	39.14	40.00
Mulberry	25.93	9.55	6.74	21.64	95.00
Ivy	6.92	−0.41	1.48	—	15.00
Algae (*Ulva lactuca*)	58.65	19.00	10.74	21.35	85.00

[a] Feed coefficient: food amount given for body weight increment/body weight increment.

From Rho, S., Master's thesis, University of Pusan, 1982, 1. With permission.

FIGURE 4.   Plastic culture baskets used for rearing abalones in the sea by hanging method.

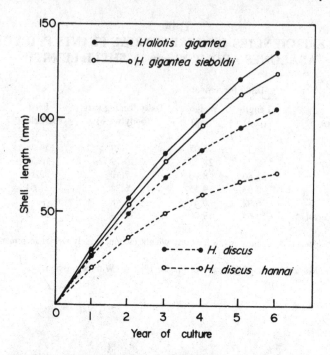

FIGURE 5.   The growth of four species of abalones for 6 years in three different places; *H. gigantea* and *H. gigantea sieboldii* reared in Jeju where water temperature ranged from 14.15 to 25.87°C; *H. discus* in Ilkwang where water temperature ranged from 11.30 to 22.29°C; and *H. discus hannai* in Ullungdo where water temperature varied from 9.73 to 23.82°C.

In order to provide shelters for these seed abalones about 2 cm in length against predators such as starfish, crabs, carnivorous fishes, and predatory molluscs, they are dropping lots of stones, waste ships, and various types of concrete materials shaped like apartments in the sea where natural foods are abundant. Fisheries scientists are attempting to mass produce abalones by feeding them land plants as a food source. Since marine algae as principal food are seasonally limited, using land plants will make a great contribution to the mass abalone production. Accordingly, many farmers are now interested in cage culture instead of release-recapture method.

Abalone seeds cultured in the plastic baskets and released in the sea bottom are preyed upon by various starfish, crabs, fishes, and predatory molluscs. This kind of predator is posing a great threat to abalone culture and is causing up 80% mortality. The Korean Fishery Administration is considering a policy to buy the starfish from the fishermen as an incentive to clearing farming beds. In addition to that, fucoids, calcareous sedentary polychaetes, polyzoans, ascidians, and bivalve molluscs can be a problem by causing shell damage and death if present in large numbers.

Cooperation between government bodies, fishery cooperatives, and researchers is focusing on the development of this new promising culture industry. The new awareness of genetic improvement for fast-growing abalones is rising for the future of our abalone culture industry.

REFERENCES

1. **Yoo, S. K., Park, K. Y., and Yoo, M. S.,** Biological studies on abalone culture. I. Growth of the *Haliotis discus hannai*, Bull. Natl. Fish. Univ. Pusan, 18, 95, 1978.
2. **Ino, T.,** *Biological Studies on the Propagation of Japanese Abalone (Genus Haliotis)*, Tokai Shobo, Tokyo, 1953, 108.
3. **Yoo, S. K.,** Abalone culture, in *Mariculture*, Saero, Seoul, 1979, 309.
4. **Rho, S.,** The utilization of terrestrial plants as food of young abalone *Haliotis discus hannai*, Master's thesis, University of Pusan, 1982, 1.
5. **Pyen, C. K. and Chung, S. C.,** On the rearing and growth of young stage of the abalone, *Haliotis discus.*, Bull. Mar. Resour. Res. Inst. Cheju Natl. Univ., 8, 41, 1984.

FARMING THE SMALL ABALONE, *HALIOTIS DIVERSICOLOR SUPERTEXTA,* IN TAIWAN

Hon-Cheng Chen

INTRODUCTION

Abalone, famous for its delicacy and high market value, has attracted tremendous interest in many countries and its cultivation is considered one of the most promising businesses. In 1972, there were only two nations studying abalone aquaculture.[1] Improvement of hatchery and culture technology, and expansion of culture area has continued since this time.[2,3] Consequently, abalone aquaculture has rapidly developed and production has greatly increased, particularly in Taiwan and Japan.

The species cultured in Taiwan is the native small abalone, *H. diversicolor supertexta,* which is completely different from the Japanese black abalone, *H. discus.*[4] In Taiwan, people prefer small abalone to other species of *Haliotis* because of its palatability, proper size for banquets, commercial value, and other favorite qualities. In addition, there is an unfounded belief that small abalone can be used as a medicine to cure eye disease or can be used as a tonic to increase body strength.

Aquaculture of small abalone in Taiwan started many years ago. At first, it was limited to small-scale operations or served as short-term stocking to wait for a better market price.[5] Knowledge of the biology and culture technology of small abalone was scarce at the time. Total annual production did not increase significantly and even decreased before 1976. Most ot the production at this time came from fishing.

Due to the increased demand for small abalone and the soaring market price, studies were begun to develop technology for artificial propagation and mass seed production.[6] Many commercial hatcheries were established to produce millions of seed abalone for grow-out culture (Figure 1). Scientific research, along with valuable culture experience, contributed tremendously to the development of advanced culture technology. This greatly improved survival and growth of the abalone, and resulted in a marked increase of total production in recent years.

This chapter will describe the techniques developed to culture *H. diversicolor supertexta,* because the cultivation of small abalone is unique to Taiwan and the culture techniques are quite different from those used to culture other species of *Haliotis.* This chapter will present culture techniques, hatchery operation, general biology, growth requirements, problems and constraints, and developed strategies for culture. The techniques presented may also be useful to culture similar species in other tropical countries.

HISTORY OF DEVELOPMENT, PRESENT STATUS, AND PRODUCTION

Culture of small abalone began at a very small scale 20 years ago, when the price increased very dramatically and fish farmers realized that small abalone could survive for a period of time in captivity. Adult or juvenile abalone were caught from the littoral zone and stocked in enclosed rocky pools with limited food to wait for a better market price or for further growth. However, there was poor survival and slow growth rate due to a lack of rearing techniques. In addition, there was a shortage of seed. These factors caused very little progress in pond culture for more than 10 years. The gradually increasing annual production of small abalone before 1971 was simply due to fishing in natural coastal waters (Figure 2). In 1977, seed production and rearing methods were developed in some commercial farms.[6] The demand for abalone seed was solved and the pond culture area greatly expanded. Since 1977, harvest in coastal waters has become unimportant to the total annual production.

FIGURE 1. Commercial hatchery at Fu-Long producing 2 million seed abalone annually.

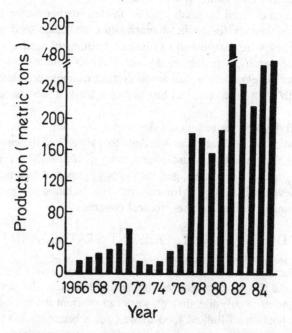

FIGURE 2. Total annual production of small abalone *H. diversicolor supertexta* in Taiwan.

A rapid increase in production occurred in 1981 but it decreased again in 1982. This unusual event was caused by many fish farmers, after several years of experience and making large profits, increasing the number of culture ponds along the rocky coast, and stocking more juveniles. However, most of the new culture ponds were built without permission from the government. These ponds were banned from use and some were even destroyed. This caused the production to fall the following year. Aside from this unusual increase, total production in the last 2 years has been higher than in 1977 to 1980. The main reasons for this higher production are advanced propagation operations, rising market price, and improvement of culture techniques. At present, there are more than 10 ha for pond culture and at least 25 hatcheries producing a total of 35 million seed abalone.

Experimental or small-scale sea ranching of small abalone was also carried out along the rocky coast of eastern Taiwan.[7] Although sea ranching produces a slower growth rate than pond culture, it is still cheaper and easier. Pond culture is hindered by construction costs and limited intertidal area suitable for use.

Fish culturists in southern Taiwan have tried to raise this species in land-based cement ponds with pumped sea water. Water temperatures in the summer afternoon are usually very high. Survival and growth are fairly satisfactory, although not as good as those in intertidal rocky ponds. Small abalone are reared in 32°C water with no ill effects or reduction in growth.[8] It is still worth expanding the onshore culture system in southern Taiwan because of limited available area and government regulations on use of the rocky coast.

Another possibility to increase production is cage culture in sheltered areas where damage from typhoons and monsoons would be minimal. The feasibility of cage culture is still being investigated since there are still problems with predator control and proper substratum.

BIOLOGY AND ECOLOGY

H. diversicolor supertexta, a subtropical or tropical species, is the most abundant abalone species found in Taiwan. The distribution is not very wide — southern Japan, Korea, and Taiwan.[9] Small abalone prefer behind, below, and crevices of rocks in the sublittoral zone (<10 m) on exposed coasts where a variety of seaweeds flourish. Adult abalone, especially ripe females, will aggregate in shallow water during the spawning season. This is thought to increase fertilization, normal development, and larval survival. Small abalone, like other species of *Haliotis*, are nocturnal and often move to deeper water in the summer.

Newly settled juveniles feed on benthic diatoms and green microalgae but switch to eating seaweeds after formation of the first respiratory pore. Small abalone prefer soft seaweeds to hard seaweeds in this order of preference; *Scytosiphon* spp., *Ulva* spp., *Gracilaria* spp., and *Sargassum* spp.[10] Shell coloration differs according to the alga eaten, green after eating Chlorophyceae and brown after eating Rhodophyceae.[11]

Almost all small abalone are dioecious. Usually, the color of the mature gonad is dark green in females and milky white in males. However, occasionally (<1%) hermaphroditic mature abalone are found. The gonad of these animals is green with a white tip at the end of the conical appendage. When induced to spawn, these animals first eject sperm, then release eggs for a while, and later release sperm again (Figure 3). Eventually, the eggs are self fertilized and develop into normal veligers. *H. diversicolor supertexta* becomes sexually mature in nature at 3.0 to 3.5 cm depending on location and abundance of food.[12]

Number of open respiratory pores increases as the animal increases in size, 7 to 8 pores are found in an individual of 6 to 7 cm shell length. Small abalone grow to a maximum length of 10 cm. Although it is not known how long small abalone live, estimates from shell bands indicate they can live at least 7 years.

Eggs are spherical, dark green, and 200 to 220 μm in diameter. Normal eggs have an outer layer of jelly. Eggs without this jelly layer are considered abnormal and discarded.

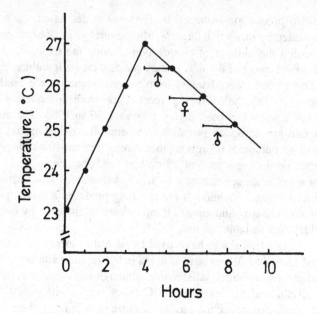

FIGURE 3.   One hermaphroditic mature abalone ejecting sperms first and then releasing eggs while kept individually and induced to spawn.

Table 1
EMBRYONIC DEVELOPMENT OF SMALL ABALONE AT DIFFERENT TEMPERATURES

Embryonic stages	Water temperature °C				
	19	**21**	**24**	**27.5**	**30.5**
2-Cell	20 min	15 min	10 min	5 min	5 min
4-Cell	50 min	45 min	30 min	25 min	20 min
8-Cell	2 hr 20 min	2 hr	1 hr 50 min	1 hr 30 min	1 hr 10 min
16-Cell	3 hr 50 min	2 hr 15 min	2 hr 40 min	2 hr 10 min	2 hr
Morula	6 hr 20 min	4 hr	3 hr 50 min	3 hr 20 min	2 hr 30 min
Blastula	8 hr 10 min	6 hr 40 min	5 hr 20 min	4 hr 50 min	3 hr 20 min
Gastrula	10 hr 50 min	8 hr 20 min	7 hr 50 min	7 hr 10 min	4 hr
Early trochophore	12 hr	10 hr 25 min	9 hr 20 min	7 hr 45 min	6 hr
Late trochophore	13 hr	11 hr 50 min	10 hr	8 hr 40 min	6 hr 50 min
Hatching	14 hr	12 hr 20 min	11 hr	9 hr 10 min	7 hr 30 min

Egg cleavage and embryonic development is classified according to the larval stages described by Oba.[13] Larval development is greatly affected by salinity and water temperature.[14] Hatch out occurs 11 hr after fertilization at 24°C (Table 1). Planktonic larvae pass through trochophore and veliger stages, and are ready for settlement after 4 days. At 24°C, the first respiratory pore is formed at 1.8 mm (30 days) and the fifth pore appears at 2.5 mm (1.5 months).[15]

There are many predators on the small abalone. Planktonic larvae and newly settled juveniles are often eaten by various fishes (*Gobius* sp., *Awaous* sp.) and crustaceans (*Tisbe* sp., *Gammarus* sp.). Larger individuals are preyed upon by *Panulirus* sp., *Thalamita* sp., *Octopus* sp., black porgy (*Acanthopagrus* sp.) and sea stars.[6] Limpets, chitons, oysters, mussels, polychaetes, and serpulid worms are commonly found in the same habitat with small abalone, and compete for food and space. So far, there are no effective methods to prevent these co-occurring organisms.

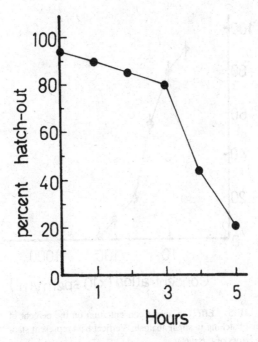

FIGURE 4. Effect of preservation of sperm on the percent hatch out of small abalone.

HATCHERY TECHNIQUES

Suitable spawning conditions occur from October to January when the water temperature is between 20 to 25°C. However, the best water temperature (22 to 24°C) occurs during October and November. These water temperatures reduce abnormality, produce healthy larvae, allow normal metamorphosis, and enable easier handling. Fully mature individuals with swollen gonads and 5.8 to 7.0 cm shell length are selected as brood stock. The most active individuals that have grown in intertidal ponds are usually the best spawners.

First, the abalone are exposed to air under mild sunshine for more than 30 min. They are placed into tanks with flowing UV-irradiated sea water as described by Kikuchi and Uki.[16] Thermal shock combined with the above treatment usually gives the best results. They usually spawn within 4 to 6 hr after treatment. However, during the prime breeding season, simply increasing the water temperature by 1°C/hr for 4 hr and then decreasing to ambient in 3 hr is sufficient to induce spawning. Rapidly changing the water temperature up and down causes the abalone to be reluctant to spawn, especially for individuals less than 5.8 cm. It is better to spawn the animal immediately after removing it from the grow-out ponds because delay or transporting long distances always causes a low percent spawning and larval survival. Pond-reared females have larger swollen gonads and are better to use than wild-caught females. Small abalone readily release their gametes, unlike other species of *Haliotis* which are more difficult to spawn, and therefore do not require the addition of hydrogen peroxide,[17] ammonia, or gonad extract. For normal fertilization, males and females (1:5) are kept separate during spawning. The number of eggs spawned by an individual female, 6.0 to 7.5 cm shell length, is between 130,000 to 200,000.

Newly spawned eggs must be fertilized within 1 hr after spawning because delay will result in a low percent fertilization. The number of normal larvae after hatch out decreases sharply when sperm is used that is more than 3 hours old (Figure 4). Sperm concentration of 45,000/mℓ produces a high percent hatch out (Figure 5). A high sperm concentration of 720,000/mℓ usually causes polyspermy which results in only 13% normal hatch out. Similar phenomena and effects have been reported by a few workers.[11,18]

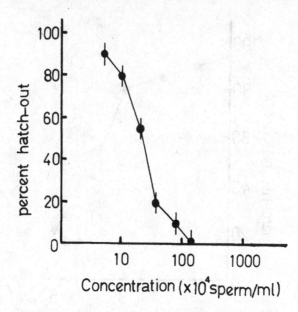

FIGURE 5.   Effect of sperm concentration on the percent of normal hatching in small abalone. Vertical bars represent standard errors of the mean.

After fertilization, the hatch-out tanks contain large numbers of sperm, dead eggs, excreted mucus, and metabolites from the spawners as well as microbes. This results in a large quantity of abnormal larvae and poor survival. Washing the fertilized with filtered sea water avoids fouling in the hatch-out tanks and is one of the most important methods to ensure a high percent hatch out. Increasing the number of washing will raise the percent of hatch out. The eggs are washed at least five times. A continuous flow of sea water or mild aeration until hatch out is even better for producing healthy larvae (Figure 6).

At 24°C, larvae become early veligers 3 hr after hatch out and begin swimming near the water surface. Swimming veligers are not fed, but take naturally occurring plankton in the water as nourishment. They are collected and placed in indoor settlement tanks (3 m × 1.5 m × 1 m) at a density of 1 to 3 larvae/m ℓ. Twelve to 18 hr later, corrugated plastic sheets (75 cm × 60 cm or 60 cm × 45 cm) that have been previously immersed in outdoor seawater ponds to grow a layer of benthic algae, are hung vertically in the settlement tanks as seed collectors. The benthic algae are usually diatoms (*Navicula* spp., *Nitzschia* spp., *Surirella* spp., *Asterionella* spp., *Amphora* spp., *Chaetoceras* spp.), green algae (*Chlorella* spp., *Enteromorpha* spp.) and blue-green algae (*Phormidium* spp., *Oscillatoria* spp.). There is about 5 cm between each sheet and 10 cm between the bottom of the sheet to the bottom of the tank. The sea water in the settlement tanks is not changed and light intensity is kept below 100 lx for the first few days to ensure an even settlement of larvae. The sheets with newly settled seed abalone are usually moved to outdoor tanks after 1 week. Addition of γ-aminobutyric acid (GABA) to induce rapid settlement, as suggested by Morse et al.,[19] is not practiced, but insulin is applied to hasten the growth of seeds.

MANAGEMENT OF NURSERY POND

Four days after hatching, all seeds are firmly attached on the sheets and cannot be washed away by the water current. Therefore, running sea water and strong aeration is begun in the tanks. After the newly settled seed are shifted to outdoor nursery ponds, running sea water, or in combination of strong aeration, is provided to ensure better growth of the seed. Water temperature and salinity are the two major factors influencing survival and growth. In northern

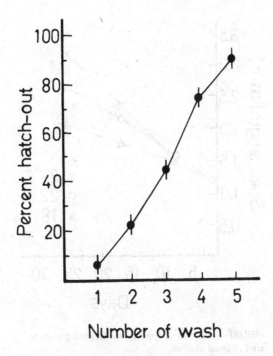

FIGURE 6.   Effect of wash frequency of fertilized eggs on normal hatching in small abalone. Vertical bars represent standard errors of the mean.

Taiwan, production of seed and juveniles (<3 mm) is carried out in the middle of winter. During this period, rain and very cold weather begins, nursery ponds are covered with plastic or PVC sheets to prevent temperature fluctuation and keep salinity constant. It is best, if possible, to maintain the water temperature at 22 to 27°C and salinity at 32 to 35‰.

The seed density on each corrugated sheet varies according to animal size, season, pond management, and culture techniques. Optimal survival and growth is obtained at densities of 350 to 1500 individuals per sheet. If too crowded, the population is thinned down by connecting a new sheet to the bottom of the original one and allowing the seed to disperse. In 2 to 3 months, the seed grow into juveniles (3 to 5 mm). The juveniles are large enough at this point to be transferred (by brushing them off the sheet) into other nursery ponds. Cut rocks or oval stones are now placed on the bottom of the nursery ponds to serve as shelters. Chopped fragments of *Gracilaria* sp. and *Ulva* sp. are supplied until the juveniles reach 15 mm (late spring) and are the correct size for stocking grow-out ponds. Juveniles (<15 mm) are stocked in nursery ponds at 3500 to 7000/m² and still show fairly good growth.

GROWTH REQUIREMENTS

Abalone growth can be affected by a number of external and internal factors.[20] Factors which are controllable should be properly maintained with the optimal ranges in order to ensure fast growth. Larvae hatch out without a shell gland and die within 1 or 2 days when the fertilized eggs are reared at 30.5°C. Shell growth of seed reared in sea water of normal salinity is retarded at 18°C and the fastest growth is found at 27°C (Figure 7). Hence, optimal temperature for veligers and seed is 22 to 27°C. Juveniles or subadults grow best at temperatures from 24 to 30°C (Figure 8). Furthermore, rapid growth of subadults and adults in pond culture occurs when the water temperature increases to or exceeds 26°C during the early summer. Therefore, the older the individual, the higher the optimal temperature. In addition, small abalone have a higher optimal temperature than other cold water species (*H. discus hannai*).[20,21]

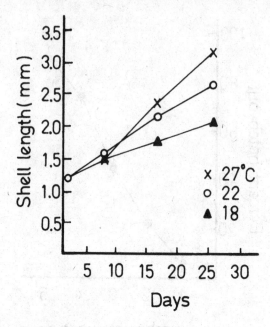

FIGURE 7.   Effect of temperature on shell growth in seed of small abalone.

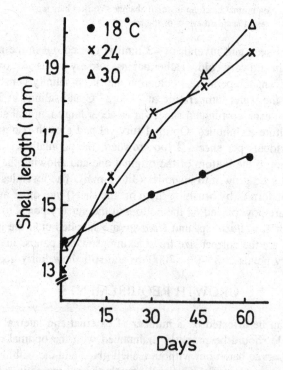

FIGURE 8.   Growth of juvenile small abalones at three different temperatures in 35‰ for 60 days.

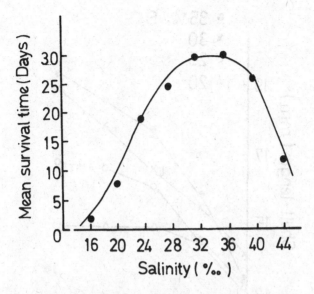

FIGURE 9.   Mean survival time (number of days required to cause 50% mortality) of seed abalone in different salinities.

Salinity has a very significant effect on growth and survival, because small abalone is a typical stenohaline species. Seed abalones can tolerate a slightly wider range of salinities than embryos,[14] but both cannot survive below 24‰ (Figure 9). Growth of juveniles is also reduced with decreasing salinity and they die after 60 days in 25‰ or 30 days in 20‰ (Figure 10). Salinities between 32 to 35‰ are considered optimum for embryo survival and pond growth. All hatcheries and ponds are located away from estuaries or river.

Small abalone are very sensitive to reduced levels of dissolved oxygen. Small abalone will crawl out of the water and be exposed to air during warm mornings when the dissolved oxygen decreases to 1.5 ppm. In many cases, small abalone shipped long distances in water usually die from lack of oxygen and careless handling. A small abalone (5.0 cm) consumes oxygen at 60 mg O_2/kg/hr at 23°C.[22] In addition, the critical oxygen level needed for life is more than 4 mg/ℓ O_2.[8,23] Pond water should be changed frequently with clean sea water and aerated for best survival.

Small abalones consume more *Ulva* spp. and *Enteromorpha* spp. than *Gracilaria* spp. when provided with a mixture of these algae.[24,25] When algae are offered separately, it is interesting to find that *Gracilaria* sp. produces the best growth, followed by *Ulva* sp. and *Enteromorpha* sp. (Figure 11). An artificial diet, a newly developed high-priced product consisting of fish meal and soybean meal, produces the least growth. The main reason for poor growth with the artificial diet is due to the diet dissolving very easily in the water and consequently polluting the water. No abalone farms in Taiwan use artificial diets at the present time. Although *Gracilaria* spp. is undoubtedly good for growth, it is not known how it affects gonad development and maturation. Experiments are in progress to study the effect of *Gracilaria* spp. on gonad maturation.

Hydrogen sulfide is notoriously toxic to aquatic organisms and is produced by the reduction of sulfate in anoxic water or decomposition of dead animals, food residues, excreted metabolites, and feces. Abalones are very sensitive to hydrogen sulfide and a concentration as low as 0.05 ppm will retard growth of juvenile abalone. Juveniles 9 to 12 mm in length and exposed to a concentration of 0.3 ppm hydrogen sulfide, are alive but show very little growth (Figure 12). At 0.5 and 1.5 ppm, juveniles survive for only 84 and 16 hr, respectively.[11] Ammonia is another toxic substance usually present in grow-out ponds when sea water is not changed. Although its toxicity to small abalone is less than that of hydrogen

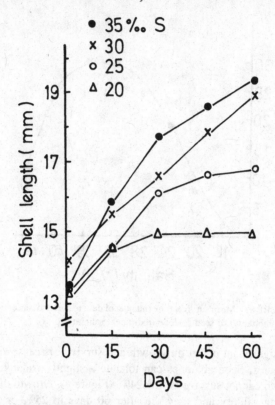

FIGURE 10. Growth of juvenile small abalones in four different salinities at 26°C for 60 days.

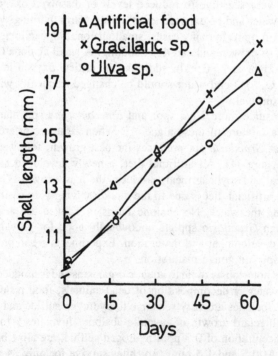

FIGURE 11. Growth of juvenile small abalones fed with different food for 60 days.

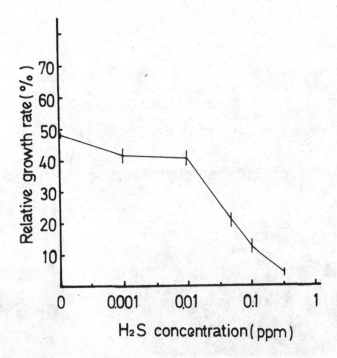

FIGURE 12. Growth of small abalone exposed to various concentrations of H₂S for 30 days. Relative growth rate means percentage of increment gained after 30 days over original shell length. Standard errors of the mean are represented by vertical bars.

sulfide, a concentration of 0.5 ppm can still retard their growth. Therefore, it is necessary to minimize chronic effects of hydrogen sulfide and ammonia on growth by regular cleaning of bottom detritus and constant exchange of sea water.

Small abalone can tolerate a wide range of pH, 6 to 9, but growth is reduced at the two pH extremes. Therefore, it is best to keep the pH of the pond water as near pH 8 as possible. Although they can still survive for more than 4 days in 0.1 ppm copper or 0.1 ppm mercury, these metals will accumulate within the viscera and body tissues. Pollutants in the water should be avoided to produce good quality small abalone, and prevent toxic concentrations for human consumption.

Different substrata were used to test their effects on growth. It was found that cut rock with a coarse surface is the best, followed by smooth-surface oval stone, and semicylindrical PVC pipe. PVC pipe can be moved or shaken when the water velocity is increased and hence, is not considered a safe shelter for use.

SITE SELECTION AND POND CONSTRUCTION

The severe effects of temperature and salinity on growth of small abalone have been mentioned previously. Thus, site selection must be done carefully to avoid areas of rapidly changing salinity or low salinity. The most suitable areas are along the rocky coast of northeast and eastern Taiwan where big rivers are not present, and sea water is always very clean and has normal salinity.

Two pond systems with different sea levels are used in the production of small abalone but most ponds are constructed in the intertidal zone (Figure 13). Nowadays, due to regulations on use of tidal areas, quite a few ponds are set up with concrete walls on shore at some distance from the sea, like nursery ponds but slightly larger in size. Onshore ponds need more electricity and equipment for pumping and delivering clean seawater, which

FIGURE 13.   Two types of grow-out ponds. Intertidal pond is in the distance with splashing waves at offshore wall, and onshore pond is installed with underwater aerator to increase the dissolved oxygen.

results in higher operating costs. Also, the daily water temperature fluctuates more in onshore ponds than in intertidal ponds. The water temperature in onshore ponds very often increases to more than 33°C during summer afternoons. Temperature fluctuations and extremely high water temperature in onshore ponds can cause retardation of growth. Therefore, for all of the above-mentioned reasons, onshore ponds are not very productive, although they are good for harvesting.

The best intertidal ponds are located on a protruding cape on a very exposed rocky shore. Incoming waves carry clean seawater through pipes in the walls from one end and drive out old pond water from the other end. This provides good water circulation that clears away food residues and toxic substances. The pond is built with the bottom slightly higher than mean low tide level of neap tide to facilitate draining during cleaning. Cut rocks and oval stones are placed neatly on the concrete bottom as shelters (Figure 14). To avoid damage from typhoon and monsoon waves, most abalone ponds have strengthened concrete offshore walls wider than 1.5 m. Several 300-cm plastic pipes are installed horizontally at various heights on the offshore walls to act as inlet or outlet and to adjust the water level in the pond. Pond sizes vary according to topography, available area, and investment but are usually in the range of 0.1 to 0.5 ha for the ease of management, cleaning, and harvesting.

GROW-OUT OPERATION

Stocking begins in spring and early summer when the sea water temperature has increased to more than 22°C and juvenile abalone have reached 15 mm or longer in the nursery ponds. In advance, intertidal ponds are cleaned by draining off the water to wash down the bottom and rocks to get rid of clay sediment, fouling organisms, predators, and toxic substances. It is best to clean the bottom every 2 months during the rearing period. A stocking density of 150 to 250/m^2 is generally practiced. Sometimes small quantities of wild-caught subadults

FIGURE 14. Cut rocks (top) and oval rocks (bottom) used as shelters on the concrete bottom of the grow-out pond.

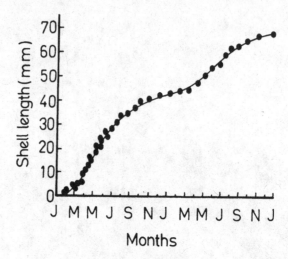

FIGURE 15.   Monthly growth curve of small abalone, *H. diversicolor supertexta*, cultured in grow out ponds in Taiwan.

of a larger size than the juvenile abalone are also planted in the same pond to utilize the carrying capacity until the juveniles grow larger. The wild subadults are harvested when the juveniles need extra rearing space. Stocking rates over 500/m² causes detrimental effects and poor survival. It takes 6 months for juveniles to reach a market size of 6 cm shell length at a high stocking density (400/m²) but only 4 months is needed at a low stocking density (200/m²).[8] Since it is not always possible to select the best sites for grow-out, intertidal ponds with very limited exchange of sea water are stocked at a low density and bottom residues are regularly cleaned.

The growth rate of small abalone in grow-out ponds depends on their initial size with small individuals growing faster than large individuals. Shell growth after 6 months, for 13-mm and 45-mm abalones were calculated to be 29 mm (223%) and 15.5 mm (34%), respectively. This indicates that larger abalone are still worth further growth in pond culture.

Percent survival depends on stocking density, animal size, and pond management, but is usually more than 70%. Small juveniles (5 mm) were previously stocked but resulted in very poor survival. It is generally recognized, the bigger the juvenile the higher the percent survival. The ideal size for stocking is 25 mm, which is the same size as wild-caught subadults. If it is not possible to get enough quantities of this size, then smaller individuals (>15 mm) are used and they still have a high survival rate.

Gracilaria sp. is one of the economically important algae cultured in southern Taiwan and can be purchased cheaply in large quantities.[8,26] Although many species of seaweeds can be used as food, *Gracilaria* sp. is the only algae which is feasible for commercial use. For example, *Ulva* sp. is harvested only in the summer and is also tedious, time-consuming work. Therefore, *Gracilaria* sp. is fed exclusively every other day by scattering it evenly into the grow-out pond. While feeding *Gracilaria* sp., crabs hiding in the fronds can be removed and killed to avoid predation. Conversion ratio for *Gracilaria* sp. calculated from experimental data, although very poor, was still 12:1. This rate has satisfied most abalone farmers because the cost of food is at most 8% of the market price of small abalone.

Small abalone grow fastest in the warmer months of April to September while slower growth occurs from October to March (Figure 15). This is also true for field-stocked abalones in ocean ranching.[7] Although this species is small in size, its growth rate is very encouraging with 4.3 cm after 1 year and 6.9 cm after 2 years in pond culture. However, it only grows to 3.9 cm after 1 year and 5.4 cm after 2 years in nature.[7] Growth of this species is even slower in Japan, growing to 2.5 to 3.0 cm after 1 year.[13] Annual yield per unit area was 3 kg/m^2 in 1981 but it has now increased to 4.0 kg/m^2 due to improved techniques and higher survival rate.[8]

After stocking, many crabs are present in the ponds and they cause heavy predation on subadult small abalones. The crabs are killed or driven out by applying dipterex or sumithion during low tide and the residue washed out thoroughly with the incoming high tide. This method still enables normal growth of the abalone. Another method usually used is to directly kill the crabs by SCUBA diving, but is is not very effective and must be carried out periodically.

Harvesting of market-sized abalones begins 4 months after stocking the large wild subadults or 6 months after planting the hatchery-produced juveniles. Market size varies depending on preference and season but it is usually larger than 4 cm. Market price fluctuates all the time depending on the amount of supply and spawning season, ranging from $24 to $45 (U.S.) per kilogram. The larger the animal, the cheaper the unit price. The abalone are harvested by either emptying the pond or SCUBA diving in large ponds. A diver can collect about 50 kg in 2 hr. Harvesting occurs when the demand is good and the price is high.

DISEASE

A high percent survival is obtained in grow-out ponds because infectious and epidemic diseases are not a serious problem. Although infection by *Vibrio parahaemolyticus* is the main cause of death in *H. discus hannai*[27] and possible pathogens, *Vibrio* spp., *Pseudomonas* spp., *Flavobacterium* spp., and *Achromobacter* spp. are present in the rearing water, small abalones inoculated with these pathogens survived.[28] This indicates that these pathogens are not infectious and *H. diversicolor supretexta* is a very hardy species.

Most mortality of small abalone occurs right after stocking into nursery and grow-out ponds. This may be due primarily to the combined effects of stress, mantle injury while removing individuals off the corrugated sheets, and lack of oxygen during air shipment.

Although no serious diseases have been found so far in pond culture, young and juvenile abalones frequently have split shells along the respiratory pores (Figure 16). The percentage of split shell individuals can be 50% of the population, but the percent varies in different ponds. Growth of split-shell abalone is much slower than normal abalone. Split-shell abalone are only 2 cm by the time when normal abalone are 5 cm. Furthermore, split-shell abalone are vulnerable to predation due to lack of protection and this causes a low percent survival of these animals.

The cause of this splitting is not completely understood. There are split shells in individuals with uninjured mantle, however, 25% of juveniles with injured mantle have split shells.[11] Carelessly brushing off juveniles from corrugated sheets injures the mantle and is probably the major cause for the split shell.[8,26] High temperatures during induced spawning and high concentrations of toxic substances in the pond may also contribute to shell splitting. Although hydrogen sulfide will retard growth of small abalone, it will not cause this abnormally.[8,11] Small splits can be healed after 2 months if rearing conditions are improved but large splits are unaffected.

Chemical content in normal and split-shell individuals were analyzed. There was a marked depression in calcium and zinc concentrations in split-shell individuals (Table 2). Chen[29] found that after hatch out, addition of calcium bicarbonate can speed up the formation of the veliger shell. Shell growth of juvenile abalone can also be increased by the addition of calcium chloride.[30]

It is hoped that application of dissolved calcium will solve the split-shell problem and, in addition, provide a better rearing environment. Brushing off seed into rearing ponds is tedious, time-consuming work, and mantle injury is probable with careless handling. Therefore, anesthetic drugs and thermal shock should be used to avoid this problem. This has now been applied in some private hatcheries.

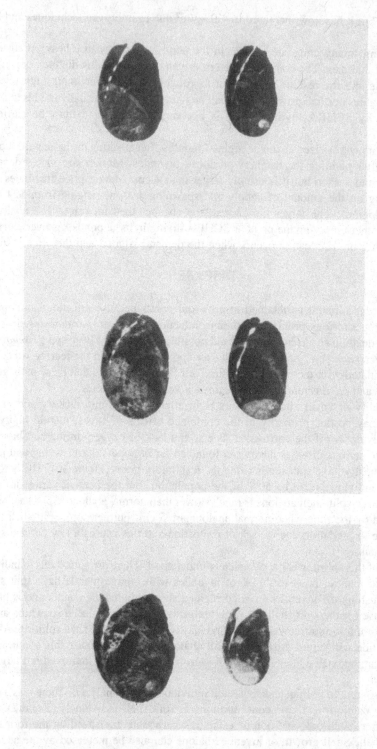

FIGURE 16.   Abalone shells split along the respiratory apertures. Minor split (top) will heal if favorable rearing conditions are provided (middle), but shells with large split (bottom) will not heal.

Table 2
THE CONTENT OF Ca, Fe, Mg, Cu, AND Zn IN SPLIT AND NORMAL SHELLS OF SMALL ABALONE FED WITH DIFFERENT FOODS

	Normal shells (ppm)			Split shell (ppm)
	Artificial food	Ulva sp.	Gracilaria sp.	Gracilaria sp.
Ca	871,00	833,600	743,800	422,100
Fe	115.9	152.3	131.3	141.4
Mg	320.1	326.1	291.5	370.2
Cu	26.0	26.3	21.3	18.4
Zn	25.5	21.0	27.0	8.9

PROSPECT, PROBLEM, AND RESTRAINT

Aquaculture in Taiwan is noted for its diversity.[31] Many species have a high market price which is especially true when total production is low and does not meet the urgent demand. If total production increases very rapidly, resulting from developed aquaculture techniques and expanded culture areas, the price has to decline due to the limited market and cultivation of this species will no longer be a profitable business. This is true for many species, such as milkfish, Chinese carps, grass shrimp, and hard clam.[32] However, small abalone, as well as a few other species do not follow this trend and the market price has remained sky high. The price has fluctuated between $30 to $40 (U.S.) per kilogram (live weight) in spite of increased total production in recent years. This price is still the highest of the species cultured in Taiwan. In addition, the annual rate of return on investment is usually 50% with a high of 161%, depending on the percent survival, duration of operation, investment cost, and sale price. This profitability makes the cultivation of small abalone one of the most promising industries and more people are expected to enter this field in the near future.

Due to the limited area of rocky shores and regulations on use of intertidal land, it is unlikely production will increase very rapidly. Therefore, scientific research on advanced technology in cage culture and onshore farming are still worthy of further study. Other possible methods to increase production include developing new strains by selective breeding or hybridization with other species to get faster growth and better tolerance to changing environments.

Besides *Gracilaria*, other algae have to be studied to ascertain their effect on gonad maturation, and production of normal eggs and healthy larvae. Further development is restrained by increasing industrial and domestic pollution from discharge of untreated effluent. Control of split shell and predators has to be intensively studied to reduce mortality and to increase growth rate. Finally, the importation of another small abalone from a foreign country into the Taiwan market has to be prohibited. It is generally recognized in order to further expand this enterprise, social and economic problems have to be solved, and more advanced culture techniques have to be developed.

SUMMARY

Culture of the small abalone, *H. diversicolor supertexta*, is conducted exclusively in Taiwan and has become one of the most successful industries in recent years. Annual production, as well as culture area has increased since 1977. Newly developed rearing techniques and mass production of 15-mm hatchery-reared juveniles are the main reasons contributing to these achievements. Intertidal ponds are stocked with juveniles at 200 to

250/m² in spring and annual yields of 4.0 kg/m², at a size of 20 to 30 g each, are reported from most ponds. Survival in grow-out ponds varies, but it is usually more than 70% because there are not serious disease problems. However, abalones in certain hatcheries quite often suffer from the shell being split along the respiratory pores that is caused by the mantle being injured when the juveniles are brushed off the plastic sheets for transfer. *Gracilaria* is the main food given to the juveniles because it is abundant, inexpensive, and has a conversion rate of 12:1. Its effect on inducing gonad maturation remains uncertain. Cut rocks or oval stones are placed on the pond bottom as shelters, and regular cleaning of the ponds reduces toxic substances and ensures fast growth. The optimal water temperature (24 to 30°C) and salinity (30 to 35‰) are required for high survival and fast growth. Small abalones with favorable conditions in grow-out ponds can reach 4.3 cm in shell length after 1 year and 6.9 cm after 2 years. Ocean ranching will produce 3.9 cm after 1 year and 5.4 cm after 2 years. Spawning occurs in the early winter when the water temperature is 22 to 24°C. Spawning is induced by a combination of exposure to air, thermal shock, and UV-irradiated sea water. Sperm concentration is maintained at 22,000 sperm/mℓ during fertilization and fertilized eggs are washed more than five times to ensure a high percent hatchout.

There are still many unsolved problems restraining the further development of abalone culture in Taiwan. The most important problems are limited suitable grow-out areas, regulations on use of intertidal land, industrial pollution, and lack of an export market. Therefore, cage culture, onshore farming, and ocean ranching are under consideration, although they are not as productive as intertidal ponds. Other culture methods are suggested, including selective breeding and development of new strains.

REFERENCES

1. **Bardach, J. E., Ryther, J. H., and McLarney, W. O.,** *Aquaculture,* Wiley-Interscience, New York, 1972, 777.
2. **Leighton, D. L.,** The influence of temperature on larval and juvenile growth in three species of southern California abalones, *Fish. Bull.,* 72(4), 1137, 1974.
3. **McCormick, T. B. and Hahn, K. O.,** Japanese abalone culture practices and estimated costs of juvenile production in the U.S.A., *J. World Maricult. Soc.,* 14, 149, 1983.
4. **Kafuku, T. and Ikenoue, H.,** Abalone *(Haliotis (Nordotis) discus)* culture, in *Modern Methods of Aquaculture in Japan, Developments in Aquaculture and Fisheries Science,* Vol 11, Kafuku, T. and Ikenoue, H., Eds., Kodansha, Tokyo, 1983, 172.
5. **Lin, S. K.,** Temporal stocking of small abalone and its dryness at low temperature, *China Fish. Mon.,* 138, 13, 1964.
6. **Chen, H. C. and Yang, H. H.,** Artificial propagation of the abalone, *Haliotis diversicolor supertexta, China Fish. Mon.,* 314, 3, 1979.
7. **Peon, S. C.,** Studies on the Age and Growth of Small Abalone in Hua Lien, Master's thesis, National Taiwan University, Taipei, 1980.
8. **Chen, H. C.,** Recent innovations in cultivation of edible molluscs in Taiwan, with special reference to the small abalone *Haliotis diversicolor* and the hard clam *Meretrix lusoria, Aquaculture,* 39, 11, 1984.
9. **Kuroda, T.,** A catalogue of molluscan shells from Taiwan, with descriptions of new species, *Mem. Fac. Sci. Agric. Taihoku Imp. Univ.,* 22(4), 71, 1941.
10. **Chiang, Y. M. and Lai, C. F.,** Studies on the feeding habit of small abalone, *China Fish. Mon.,* 284, 6, 1976.
11. **Chiu, C. C.,** Biological Studies on the Propagation and Larval Rearing of Abalone, *Haliotis diversicolor,* Masters thesis, National Taiwan University, Taipei, 1981.
12. **Tzeng, W. N.,** The biology of reproduction in the abalone, *Haliotis diversicolor supertexta* in the northeastern Taiwan, *J. Fish. Soc. Taiwan,* 5(1), 24, 1976.
13. **Oba, T.,** Studies on the propagation of an abalone, *Haliotis diversicolor supertexta.* II. On the development, *Bull. Jpn. Soc. Sci. Fish.,* 30(10), 809, 1964.
14. **Yang, H. S. and Chen, H. C.,** Effect of temperature and salinity on the embryonic development of abalone, *Haliotis diversicolor supertexta, J. Mar. Sci.,* 21, 78, 1979.

15. **Chu, L. H.,** Studies on the Development and Post-Larval Growth of Small Abalone (*Haliotis diversicolor supertexta*), Master's thesis, Chung Sun University, Kaoshiang, 1984.
16. **Kikuchi, S. and Uki, N.,** Technical study on artificial spawning of abalone, genus *Haliotis*. II. Effect of irradiated sea water with ultraviolet rays on inducing to spawn, *Bull. Tohoku Reg. Fish. Res. Lab.*, 33, 79, 1974.
17. **Morse, D. E., Duncan, H., Hooker, N., and Morse, A.,** Hydrogen peroxide induces spawning in molluscs, with activation of prostaglandin endoperoxide synthetase, *Science*, 196, 298, 1977.
18. **Kikuchi, S. and Uki, N.,** Technical study on artificial spawning of abalone, genus *Haliotis*. III. Reasonable sperm density for fertilization, *Bull. Tohoku Reg. Fish. Lab.*, 34, 67, 1974.
19. **Morse, D. E., Hooker, N., Duncan, H., and Jensen, L.,** γ-Aminobutyric acid, a neurotransmitter, induces planktonic abalone larvae to settle and begin metamorphosis, *Science*, 204, 407, 1979.
20. **Cox, K. W.,** California abalones, family Haliotidae, *Fish. Bull.*, 118, 1, 1962.
21. **Lee, T. Y., Pyen, C. K., Chin, P., and Hong, S. Y.,** Seed production and rearing of the abalone, *Haliotis discus hannai*, *Publ. Inst. Mar. Sci., Natl. Fish. Univ., Busan*, 11, 47, 1978.
22. **Chen, H. C.,** Water quality criteria for fish farming, *Fish. Ser., Cty. Agric. Proj. Dev.*, in press.
23. **Jan, R. Q.,** Studies on the Oxygen Consumption of Small Abalone, Master's thesis, National Taiwan University, Taipei, 1980.
24. **Tenore, K. R.,** Food chain dynamics of abalone in a polyculture system, *Aquaculture*, 8, 23, 1976.
25. **Tunbridge, B. R.,** Feeding habits of paua, *Fish Rep., N.Z. Mar. Dept.*, 20, 1, 1967.
26. **Chen, H. C.,** Studies on the aquaculture of small abalone, *Haliotis diversicolor* in Taiwan, in *Proc. ROC-JAPAN Symp. Maricul TML* Vol. 1, Liao, I. C. and Hirano, R., Eds., 1984, 143.
27. **Matsunaga, J.,** Studies on the diseased abalone, *Fish Pathol.*, 2(1), 11, 1967.
28. **Chen, C., Cheng, M. H., and Chen, J. D.,** The Investigation of Abalone Diseases in Northeast Taiwan, Master's thesis, National Taiwan College of Marine Science and Technology, Keelung, Taiwan, 1981.
29. **Chen, H. C.,** unpublished data, 1983.
30. **Sakai, H.,** Changes in shell size in young abalone cultured in calcium added sea water, *Aquiculture*, 22(3), 105, 1974.
31. **Chen, T. P.,** *Aquaculture Practices in Taiwan*, Fishing News Books, Surrey, 1976, 161.
32. **Li, Y. P. and Yuan, P. W.,** Status of aquaculture in Taiwan, in *Proceedings of ROC-US Coop. Sci. Sem. Fish Dis.*, Kou, G. S., Fryer, G. H., and Landolt, M. L., Eds., Natl. Sci. Counc., Taipei, ROC, 1981, 1.

CULTURE OF *HALIOTIS TUBERCULATA* AT THE ARGENTON EXPERIMENTAL STATION, FRANCE

Kirk O. Hahn

INTRODUCTION

The Centre National pour L'Exploitation des Oceans (C.N.E.X.O.) began studying the aquaculture of *Haliotis tuberculata* in 1973. The initial research was conducted at the Centre Oceanologique de Bretagne and the techniques were modified from existing research methods used in foreign countries, especially Japan. The goal of the French research project was to obtain scientific and technical information necessary for growing 1-year-old juvenile *H. tuberculata* from brood stock raised in the laboratory. The research initially concentrated on developing reliable techniques for larval rearing.

After the first experiments were successful, a pilot abalone hatchery was begun in 1976. The goals of the experimental hatchery were to have a minimum annual capacity to produce 50,000 1-year-old juveniles, test the techniques developed in the laboratory on a commercial scale, furnish animals for growth experiments, and estimate the cost of producing 1-year-old abalone (2 cm in length and 1 g in weight). This chapter summarizes the techniques described in a report on the French abalone aquaculture.[1]

CULTURE FACILITY AND EQUIPMENT

The hatchery is located in Argenton, on the northwest coast of Finistere, Brittany. The hatchery has a surface area of 165 m² and the entire culture process (spawning to grow out) occurs indoors. The facility was originally built to hold crustaceans in the 3600 m² outdoor pond. After conversion to an abalone hatchery, the pond became a holding area for adult abalone, then later a water reservoir for land-based tanks. The pond is connected to the ocean by two valves and pond level follows the tides. The average water level varies from 1.8 to 3.4 m, but decreases to between 0.5 and 1.5 m at low tide.

The water intake pipe for the hatchery is in the center of the pond. Two water intake lines allow periodic cleaning without stopping water flow to the hatchery. The water is pumped in two stages. The water is first pumped from the pond through a sand filter to a holding tank and a second pump circulates the water through the hatchery.

The main pump for the hatchery produces a flow rate of 20,000 to 35,000 ℓ/hr. Water passes through a pressure sand filter (grain size, 0.4 to 0.6 mm). The sand filter is hooked up in parallel, which allows quick deactivation when it is not needed. Water from the filter can be sent to different growing ponds, and, at the same time, to a holding tank where it is mixed with preheated water before being routed to the growing ponds. The initial sand filter isn't necessary if the water in the pond is allowed to stand undisturbed before being used for the hatchery. However, the filter on the recirculated water inside the hatchery is indispensable, especially for rearing juveniles up to 2 months old.

The main pump is very reliable and only needs normal maintenance. The recycling pump (10,000 ℓ/hr) is fairly fragile because the internal filter easily clogs. A prefilter before the pump and a removable and transparent pump cover is very useful in preventing clogging of the pump.

The water inside the hatchery is partially recycled and filtered through sand, and the used water from the hatchery is channeled back to the sea, away from the pond duct openings. The water inside the hatchery is recycled by a simple system adapted to the needs of the facility. A 1300-ℓ holding tank mixes new and used water, and acts as a safety feature in case the main pump fails. Normally, aeration of the water in the holding tank is unnecessary before recirculation, since the total water renewal time is short (3 hr), organic wastes levels are low, and oxygen consumption by abalone is low. The amount of organic wastes in the water is reduced by a pipe system, parallel to the water circulation pipes, which receives the water when the tanks are cleaned. In addition to better sanitation, the waste water pipe saves time and energy during cleaning.

Recycled water is heated in a heat exchanger with a total capacity of 40,000 kcal/hr capable of producing temperatures between 40 to 80°C. A total flow rate of 10,000 ℓ of thermoregulated water per hour is ensured by hooking up the heat exchanger in parallel to the ambient water line. Thermoregulation is controlled by a three-way valve linked to a thermostat and guarantees a temperature fluctuation of less than ±1°C. (The variation in temperature at the facility is actually a lot less than this level.)

The water recirculation system has several advantages. When the water volume needed for the hatchery is below 10,000 ℓ/hr, the unused heated water is recycled to the holding tank which ensures the water is oxygenated and eventually degased. However, when water needs are greater, water is either partially or completely recycled. This arrangement also guarantees that new water is used before recycled water.

Water quality is very good at the hatchery. The pH varies from 7.98 (metamorphosis tank) to 8.39 (recycling tanks); dissolved oxygen concentration is about 9.3 mg/ℓ as it enters the rearing tank; and nitrite NO_2 content is 0.001 to 0.004 mg/ℓ. Nitrate NO_2 content in the culture tanks is 0.073 mg/ℓ in tanks with vertical settlement plates and 0.040 mg/ℓ in tanks without vertical settlement plates. Water quality analysis at different points along the water circulation path indicates some recirculation is possible without using a biological filter.[2]

An air compressor with a 300-ℓ reserve tank ensures a constant supply of air at an air pressure of two bars and a flow rate of 15,000 ℓ/hr, regardless of the water volume aerated. The compressor runs approximately 20% of the time to supply air for the facility. The entire network (tubing and valves) for air circulation is made with 16 to 20 mm PVC pipes. Air is distributed to the tanks by thick-walled (20 mm exterior, 8 mm interior) rubber tubing closed at the end. The volume of air delivered to each tank is determined by the diameter of a syringe needle inserted through the tube wall. This method avoids the use of individual petcocks and the system can be dismantled by simply pulling the needles out of the tube, without any loss of pressure. Filtration and removal of oil from the air is an inconvenience of using pressurized air. However, pressurized air provides uniform aeration, independent of water height, that cannot be produced by a standard air generator.

HATCHERY TECHNIQUES

Conditioning

Adult abalone used for brood stock are conditioned with water temperature. The brood stock is separated by sex and kept for 6 to 12 months in tanks at a density of 10 to 20 adults (80 to 120 g) per square meter. The conditioning tanks are kept dark with covers and are slightly lighted at night. Water temperature is increased 1°C every 4 days from the ambient temperature until the water in the pond reaches 18°C.

Determining the level of gonad maturation by visual examination of gonad volume is imprecise since ovaries and testes increase in size independently of maturation state, especially during periods of stress. Measuring gonad maturation by calculating the effective accumulative water temperature (EAT) is more efficient than visual examination alone. (See Chapter 3 — Conditioning.) The EAT method is more informative if the level of gonad growth is determined before beginning conditioning. Groups may start at different gonad maturation levels, and will require different lengths of time before becoming fully mature. The average gonad maturation level of each group of brood stock is measured every 2 months.

Individuals are fully ripe at about 1500°C-day (e.g. 89 days at 18°C). Conditioning is possible throughout the year if only small quantities (~1 to 2 × 10⁵) of eggs are needed at each spawn induction, however conditioning is more difficult when large quantities (~1 to 2 × 10⁶) of eggs are needed.

Induction of Spawning and Fertilization

To induce spawning, mature adults are first air dried for 0.5 to 1 hr, placed into bags (individually or pairs of same sex) and suspended in tanks for spawning. Adults are spawned in 20-ℓ cylindrical, conical tanks with well aerated, stagnant (nonflowing) 1-μm filtered sea water. The water temperature in the tank is increased 4°C above the water temperture in the conditioning pond (18°C). These conditions will cause spawning in fully mature adults within 5 to 24 hr.

The thermal shock method of spawning induction is efficient with naturally matured wild animals; however, it is less efficient when the brood stock has been artificially conditioned. UV-irradiated sea water and hydrogen peroxide do not effectively induce spawning in *H. tuberculata*. Thermal shock is the only technique used at the present time.

The quantity of eggs spawned is estimated by counting the eggs in three 1-mℓ samples collected with an automatic pipette. Sperm density is measured with a hemacytometer. Using the methods developed for *H. tuberculata* (small water volume with aeration and temperature regulation, 20-min fertilization), the quantity of sperm needed for complete fertilization is inversely proportional to ova density.

Sperm is quickly added to a known concentration of eggs in the spawning tank. Aeration is stopped as soon as the fertilized eggs form polar bodies. The eggs are allowed to settle to the tank bottom and 80% of the water is decanted. The container is refilled/decanted twice to eliminate excess sperm.

Larval Culture and Metamorphosis

The fertilized eggs either develop in the tank used for spawning or are transferred to flat-bottom tanks (90 cm × 62 cm × 50 cm). The maximum density of fertilized eggs is 3 × 10⁵ eggs per 20 ℓ in cylindrical spawning tanks and 1.5 to 2 × 10⁶ in the flat-bottom tank. The fertilized eggs in the flat-bottom tank settle uniformly on the tank bottom and are raised in 1-μm filtered sea water with antibiotics (streptomycin sulfate, 50 mg/ℓ and penicillin G, 30 mg/ℓ, or chloramphenicol, 8 mg/ℓ).

Hatch out occurs about 13 hr after fertilization at 20°C. After hatch out, newly hatched trochophore larvae swim to the surface and the water in the bottom of the 20-ℓ tank (containing unfertilized eggs and empty egg cases) is siphoned out and replaced by clean water. The larvae remain in the tank for 20 to 30 hr until operculum formation and the veliger larvae are capable of retracting into the shell. Larvae are less fragile after this stage and can be concentrated. The larvae on the tank bottom are separated from the actively swimming larvae at the surface. Each day the larvae are rinsed and transferred to clean containers. The water is completely replaced each morning, and 75% is replaced at the end of the day.

In the flat-bottomed tanks, newly hatched trochophore larvae are separated from the unfertilized eggs, abnormal eggs, and empty egg cases by a stream of water between a pair of tanks. The initial tank is cleaned and refilled with water after the larvae are transferred. The larvae are transferred between the pair of tanks for the first 3 days of larval culture.

On the 4th day after hatch out, larvae (density varies from 200 to 300 larvae per liter) are placed into polyester settlement tanks (2 cm × 0.5 cm × 0.35 cm) with a white-gel coating for metamorphosis. The juveniles remain in these tanks until they are 8 months old.

Grow Out

The initial grow-out method, developed from the Japanese methods, generated a satisfactory production (50,000/year), but the variability was very large. A second grow-out method was developed in 1977 with modifications added in 1979. The differences between the two methods occur during the first 3 months of benthic life, the most critical culture period.

The initial method used vertical settlement plates for increasing the usable surface area in the rearing tanks for both juvenile habitat and diatom growth. Groups of ten plates hung from hooks connected to PVC tubing, 2 cm in diameter. This method required low maintenance (once a week per unit) but it was very expensive (high operating and labor costs to set up and clean the plates). In addition, the production of juveniles varied between tanks. Maintaining favorable culture conditions in the grow-out tanks was difficult during the critical period (juveniles at 500 to 1000 μm) when the balance of available food was very unstable. The difficulty with variability might have been due to the grow-out method, or the general conditions found at the hatchery, i.e., air and water temperature.

Due to the variability, a second method was developed at the hatchery. Larval culture is the same as in the first method, but the grow-out tanks do not have vertical settlement plates and larvae settle directly onto the tank walls and bottom. This method also requires the production of high quality diatoms, an artificial diet, and harvest of macroalgae from nature for feeding the juvenile abalone.

Juveniles are raised in the settlement tanks for 6 to 8 months. After this time, juveniles are transferred to 1 m $\times$ 1 m $\times$ 0.13 m tanks for continued growth to 1 year old. The tanks are stacked in racks of four tanks. This arrangement allows raising juveniles in four different water temperatures. Also, the water level in each tank can be varied by adjusting the drain pipe.

FOOD

Diatoms

The diatom culture room has a surface area of 10 m^2. The room is supplied with 1-μm filtered sea water (additional filtration is possible to less than 1 μm). Carbon dioxide (1%) is added to the air going to the diatom cultures. The aeration to the diatom cultures is independent from the rest of the hatchery, and regulated with flow meters. Natural light enters through two windows facing east and artificial light is supplied by six 80-W fluorescent tubes (daylight type).

The diatom culture is either done in 20-ℓ carboys or polyethylene bags. Twenty-five percent of the carboy volume is taken daily. The culture technique with polyethylene bags is different because it requires "blooming" of the culture. The cultures start with an initial concentration $\geq$ 3 $\times$ 10^5 cells per milliliter. The densities vary from 2 to 3 $\times$ 10^6 cells per milliliter for *Platymonas suecica*, and 5 to 10 $\times$ 10^6 cells per milliliter for *Pavlova lutheri*. *Prasinocladus marinus*, a species similar to *Platymonas suecica*, was isolated at the Argenton facility and does not deteriorate when benthic for more than 1 week, in contrast to *Platymonas suecica*. Diatoms are cultured in Conwy medium (1 mℓ/ℓ of inorganic nutrients and 0.1 mℓ/ℓ of vitamin solution).

Inorganic and vitamin solutions can be added to the diatom cultures using standard flame sterilization and opening the containers for inoculation. This method produces satisfactory cultures for a short period, but there is usually rapid contamination from bacteria, ciliates, molds, or other diatoms, thus limiting the duration of cultures to about 1 week.

Contamination problems were solved by developing a second culture method. The nutrient solutions are injected by syringe into vacuum tubing with a small interior diameter (10 mm) and thick walls (10 mm), as filtered sea water enters the container. The thick walls of the tubing ensure that the needle hole closes automatically after removal of the syringe. The tubing is also rinsed with 90% ethanol after injecting the nutrient solutions.

Diatoms are cultured in 50 to 80 ℓ polyethylene bags when there is a large demand (e.g., greater than 20 ℓ/day).[3] The polyethylene bags are 350 mm wide and 0.15 mm thick. Each bag actually consists of two envelopes: the external envelope is reused until it is warn out (several months), and the internal envelope is used for up to four cycles of production. Each cycle lasts from 5 to 7 days with initial concentrations varying from 4×10^5 cells per milliliter (*Platymonas suecica*) to 1×10^6 cells per milliliter (*Pavlova lutheri*).

Artificial Diet

An artificial diet has been used at the hatchery since 1977. The artificial diet simplifies juvenile rearing during the critical 1- to 3-month-old period. The diet is made from soybean meal, wheat, lactoserum, vegetable oil, minerals, vitamin complex, and antioxidants. The diet is 21% protein, 3.5% lipids, 3.5% cellulose, 17% minerals, and has a 9 to 10% moisture content. The artificial diet is stored in a dark dry place. The particle size for feeding is determined by filtering. The artificial diet is used between days 30 and 40 following larval settlement. The artificial diet avoids using the constantly changing natural diatom growth on the tank walls which causes increased variation of growth. A sudden change in the natural diatom growth can cause heavy mortality.

Macroalgae

The red alga, *Palmaria palmata (Rhodymenia palmata)*, produces the best growth rate and occurs naturally in large quantities along the Brittany coast. The algae is collected at low tide and can be stored for about 10 days in 500-ℓ cylindrical, conical tanks with aerated seawater flowing into the bottom of the tank. One tank is needed to feed 50,000 1-year-old juveniles.

When the tanks used for red algae are connected to the water-circulation system, plastic grids inside the tubing prevent algae from clogging the pipes. The algae is cut mechanically and separated by size before being fed to the juveniles. The size of the cut algae corresponds to the juvenile size. To prevent floating, the algae is rinsed in cold fresh water and then placed in a well aerated sea water tank for 1 hr. This procedure degases the algae and prevents it from floating. This procedure is especially necessary during periods of active photosynthesis.

IMPORTANT CULTURE FACTORS

Light

Light primarily influences the growth and quality of the diatoms and macroalgae. The light intensity in the settlement tanks is critical for maintaining the proper balance between growth and quality of the diatoms used as food for the newly settled juveniles. The light intensity necessary for good diatom quality (2500 to 3000 lx in white-bottom tanks, 3000 to 3500 lx in gray-bottom tanks) also promotes growth of the diatoms. The proliferation of diatoms can become a mechanical obstacle for the juveniles and may result in heavy mortality as early as the third day after settlement. Reducing the lighting slows diatom growth, but also lowers the quality of the diatoms. Diatom quality quickly deteriorates and heavy mortality occurs in the tank. Additionally, variations in light intensity can cause sudden changes in pH and oxygen concentration, which influence early mortality. The best solution to this dilemma is to provide the optimal light intensity for rapid diatom growth with the addition of inorganic nutrients to the water in the settlement tank. As soon as the juveniles settle and have completed metamorphosis, the water flow is adjusted to "rinse" away excess diatoms. The light intensity must be maintained above 300 lx when feeding macroalgae. Oxygen concentration can decline sharply during periods of low light intensity and reduced water flow.

Water Height

During the first years of culture at the hatchery, the water depth was 0.5 to 1 m in the rearing tanks used for grow out. This water depth was necessary for using the settlement plates. After the change in the rearing methods, which eliminated settlement plates, the water depth was excessive and was reduced without affecting the growth or production of juveniles. The macroalgae remained in better condition for longer periods since the water was more homogeneous. Water height never exceeds 15 cm with present rearing methods. The change in rearing methods (settlement on tank surfaces without settlement plates and reduced water depths) had a significant effect on the efficiency of the hatchery. The rearing tanks are now shallower, lighter, take less space, and are easily stacked. Each tank still holds the same number of abalone as the previous tank, but the rearing capacity per floor surface area of the hatchery has greatly increased.

Aeration

Aeration is supplied to the rearing tanks from settlement until the juveniles are 6 months old. Aeration increases mixing and renewal of water for benthic diatoms. Without aeration, there is rapid degradation of the macroalage in the tank and the juveniles move toward the water surface and risk a greater chance of desiccation. Increasing the water flow just pushes the algae toward the drain and does not solve the problem. Increased water flow also causes higher pumping costs, especially during winter. Aeration is not supplied to rearing tanks with juveniles 6 to 12 months old, except during periods when the water pumps fail.

Food

Cultured diatoms are used for food until day 40 after settlement. The cultured diatoms must be in the exponential growth phase when fed to the juveniles. The light intensity in the diatom cultures must be sufficient for the species being cultured while, at the same time, limiting its growth. A rapid diatom proliferation hinders the movement of juveniles by settling on the shell. In addition, the successive settlement of diatoms induces death of the lower layers, resulting in rapid mortality of the juveniles.

The feeding schedule (food type and amount) changes as the juveniles grow: day 0 to day 15 — *Platymonas suecica* at 2.5 cells/$\mu\ell$, *Pavlova lutheri* at 30 cells/$\mu\ell$, 0.5 mℓ/ℓ inorganic nutrient solution, and lighting 2000 to 3000 lx; day 20 to day 30 — *Prasinocladus marinus* at 5 to 20 cells/$\mu\ell$, 0.5 mℓ/ℓ inorganic nutrient solution, and lighting 2000 to 3000 lx; day 30 to day 40 — the unicellular diet is replaced with an artificial diet. Red algae is given to the juveniles after day 40. Feeding young fronds of *Palmaria plamata*, whenever possible, is very important. In general, daily distribution of food is best for ensuring good juvenile growth.

Problems Caused by a Massive Oil Spill

The oil tanker, *Amoco Cadiz*, wrecked several kilometers from the hatchery in 1978. The outside pond wall served as an efficient protection against the floating oil. However, dissolved hydrocarbons in the sea water entered the grow-out tanks. The levels of dissolved hydrocarbons were 4.9 μg/ℓ at the main pump and 4.4 μg/ℓ in the recirculation holding tank. Water analyses showed the hydrocarbon chain length was shorter in the holding tank. The concentration peak was at C_{30} chains (336 ng/ℓ) in the outdoor pond and C_{25} chains (320 ng/ℓ) in the holding tank. No mortality was immediately caused by the oil pollution. The construction of the hatchery facility allowed time during the accident for consideration of alternative solutions to the problem without rush.

EVALUATION OF CULTURE TECHNIQUES

Seed Production

The Argenton hatchery raised 57,800 1-year-old juveniles (16 to 25 mm) in 1976, the first year of production. The average production rate was 3320 juveniles per square meter of rearing surface, with a range from 0 to 11,100 juvenile per square meter. The researchers believed the variation in production in different tanks was caused by the use of vertical settlement plates and naturally occurring diatoms as food. Production at the hatchery decreased slightly in the second year to approximately 51,000 1-year-old juveniles ($2200/m^2$, but the variation ($1800/m^2$ to $6600/m^2$) between tanks was reduced by using an artificial diet. The 1-year-old juveniles ranged in size from 15 to 27 mm.

In 1978 several rearing methods were tested and while total production (36,000 juveniles) decreased, the average production rate ($4200/m^2$) was increased by some of the newly developed methods. The artificial diet greatly reduced the variability between tanks and increased the average production per tank by 52%. The best results were obtained with the use of an artificial diet during a short period of the juvenile stage: 3900 juveniles per square meter for 14 m^2 (i.e., 14 tanks). The production rate was very low with vertical settlement plates ($2552/m^2$ for 35 tanks).

Mortality

The simplest estimate of survival is comparing the number of surviving juveniles to the number of larvae introduced for settlement. The average survival for juveniles was 4.1% (ranging from 0 to 13%) in 1976, and 1.9% (0.5 to 15%) with vertical settlement plates and 7.3% (2.1 to 15%) with the artificial diet-macroalgae method in 1977.

Mortality (after day 15) was monitored more closely by estimating the number of dead individuals in the tanks at each cleaning. Although this method was imprecise, it showed periods of important mortality loss. A sharp increase in mortality was seen from day 50 until day 70 with the vertical settlement plate method. The plates hindered water flow and aeration, and caused difficulties in controlling the diatom growth. Mortality was caused primarily by deterioration of food quality caused by the growth of filamentous algae, which killed the diatoms on the plates. Mortality was also caused by the facts that cultured diatoms did not supply the necessary energy requirements for the juveniles at this developmental stage and the size of the diatoms was disproportionate to the size of the juveniles.

Juvenile mortality is reduced when using an artificial diet. The artificial diet allows more control over the culture and mortality is evenly distributed over time with no sudden deaths. Mortality is also reduced after the juveniles are 6 months old. Mortality is approximately 1% from 6 months old to 12 months old. Mortality is a function of density and becomes significant at densities above 2500 juveniles per square meter at 1 year old. A light intensity lower than 300 1x in the rearing tanks causes quick deterioration of the diatoms, regardless of the culture method being used.

Growth

Generally, growth is slow for the first 90 days after settlement. Average juvenile size is about 4.5 mm at 100 days and 9.5 mm at 200 days. The growth rate is independent of the rearing method used for culture. Monthly growth rates vary between 1 and 1.5 mm with the fastest growth in animals raised at low density.

Growth is greatest during the first 90 days with vertical settlement plates (1.5 mm/month with vertical plates, 1.0 mm/month with an artificial diet). However, the growth with either method is identical after 1 year. An advantage of the artificial diet method is the ability to maintain high rearing density without reducing growth rates. The average size of food particles given to juveniles is very important for maintaining optimum growth. Growth is reduced if the food particles are too large compared to the juvenile size. The food particle size range should be narrow and the average size should increase as the juveniles grow (e.g., 200 to 250, 250 to 400, 400 to 500 μm).

In 1976, juveniles were raised in densities from 1500/m² to 3000/m² and averaged less than 2 mm/month from 6 to 12 months old. The average length of animals changed from 11.4 mm at 6 months to 19 mm at 12 months, and total biomass changed from 15 kg to 61 kg. Rearing density has a significant effect on the growth. Growth is about 3 mm/month at 1000/m² and only 2.1 to 2.6 mm/month at 3000/m². The growth rate was slowly reduced as the rearing density was increased from 80/m² to 5000/m². However, individuals at the two highest densities tested (3750/m² and 5000/m²) showed a rapid decrease in growth when they reached a critical size. Both densities showed rapid declines when juveniles reached 12.9 mm in length.

Rearing juveniles in the dark does not cause any noticeable effects and growth is equal to or slightly faster than in normal lighting. No significant differences are found between densities of 2000/m² and 2500/m² reared in ambient light or the dark. The cost savings of rearing juveniles in the dark are not negligible.

OCEAN RANCHING

The artificial production of abalone seed is only one step in the aquaculture of this species. The juveniles from the hatchery are placed in artificial habitats below the intertidal zone in relatively strong wave-action areas. This zone is extremely rich in macroalgae but exposed to storms. Juvenile abalone in the wild naturally seek shelter for protection and food. Therefore, the juvenile is very mobile for the first 2 years after release. Several types of habitats were tested for juvenile survival in the wild and recovery of adults after grow out. The efficiency of a habitat is judged by the recovery rate and growth. The first trials in France used mounds of rocks from the planting site as habitats. The juveniles had good growth (1.7 to 2 mm/month) but there was a poor recapture rate (4%). Habitats made of plastic tubes also proved unsatisfactory.

Major progress was made in ocean ranching by building concrete habitats in the intertidal zone. Recapture rates varied from 40 to 50% after 2.5 years, depending on initial reseeding density. The animals had a growth rate of 1 to 2 mm/month. These results were obtained without any management or food supplementation. Research efforts are now being concentrated on designing habitats that allow entry of food while restricting abalone movements away from the area. Juvenile abalone catch free macroalgae during ebb and flow of the water into the habitats. The goal is to use this "free" food by taking advantage of natural water flow to supply algae and proper rearing conditions, without labor or hatchery costs.

FINANCIAL EVALUATION OF CULTURE TECHNIQUES

Hatchery

The costs associated with hatchery operations are related to the outside air and water temperatures, and water pumping requirements. During the 2 years from 1976 to 1978, 0.34 ℓ of fuel and 1 kWhr per juvenile were needed to produce 113,200 1-year-old juveniles. The energy costs were greatly reduced (one third the energy of previous years) in 1979 to 1980, 0.1 ℓ of fuel and 0.3 kWhr per juveniles were needed to produce 145,000 juveniles. The costs were almost identical in 1980 to 1981, 0.1 ℓ of fuel and 0.5 kWhr per juvenile.

Ocean Ranching

Current recapture rates are approximately 50% after 2.5 years with prototype habitats and higher recovery rates expected in the future with improved habitats. At the beginning of

Table 1
ESTIMATED PROFIT WITH OCEAN RANCHING OF JUVENILE ABALONE IN FRANCE. 100,000 JUVENILES REARED AT 1000/HABITAT. PRICE OF ADULT ABALONE IS $4.00/kg @ 10 INDIVIDUALS PER KILOGRAM

Costs	Cost per juvenile		
	$0.08	$0.12	$0.16
Juveniles	8,000	12,000	16,000
Habitats	8,000	8,000	8,000
Labor			
Placing habitats in field	800	800	800
Upkeep of habitats	1,000	1,000	1,000
Harvesting of adults	300	300	300
Total	18,100	22,100	26,100
50% Recovery			
Total income	20,000	20,000	20,000
Profit from first production	1,900	−2,100	−6,100
Profit from subsequent productions	10,700	6,700	2,700
60% Recovery			
Total income	24,000	24,000	24,000
Profit from first production	5,900	1,900	−2,100
Profit from subsequent productions	14,700	10,700	6,700
70% Recovery			
Total income	28,000	28,000	28,000
Profit from first production	9,900	5,900	1,900
Profit from subsequent productions	18,700	14,700	10,700
80% Recovery			
Total income	32,000	32,000	32,000
Profit from first production	13,900	9,900	5,900
Profit from subsequent productions	22,700	18,700	14,700

Note: Prices have been converted to U.S. $.

(From Flassch, J. and Aveline, C., *Pub. C.N.E.X.O.*, *Rapp. Sci. Tech.*, 50, 1, 1984. With permission.)

1982, after 9 months in the habitats, the recovery rate was 70 to 90%, and abalone showed good growth without supplemental food or labor.

Table 1 shows the expected profit with various costs of hatchery-reared juveniles and recovery rates after 2.5 years of ocean ranching. The cost of labor is over estimated since it assumes food would be manually supplied to the habitats. (Labor costs would be considerably less if algae were supplied naturally.) The calculations are simplified by assuming that the cost of the habitats is paid after the first harvest (3 years of culture). The life span of a habitat should be longer than three cycles of production (9 years). A recapture rate of 60% for the first production and 70% for the following productions is expected. These recapture rates count only the abalone found inside the habitats. Abalone found outside the habitats could significantly increase the number of harvested abalone and increase the actual recapture rate. The return on the investment figures corresponding to the recapture rates are relatively stable from year to year with a constant retail price. Food and labor costs, which usually increase faster than other costs and contribute most to the final cost, are nonexistent.

The expected harvest weight for 100,000 seeded juveniles with 60% recapture is about 6000 kg. The minimum production for a successful aquaculture business should be about 18,000 kg/year or approximately 300,000 seeded juveniles.

SUMMARY

Hatchery production of juvenile abalone in France will mainly provide animals for ocean ranching in intertidal zones rich in macroalgae. The potential of a hatchery depends on two factors: (1) the capacity of the facility (e.g., size of tanks, pumping, and filtration), and (2) productivity of the personnel. The productivity is a function of both the qualifications of each individual and the total number of employees. For example, a competent technician working alone would be less productive than if he were assisted.

From 1976 to 1979 at Argenton, the average number of juveniles produced per culture surface area varied from 2600 to 3900/m^2. Since 1980, average production is 5000/m^2 or 120,000 juveniles for an entire culture cycle. The objective of the Argenton hatchery is 200,000 1-year-old juveniles over two cycles. Future commercial facilities should expect productions of 5000/m^2 or better.

Significant improvement in juvenile growth can also be expected.[4] Obtaining a growth rate of several millimeters a month is possible, producing 20-mm juveniles in approximately 8 months. This saving in hatchery culture time has a direct effect on production capacity.

The Argenton hatchery has determined the production capacity per unit surface, but the facility is too small for economic success. Energy costs for this hatchery average 0.1 ℓ of fuel and 0.5 kWhr per juvenile. Estimated costs for raising juveniles at a commercial hatchery would be between \$0.07 to \$0.12 (U.S.) per animal for a hatchery producing 1 million juveniles. These estimated costs are compatible with a profitable production (Table 1).

Commercial hatcheries will supply juveniles to aquaculturists for ocean ranching in artificial habitats. The probable area for this type of culture would be along the coasts of Brittany where the littoral zone is rough and the tides high. The present methods ensure more than a 50% recovery rate after 2.5 years in the ocean.

ACKNOWLEDGMENTS

I would like to thank Teva Siu for translating the French scientific papers.

REFERENCES

1. **Flassch, J. and Aveline, C.,** Production de Jeunes Ormeaux a la Station Experimentale d'Argenton, *Pub. C.N.E.X.O., Rapp. Sci. Tech.,* 50, 1, 1984.
2. **Couteaux, B.,** *Contrôle de l'eau dans les Élevages Marins á Caractére Intensif — Recyclage,* Memoire ENSAR, Rennes, 1976, 1.
3. **Flassch, J. P.,** Production d'algues unicellulaires á des fins d'aquaculture, *Oceanis,* 4(1), 1, 1978.
4. **Cochard, J.,** Recherches sur les Facteurs Determinant la Sexualite et la Reproduction Chez *Haliotis tuberculata* L., Ph.D. thesis, Universite de Bretagne Occidentale, 1980.

ABALONE AQUACULTURE IN NEW ZEALAND, AUSTRALIA, AND IRELAND

Kirk O. Hahn

NEW ZEALAND

Haliotis iris

The natural black color of the meat is one of the major hindrances to *H. iris,* paua or black paua becoming a major economically important abalone species. Consumers have negative connotations to the black color, associating it with spoilage or decay, especially since other abalone species have creamy-white meat. Therefore, the meat must be bleached before it is acceptable to the public for sale.[1] The cost of fresh, unbleached paua is $8 (N.Z.) per kilogram in local markets.[2]

Although the meat is aesthetically of low quality, the paua fishery is still valuable and unique because the salable product is the shell rather than the meat. The value of raw shell is $5 (N.Z.) per kilogram for superior grade and $2.50/kg for third grade. Approximately $2 to $3 million worth of manufactured shell jewelry is exported annually to the U.S., Australia, Singapore, Malaysia, and West Germany. The world-wide demand for paua shell is very high, but there are severe restrictions in New Zealand on the export of the raw shell. This has caused an illegal industry of selling blue-dyed abalone shell (species unknown) embedded in resin. This product is almost indistinguishable from paua.[3]

Hatchery Techniques

Tong and co-workers from the New Zealand Ministry of Agriculture and Fisheries have developed several unique techniques to culture abalone in New Zealand. The research began in 1980 at the Mahanga Bay Shellfish Hatchery, near Wellington.[4]

Brood stock is collected from wild populations close to the hatchery. Paua collected from the wild in the Wellington area can be induced to spawn from May through December.[5,6] Paua can only be artificially induced to spawn when the water temperature in nature is below 16°C.[5] The drop in water temperature is needed before individuals can attain full ripeness. Paua along the South Island can be induced to spawn from late summer to fall.[6]

Ripe paua (two individuals of the same sex in each bucket) are placed in single 12-ℓ buckets. The pH of the water in the bucket is raised to 9.1 by adding 13 mℓ, 1 M NaOH. Spawning is induced by adding 40 mℓ of 6% hydrogen peroxide. The brood stock are left undisturbed for 3 hr and then the buckets are thoroughly rinsed for 15 min with clean water. Spawning begins 1 to 2 hr after rinsing out the bucket, and may continue for 4 hr.[7]

A 125-mm adult female paua can spawn up to 3 million eggs.[7] The eggs are thoroughly washed in a 100-µm mesh sieve and fertilized in a bucket with a screen on the bottom, which sits inside a bigger bucket with holes near the top, allowing the water to overflow.[4] Sperm concentration for fertilization is 25,000/mℓ (final concentration). This concentration assures 95% fertilization. After 15 min, the eggs are washed free of excess sperm.[7] The fertilized eggs are maintained in a continuous flow system until hatch out.[4]

After hatch out, any trochophore larvae on the bottom of the bucket or egg debris is siphoned off and discharged. The remaining trochophore larvae are drained through nylon sieves, washed carefully, and resuspended in 20-ℓ buckets. After 15 min, any larvae on or near the bottom are again siphoned off and discarded. Healthy larvae are resuspended at a concentration of five larvae per milliliter in 70-ℓ containers with gentle aeration.[7] The container is specially designed to allow continuous flow without loss of the planktonic larvae. A 3-cm-wide large-diameter cylinder, covered on both sides with 100 µm nitex screen is connected to an overflow pipe exiting the container. A slow water flow is begun into the container once the larvae have reached the veliger stage. The screen will not clog if the water flow is slow.[4]

Every day, the water is completely changed (the larvae are retained on a sieve and washed), and the tanks are cleaned. Larvae are culled very heavily at each washing to leave only the most healthy and actively swimming trochophores.[7] Survival to settlement, previously 30 to 40% at 16°C, has been increased to 75% by improving the system and reducing the handling of the eggs and larvae.[7,8]

Tong and his co-workers have developed an unique method to determine when the larvae reach the developmental stage necessary for settlement. About 100 larvae are placed into a test tube and five drops of concentrated HCl is added, dissolving the shell. The larvae are dehydrated with an alcohol series up to 100% ethanol. The larvae are then squashed under a cover slip, and the rows of teeth in the radula are counted. Larvae are introduced into settlement tanks when the veligers have developed a radula with ~5 to 6 rows of "teeth".[4]

At 16°C, the larvae are ready to settle in 7 days.[6] Larvae can be induced to settle onto any surface with γ-aminobutyric acid (GABA), but will die or reject the surface within 24 hr if the surface is unsuitable.[7] Surface texture may be important during settlement induction.[4] Tong is experimenting with a variety of materials for settlement substrata.[6,7]

The larval settlement and juvenile grow-out tanks are V-shaped with vertical boards interspaced along the sides. The tanks are painted with antifouling paint (unknown whether the paint contains copper or organo-tin as the active ingredient). The settling larvae probably like the surface texture.[4] The V-shaped settlement tanks produce up to 10% survival, or approximately 10,000 to 15,000 juveniles per tank.[9] Tong tried putting GABA into the paint to help induce settlement, but is did not work.[4]

The settlement plates are inoculated with two diatom species that promote good juvenile growth and survival. The diatoms were isolated by dissecting out the stomach from 2-day-old juveniles that had settled on surfaces covered with naturally occurring diatoms. It is believed that newly settled juveniles will actively select the good diatoms from the mixed array on the plates. Good diatoms are defined as ones which produce a large amount of secretion and form sheets.[4]

The isolated diatoms species are *Navicula minimata* and "Peanut". "Peanut" has not been identified but the shape of the diatom resembles a peanut. The isolated diatom species are mass cultured in polyethylene bags, using normal algal growing techniques similar to methods used to raise phytoplankton for oysters. As the diatom culture solution is removed from the polyethylene bag, the solution is passed through a filter, breaking up the sheets of diatoms. This procedure creates individual cells which will quickly stick to the settlement tank surface.[4]

Juvenile growth rate is approximately 30 μm/day for the first 50 days after settlement and then decreases to 20 μm/day. If the food source (diatoms) is increased, growth will increase rapidly to approximately 64 μm/day. *Phaeodactylum* sp., a fast-growing diatom, increased juvenile growth to 70 μm/day. The growth rate slowly decreases as the diatoms are grazed off the substratum. The juveniles also feed on macroalgae after reaching 2 to 3 mm in size. After 87 days postsettlement, the mean shell length is 3.3 mm and the growth rate is 40 μm/day. Macroalgae are given to the juveniles at this time to increase the growth rate. *Macrocystis pyrifera* and *Pterocladia* sp. (a red seaweed) produce growth rates of 84 μm/day, and *Gracilaria* sp. produces a gradual increase in the growth rate up to 105 μm/day. At 153 days, the mean length was 9.1 mm.[10]

Five species of macroalgae were tested for suitability of use as food for *H. iris* (10 to 60 mm). After 10 weeks of culture, *Lessonia variegata* had the greatest mean increase in weight (51.2%), followed by *Pterocladia* sp. (37.8%), *M. pyrifera* (28.6%), *Ulva lactuca* (4.3%), and *Champia* sp. (0.5%). The order of food preference (based on wet weight consumed) for *H. iris* was *Glossophora* sp., *M. pyrifera*, *L. variegata*, *Champia* sp., *U. lactuca*, and *Pterocladia* sp. *Glossophora* sp. was not available during the growth experiments. The total quantity of algae consumed in 30 days ranged from 459 g of *Glossophora* sp. to 201.5 g of *Pterocladia* sp.[11] Paua will eat almost any seaweed, but normally eat more brown algae since they are the most abundant in nature.[7]

After further experiments, *Gracilaria* sp. was determined to be the best food for juvenile and adult paua, even though individuals would not normally eat it in any abundance in nature. The feasibility of culturing *Gracilaria* sp., a red seaweed which is also a valuable source of agar, as a food source for paua has been studied by Karl Johnson at the Manukau Sewerage Treatment Station run by the Auckland Regional Authority. Johnson's work showed that *Gracilaria* sp. could be cultivated in a simple pond arrangement with growth rates capable of achieving a doubling of wet weight within 7 days.[12]

At the Mahanga Bay Shellfish Hatchery, adult and juvenile (>3 mm) paua are fed *Gracilaria* sp., producing an average growth rate of ~100 μm/day and juvenile growth from 3 mm to 10 mm at a rate of 126 μm/day.[4,12] The growth rate produced by *Gracilaria* sp. is more than double the rate produced by *M. pyrifera* (~40 μm/day) which is usually used as the major food for cultured abalone.[4] The fastest growth rate is obtained by raising juvenile paua at 19°C water temperature with *Gracilaria* sp. as the food source.[7] The development of a technique for weaning paua onto macroalgae at a small seed size was equally important in obtaining faster growth.[12]

Haliotis virginia

The virgin paua or white-footed paua has recently stimulated interest as a potential culture species. The meat of the virgin paua is white and more appealing to consumers than the other species of paua in New Zealand (*H. australis* and *H. iris*) which concentrate black pigment in the foot.[13] The mantle of the virgin paua is a beautiful purple color and the shell is more rounded than *H. iris*. Virgin paua behavior and food requirements are identical to those of black paua. The virgin paua can be induced to spawn from May to November. The fecundity is very low (a few thousand) due to the small size of the species.[4]

AUSTRALIA

The abalone industry is very important to Australia but over fishing is causing a decline in harvest. In recent years Australia has attempted development of abalone aquaculture as a means of stabilizing the abalone industry. There are two commercially important abalone species, black-lip abalone (*H. ruber*) and green-lip abalone (*H. laevigata*). Australia has several advantages that could make the establishment of abalone culture much easier than in other countries. Presently, the availability of and access to suitable waterfront property is unhindered by government regulations or a large population, but this will probably gradually change in the future.[14]

Preliminary experiments were conducted at the Planet Fisheries Factory at Dunalley to determine the feasibility of abalone culture in Australia.[15] Colin Sumner and John Grant of the Tasmanian Fisheries Development Authority began a research project in 1980 to develop techniques to culture *H. ruber* in Tasmania. The objectives of the project were to develop conditioning techniques for adult *H. ruber* to produce brood stock, develop techniques to reliably induce spawning in native abalone, study larval development and metamorphosis, and find the optimal diet for juvenile abalone.[16]

The hatchery took 3 months to build, and the design was a combination of Japanese and American techniques. The researchers initially used techniques developed in other countries to begin the abalone aquaculture project. Brood stock (>60 mm) for spawning was collected from natural populations in Tasmania. Animals with suitable gonad development were originally maintained in a 5,000-ℓ holding tank with running ambient sea water; however, this method was unsatisfactory because it caused massive spawning during the night. The animals collected for brood stock are now brought to the hatchery, and placed directly into spawning tanks. Artificial induction of spawning is successful from December to April, indicating that it is possible to obtain spawnable animals directly from natural populations without prior conditioning. During the initial project, two batches of *H. ruber* larvae were raised.[17] Both *H. ruber* and *H. laevigata* show promise for culture because their growth rate is faster than reported for any other species.[18] The research project in Tasmania ceased in early 1985 due to a reorganization of the Tasmanian Fisheries Development Authority.[19]

IRELAND

The feasibility of raising ormer, *H. tuberculata,* in Ireland was investigated by the Shellfish Research Laboratory in Carna, although ormer is absent from the coasts of Britain and Ireland. In 1976 and 1977, 80 ormers were provided for breeding by the States of Guernsey Sea Fisheries Committee. The Shellfish Research Laboratory now has 600 sexually mature first generation ormers (4 to 6 years old) and a large number of second generation (1 to 3 years old). The growth rate at the laboratory (60 mm in 4 to 5 years) is comparable with ormers from nature. The coast along Ireland is similar to the Channel Islands, and *H. tuberculata* is suitable for transplantation along the deeper coastline areas of the Irish west coast (e.g., Cork and Kerry). In these areas, winter sea temperatures rarely fall below 6°C. Prolonged exposure to water temperatures less than 5°C will cause mortality. Sites near Connaught and Ulster border on unacceptability as grow-out areas but are not excluded as potential sites.[20]

Since 1979 an emphasis has been placed on the development of economic and simple hatchery/nursery techniques. In 1982, the researchers developed a prototype hatchery with an approximate annual production of 80,000 juveniles (5 mm shell length). The Shellfish Research Laboratory can reliably produce good quality ormer seed for grow out. In 1984, the States of Guernsey Sea Fisheries Committee signed a 3-year contract to purchase 10,000 to 20,000 (15 mm shell length) seed per year. The juvenile ormer will be used for grow-out experiments in Guernsey. Also, two companies have made inquires about grow-out projects in Ireland.[20]

REFERENCES

1. **France, G. U.,** The value of canned paua to New Zealand, in *Proc. Paua Fish. Workshop,* Fish. Res. Div. Pub. no. 41, Akroyd, J. M., Murray, T. E., and Tayler, J. L., Eds., N. Z. Min. Agric. Fish., Wellington, New Zealand, 1982, 18.
2. **Hahn, K. O.,** personal observations, 1984.
3. **Wright, R. R.,** Export value of paua shell to New Zealand, in *Proc. Paua Fish. Workshop,* Fish. Res. Div. Pub. no. 41, Akroyd, J. M., Murray, T. E., and Tayler, J. L., Eds., N. Z. Min. Agric. Fish., Wellington, New Zealand, 1982, 24.
4. **Tong, L. J.,** personal communication, 1984.
5. **Tong, L.,** Spawning and rearing of paua, *Catch '82,* June, 19, 1982.
6. **Tong, L., Dutton, S., and Swindlehurst, R.,** Paua research update, *Catch '81,* December, 17, 1981.
7. **Tong, L. J.,** The potential for aquaculture of paua in New Zealand, in *Proc. Paua Fish. Workshop,* Fish. Res. Div. Pub. no. 41, Akroyd, J. M., Murray, T. E., and Tayler, J. L., Eds., N. Z. Min. Agric. Fish., Wellington, N.2., 1982, 36.
8. **Tong, L.,** Paua research in new phase, *Catch '83,* March, 16, 1983.
9. **Tong, L. J.,** personal communication, 1985.
10. **Tong, L. and Dutton, S.,** Growth of post-larval paua, *Catch '81,* June, 1981.

11. **Dutton, S. and Tong, L.,** Food preferences of paua, *Catch '81,* March, 15, 1981.
12. **Tong, L. J.,** Paua research shows progress, *Catch '83,* September, 18, 1983.
13. **Murray, T.,** Could the virgin paua appeal?, *Catch '83,* June, 13, 1983.
14. **Cuthbertson, A.,** *The Abalone Culture Handbook,* Tasmanian Fisheries Development Authority, Hobart, Australia, 1985, 1.
15. **Harrison, A. J.,** Abalone culturing experiment, *Tasmanian Fish. Res.,* 1(3), 1, 1967.
16. **Anon.** Abalone breeding project begins in Tasmania, *Aust. Fish.,* September, 15, 1980.
17. **Sumner, C., and Grant, J.,** TFDA experimental abalone hatchery making progress, *Aust. Fish.,* June, 14, 1981.
18. **Shepherd, S. A.,** Breeding, larval development and culture of abalone, *Aust. Fish.,* April, 7, 1976.
19. **Grant, J.,** personal communication, 1985.
20. **LaTouche, B. and Moylan, K.,** Abalone farming in Ireland, *Aquacult. Ireland,* 16, 12, 1984.

CULTURE OF THE TROPICAL TOP SHELL, *TROCHUS NILOTICUS*

Kirk O. Hahn

INTRODUCTION

The tropical top shell, *Trochus niloticus*, (also called *Trochus* or troca) is indigenous only to Indo-Malaysia, Melanesia, and Yap and Palau in Micronesia, but has been introduced in the tropical Pacific as far east as the Tuomotus of French Polynesia.[1] Commercial top shell fisheries exist in the Marshall, Mariana, Caroline, Solomon, and Cook Islands; and the Philippines, Indonesia, Papua New Guinea, Australia, New Caledonia, Vanuatu, Fiji, and French Polynesia.[2] Also, artisan fisheries for meat and shells exist on many other islands.

T. niloticus is the most economically important gastropod in the tropical Pacific. Top shell is both an important traditional food, and a leading export item which generates sale revenues that reach a broad cross-section of the local populace and is the principal source of money for many islanders. In addition, the skill, investment, and technology required for harvesting top shell is minimal, making this resource easily accessible to islanders. In north and northeast New Caledonia, for example, harvested top shell is usually the only source of money for islanders' living.[3]

"The importance of the shell trade to the coastal villager or to the nation cannot be measured solely in terms of cash earnings or Gross National Product. It is an industry ideally suited to coastal villages in that: the harvest of shell does not require investment in expensive equipment or vessels; the reefs on which it is found are often contiguous with small population centers (a villager need not leave home to enter the cash economy); the salable product requires no preservation and is easily packed and stored; and the meat (foot) of the shell is easily processed (salted and/or smoked) to provide a source of locally produced and preserved high quality protein."[4]

Shells are collected by hand, either by walking along the reef at low tide or by skin diving. There is 1 kg of edible meat in 10 kg of live *Trochus*.[5] The meat is either dried, cooked, or canned, and the shell (77% of the animal's live weight[6]) is exported to Europe and Asia for the production of mother-of-pearl buttons.[2] The shell is thick, with a very beautiful interior, and the outside is covered with oblique or radiating red bands or green bands with zones of rose (Figure 1).[7] Blank buttons are cut from the shell following the shell whorl.[8] *Trochus* buttons can be recognized by the red, brown or greenish markings on the underside.[9]

The annual world *Trochus* harvest is between 5 to 6 million kg with a wholesale value of $4 million (U.S.).[3] The price islanders receive for cleaned and dried shell ranges from approximately $356 (New Caledonia) to $660 (Palau) per 1000 kg. The shells are graded and cleaned again before export to Europe (54%) or Japan (41%), and sold for approximately $616 (New Caledonia) to $1500 (Palau) per 1000 kg.[3,6]

The final retail value of the shell is many times the value of the raw product. An individual *Trochus* shell costs $0.15 (U.S.) ($0.88/kg) but will produce about 35 buttons, worth $0.30 each or $10.50 per shell.[10] Also, the shell waste after button production can either be processed to produce mother-of-pearl chips or crushed into powder for use in making paint and nail polish. Thus, the total retail value of *Trochus* shell is approximately $300 million. Although plastic buttons are now common (only 0.5% of the buttons in the world are made from mother-of-pearl), the world demand for top shell has remained high.[3]

Due to severe overfishing, the International Union for the Conservation of Nature has place *T. niloticus* on its list of "commercially threatened invertebrates". New Caledonia

FIGURE 1.   Adult *T. niloticus*. Right — natural appearance of top shell. Left — polished top shell, exposing the pearly nacre underneath the top layer of shell.

and Philippines each harvested over 1 million kg per year during the pre World War II period; the harvest was reduced to less than 300,000 kg/year each in 1948.[11] The harvest of *Trochus* was very low (<100,000 kg) from 1958 to 1975, due to a mining boom on New Caledonia.[3] In 1978 the harvest rapidly increased to 1.9 million kg but dropped to 1 million kg in 1979.[3] There are several factors that have contributed to the decline of top shell populations. The wholesale price of top shell has increased more than 500% in the last decade which has caused intense fishing pressure near population centers.[6] Also, top shell has been over exploited due to its large size, close proximity to shore on shallow reefs, and its relative immobility.[12]

Small individuals (6 to 10 cm) with thin shells are most valuable, so individuals are usually harvested before they are 10 cm.[3] It is probable that individuals harvested at the usual legal minimum size limit of 7.6 cm are mature but have contributed very little to annual recruitment. Large individuals (>12 cm) have a low value because the shell is too thick and encrusted with calcareous algae.[6] Parasitism also contributes to the trend toward harvest of small individuals. *Trochus* shells are often infested with an organism, probably a marine fungus,[13] that bores into the shell at the tip and continues eroding the shell downward along the spire. Examination of the fine structure of the shell shows that the apex is attacked by boring organisms within a few weeks after settlement. This erosion lowers the price of the shell. Therefore, only a small amount of erosion at the tip is tolerated by buyers. Fishermen try to harvest animals having very little damage, but the animals with the least amount of damage are also the smallest and youngest.[14] These factors cause almost all top shell to be harvested once they reach the minimum size limit and removes most reproductively mature individuals from the population.

Top shell has been overfished in Palau, Yap, Truk, Australia, New Caledonia, Vanuatu, and the Philippines, which has required the establishment of stringent conservation measures.[15] For example, Aitutaki, Cook Islands, had a harvest quota of 20,000 kg in September

1984, but the fishery was quickly closed after 45,000 kg were taken in the first 2 weeks. Additionally, most of the individuals were very small and immature.[14] In Palau, about 20% of the top shells collected in 1979 were below the legal size. Even though the harvest season in Palau is only 1 month long, many fishermen collect top shell out of season and hide the shells until the season is open.[15]

Palau's top shell fishery has shown a decline in harvest since it began in 1915, and is typical of other countries in the Pacific. Palau has adopted several management policies to help reverse the decline in the top shell abundance. The minimum size of harvest is 7.6 cm (base diameter), the fishing season is only 1 month each year (June), sanctuaries from harvest have been established, and moratoriums from harvesting for a year or more have been established in some regions in Palau.[12]

Sanctuaries were established to create a resident reproductive pool of animals that could produce larvae to repopulate harvested areas. Top shell is conducive to this type of management due to its short larval period and settlement within 5 to 7 days after spawning. However, the sanctuaries first established in Palau were not successful and actually had lower densities of top shell than the exploited areas. This was due to improper selection of these sites as prime top shell habitat. The sanctuary sites in Palau were selected without taking into consideration the proper substratum for top shell or the ease of enforcing prohibition of harvesting. In view of these findings, the Palauan government has now created new sanctuaries in prime habitat areas. If harvest prohibitions can be enforced, these sanctuaries may prove invaluable to rescuing Palau's declining fishery.[12]

Some countries restrict the harvest of top shell to individuals between 6 to 12 cm, assuring the presence of large individuals to contribute to the reproductive effort. The reduced harvest is offset by the increased value of smaller individuals. This management method is not useful on all reefs. If the reef has already been heavily fished, there will be no large individuals (>12 cm) present. In these areas, all reproductively mature animals will be taken when the harvest size is reduced. The population will then require the transplanting of large individuals to maintain the fishery.[3]

ECOLOGY

T. niloticus typically inhabits shallow, high-energy portions of barrier and fringing reefs. The highest densities of top shell occur in areas having a wide reef flat that is exposed at spring low tides, an unobstructed exposure to high-energy surf (waves, currents), a gently sloping bottom of uniform pavement substratum, and an abundance of coralline and low filamentous algae at 1 to 3 mm.[12] In New Caledonia, for instance, maximum densities occur in the boulder zone of shallow reef flats.[3] McGowan[11] observed the greatest densities of top shell on the seaward side of high island barrier reefs; however on atolls, the highest densities were on the lagoon side of the reef. The physical and biological characteristics of these two reef habitats are very similar and provide optimal conditions for *Trochus*.[11]

Low energy portions of tropical reefs usually have live coral and sand, which offer several hazards to top shell. Settling larvae and juveniles can be consumed by live coral or buried under shifting sand. Juveniles and adults probably avoid sand because it inhibits locomotion and prevents adhesion of the foot to the substratum. Movement across live coral is probably avoided because it exposes the foot to stinging nematocysts. Also, live coral and sand do not promote growth of the low filamentous algae that are the principal food for top shell. Low-energy habitats, therefore, generally do not support top shell populations.[12]

There is a significant negative correlation of top shell density with depth. Shallow sites have higher densities of small individuals; however, as depth increases, the top shell density decreases while the mean size of each individual increases (Figure 2).[1,9,16] These large individuals, which are often encrusted with coralline algae, are most often found on the reef

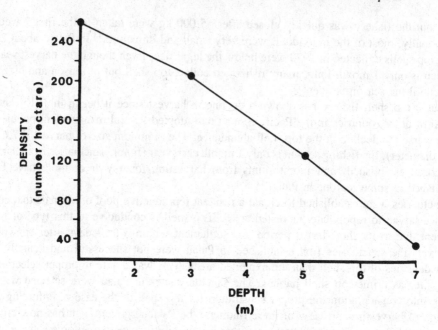

FIGURE 2. Relationship between depth and mean density of *T. niloticus* on the seaward barrier reefs of Koror State, Palau. A significant negative correlation ($r = -0.99; p < 0.01$) between depth and density was found between 1- and 7-m depth in exploited area. (From Heslinga, G. A., Orak, O., and Ngiramengior, M., *Mar. Fish. Rev.*, 46(4), 73, 1984. With permission.)

slope at 0.5- to 10-m depth.[7,9,11,17] Sometimes, they live among blocks of dead coral on the reef flat and are uncovered at low tide.[9,11,17] There are some rare individuals, usually large in size, found in deep water, but *T. niloticus* becomes very rare below 10 m.[7,12]

This density pattern suggests that settlement and/or early survival is highest on the outer reef flat, and individuals migrate outward as the grow.[1] Direct evidence has not been obtained for this hypothesized outward migration, although several mark and recapture studies have been performed. Van Pel[5] found individuals that had moved 50 m or more in one night, while Moorhouse[9] found individuals 2 years after release very close to the original release site.

The shallow reef flat inhabited by juvenile top shell typically consists of cobble and rubble that provide abundant crevices for shelter and algae for food. Very small individuals (<10 mm) have rarely been found, but they presumably occur chiefly in this shallow intertidal reef flat habitat (Figure 3). Juvenile top shells can sometimes be found near the shore, isolated far from the reef. On Moorea, French Polynesia, small (3 to 20 mm) top shell (mean density — 0.7/m²) were found several kilometers from the reef, in the mouth of the tidal inlet to Lake Temae, a brackish lake on the northeast coast. There is an unusually high amount of algal growth in the inlet, which may help explain the abundance of juvenile top shell. The substratum near the inlet consists mostly of rocks less than 10 cm in diameter, but the juvenile top shell in the inlet occur almost exclusively on rocks larger than 10 cm. These larger and more stable rocks probably have more algal growth and less accumulation of silt.[18]

In addition to the top-shell density being correlated with depth, its density is also correlated with the orientation of the reef (Figure 4). The highest densities in Palau are on reefs facing northeast and southeast; east- and west-facing reefs have very few animals. The orientation of the reef determines the amount of wave exposure and type of substratum.[12]

Although top shell density is frequently correlated with wave exposure, not all exposed

FIGURE 3. Juvenile *T. niloticus*. The shells have small spikes along the shell whorl and oblique reddish stripes.

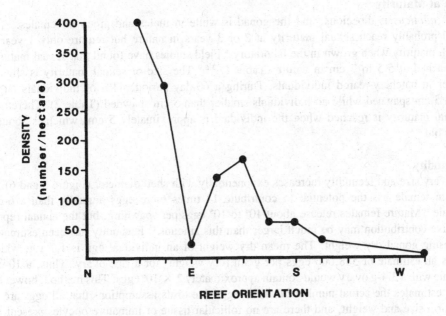

FIGURE 4. Relationship between reef orientation and mean density of *T. niloticus* at surveyed sites in Koror State, Palau. Density was positively correlated with degree of exposure to surf, and generally declines along a NE, E, S, W gradient. (From Heslinga, G. A., Orak, O., and Ngiramengior, M., *Mar. Fish. Rev.*, 46(4), 73, 1984. With permission.)

Table 1
SIZE AT MATURITY IN *TROCHUS NILOTICUS*

Country	Size at maturity (cm) Male	Size at maturity (cm) Female	Method	Ref.
Andaman Islands	6—7	6—7	Histology	23
	6—7	9	Histology	17
Australia	5—6	5—6	Histology/spawning	9
New Caledonia	7.2	7.2	Histology	20
	6.5—7	6.5—7	Histology	3
	8	8	Histology	24
Palau	5.5	5.5	Histology	25
	5.8	6.5	Histology	11
	5.0	5.0	Spawning	6

(From Heslinga, G. A., *Proc. Fourth Int. Coral Reef Symp.*, 1, 39, 1981. With permission.)

sites will support dense top shell populations. For instance, on Guam, the windward side of the island has very low top shell densities. This may be related to the substratum having a low relief topography that lacks shelter *Trochus* need for protection against direct wave exposure. Densities are higher (ranging from 0.05 to 2.13/20 m²) on the leeward side.[16]

REPRODUCTION

Size at Maturity

T. niloticus is dioecious, and the gonad is white in males and green in females.[7] Top shell probably reach sexual maturity at 2 or 3 years in nature but require only 1 year to reach maturity when grown in the laboratory.[6] Field studies have found that sexual maturity is reached at 5.5 to 7 cm in nature (Table 1).[3,6,9] The size of sexual maturity is slightly lower in hatchery-reared individuals. During a 60-day period, 44% of individuals larger than 5 cm spawned while no individuals smaller than 5 cm spawned (Table 2). Therefore, sexual maturity is reached when the individual is approximately 5 cm, which is about 1 year old.[6]

Fecundity

Ovary size and fecundity increases exponentially with shell diameter (Figure 5 and 6). A 12-cm female has the potential to contribute 14 times more eggs annually than a 6-cm female.[6] Mature females release about 10^4 to 10^5 eggs per spawning, but the annual reproductive contribution may be much larger than this amount.[15] Fecundity has been estimated by using gonad dry weight. The mean dry weight of an individual egg is 1.7 μg, which gives an estimate of 588,000 eggs per gram (dry weight) for a ripe ovary. Thus, a 10-cm female with a 3.4-g ovary would contain approximately 2×10^6 eggs. This method, however, over estimates the actual number of viable eggs due to its assumptions that all eggs are of uniform size and weight, and there are no follicular tissue or immature oocytes present.[6]

Spawning and Fertilization

Spontaneous natural spawning occurs throughout the year, indicating that there are at least some ripe individuals in the population at all times.[7] Brood stock (>6.0 cm in basal diameter) can be maintained in large (e.g., 5000 ℓ) cement tanks (30 to 50 animals per tank) with ambient unfiltered sea water (27 to 30°C) flowing at 30 ℓ/min. Fifty adults can be held in a 5000-ℓ tank and up to 200 animals in a 100,000-ℓ tank.[19] Adult top shell graze on the

Table 2
SIZE FREQUENCY DISTRIBUTION OF LABORATORY-CULTURED *TROCHUS NILOTICUS* SPECIMENS OBSERVED SPAWNING DURING JUNE AND JULY, 1980.

| Size class (mm) | N | Number observed spawning | | % Spawning |
		Male	Female	
40—44	12	0	0	0
45—49	6	0	0	0
50—54	8	1	1	25
55—59	6	4	1	83
60—64	2	0	1	50
65—69	9	5	0	56
70—74	2	2	0	100

Note: All specimens were members of the same F_1 cohort, reared on an *ad libitum* feeding regime for 12 months.

(From Heslinga, G. A., *Proc. Fourth Int. Coral Reef Symp.*, 1, 39, 1981. With permission.)

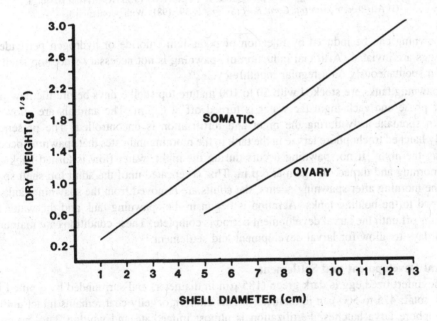

FIGURE 5. Relationship between shell diameter, and somatic-tissue dry weight and ripe-ovary dry weight in Palauan *T. niloticus*. Cube root transformation applied to weight data. Soma — Y = 0.0393 + 0.2305*X; r = 0.998; n = 26. Ovary — Y = −0.3184 + 0.1816*X; r = 0.995; n = 10. (From Heslinga, G. A., *Proc. Fourth Int. Coral Reef Symp.*, 1, 39, 1981. With permission.)

algae growing on the floors and walls of the tank. Survival is about 97% per month with these conditions.[15]

Brood stock will spontaneously spawn during the evening of the new moon or one of the three successive evenings. They exhibit a lunar spawning cycle superimposed on a nocturnal foraging cycle. Spawning occurs between 8:00 p.m. and 10:30 p.m., with males spawning first.[15] Females release a few eggs at a time as the body is retracted into the shell.[9]

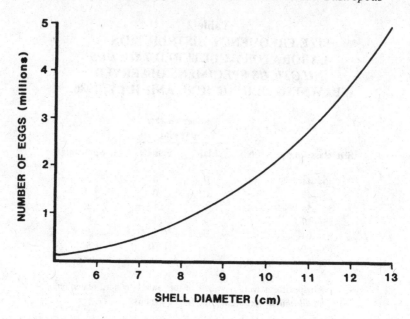

FIGURE 6.   Relationship between shell diameter and potential fecundity in Palauan *T. niloticus*. Based on size-specific dry weight of ripe ovaries (Figure 5) and mean dry weight of individual eggs (1.7 µg/egg). $Y = (-26.70 + 15.22*X)^3$. (From Heslinga, G. A., *Proc. Fourth Int. Coral Reef (Symp.*, 1, 39, 1981. With permission.)

Spawning can be induced by injection of potassium chloride or hydrogen peroxide but the eggs are inviable. Artificial induction of spawning is not necessary since top shell will spawn spontaneously on a regular monthly cycle.[19]

Spawning tanks are stocked with 50 to 100 mature top shell 3 days before the new-moon lunar phase and each night the water is turned off at 6 p.m. The animals are allowed to spawn spontaneously during the night and fertilization is uncontrolled. The presence of newly hatched trochophore larvae in the tank in the morning indicates that spawning occurred during the night. If no spawning occurs during the night, water flow is turned back on in the morning and turned off again at 6 p.m. This is repeated until the adult top shell spawn. On the morning after spawning occurs, the adults are removed from the spawning tanks and returned to the holding tanks. Aeration is begun in the spawning tank and the water flow remains off until the larval development period is complete. These conditions are maintained for 8 days to allow for larval development and settlement.[15]

Larval Development and Settlement

The unfertilized egg is dark green (185 µm in diameter) and surrounded by a pitted jelly layer (total, 475 to 500 µm in diameter). This chorion or jelly coat remains intact until the trochophore larva hatches. Fertilization is almost immediate and whithin 2 to 5 min the vitelline membrane rises to about 225 µm in diameter.[15]

The first spiral cleavage occurs after 30 min. Gastrulation is by epiboly (growth of one part over another in embryonic stages) and occurs in 5 to 6 hr, trochoblast cells are ciliated by 8 hr, gastrula begins rotating within the egg membrane after 9 hr, and the trochophore larva hatches out at 12 hr.[1,15] Newly hatched trochophores swim near the top centimeter of water, and alternately rise to the surface and then drop through the water column. The larvae are lecithotrophic and nonfeeding during the planktonic period. But examination indicates that there is no feeding prior to settlement. The larval shell begins to form at the vegetal pole about 1 hr after hatch out (13 hr) and spreads rapidly toward the anterior. Torsion occurs between 14 and 18 hr. The larval shell is completed by 20 hr (290 µm in diameter) and does not grow in size during larval development. The veliger is able to fully retract inside the shell after 24 hr.[15] Veligers continue to swim near the surface of the water but begin to show negative phototaxis (swimming to the opposite side from the light source) at about this time.[1] They will swim near the surface of the water for a minimum of 40 hr before settlement.[15]

Veligers settle and begin metamorphosis on live algal film at 50 to 60 hr. Newly settled larvae test the substratum with the tip of the propodium, flex the anterior foot edge, and pass the propodium over the mouth. The velum develops a midventral cleft that forms two lobes. The lobes are gradually resorbed, and two rudimentary cephalic tentacles and two rudimentary black eyes form near the center of the velum. The velar lobes slowly reduce in size and cilia are cast off until only a few remain beating. All cilia are cast off from the vestigial velum and metamorphosis is completed at about 70 hr.[1] It must be remembered that although the approximate timing of larval stages is repeatable, the exact timing of larval stages will vary among individuals and rearing conditions.

The duration a larva remains in the water column before settlement is highly variable with some remaining planktonic for 8 or more days before settlement.[15] Metamorphosis can be experimentally induced by algal films, red coralline algae (*Porolithon*), or γ-aminobutyric acid (GABA).[1,15] The quick response to algal films suggests that top shell are not narrowly restricted in physical substratum preference, although they show a preference for hard substratum over sand. The minimum time for larvae to metamorphose is 3 days after fertilization and all are metamorphosed in the presence of algal films by 8 days.[1]

The large tanks used for spawning are preconditioned for larval settlement by introducing unfiltered sea water (30 ℓ/min) for 2 to 3 weeks before adults are introduced for spawning. After this time there is a noticeable green algal film layer on the tank walls. Eight days after spawning, water flow is resumed in the tank and the juveniles are left to develop for 2 to 3 months. Survival from egg to metamorphosed juvenile exceeds 50% in large outdoor tanks and 85% in small laboratory culture. Using these techniques, batches of up to 500,000 juveniles have been reared.[19]

GROW OUT

The density of each culture tank is estimated every 30 days by taking the mean of twenty 100 cm² quadrat samples of the walls and floor.[15] Most individuals remain quiescent and clumped together in the corners of the tanks during daylight hours. Feeding occurs primarily during the night.[6] As the juveniles grow and begin to require more algae, the density is reduced by sweeping the walls of the settlement tank, siphoning up loosened juveniles, and transferring them to other tanks. The relationship between shell diameter and dry body weight in juvenile *T. niloticus* (<10 mm) is shown in Figure 7.[6]

The growth rate and survival of juveniles varies depending on the size of the tank used for grow out (Figure 8). Heslinga and Hillmann[15] reported that a 5,000-ℓ tank stocked with 8×10^4 larvae contained 1.5×10^4 juveniles after 2.2 months (18.7% survival) and 7000 juveniles after 4 months (8.7% survival). Mean juvenile size was 7.8 mm after 4 months with a maximum size of 15.5 mm. However, a 100,000-ℓ tank stocked with 1.5×10^6 larvae, contained 5×10^5 juveniles after 1.3 months (33% survival) and 3×10^5 after 2.2 months (20% survival). Mean juvenile size was 2.1 mm after 2.2 months. The 5000-ℓ tank produced a juvenile growth rate twice as fast as the 100,000-ℓ tank during the first 2 months, with a comparable survival rate. The slower growth in the larger tank was probably due to lower algal production on the walls and floor, and a lower water exchange rate. The 100,000-ℓ tank was 2 m deep and 8 m wide, which caused shading on a portion of the walls and floor during daylight hours. Also, less light reached the tank floor through the water and the water exchange rate was 20 times lower than in the 5000-ℓ tank.[15]

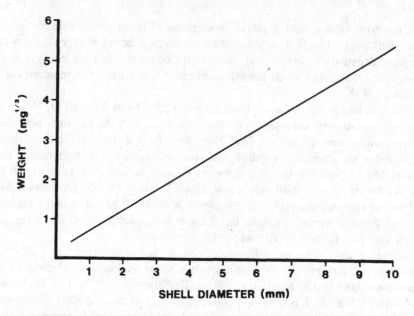

FIGURE 7.   Relationship between shell diameter and shell dry weight in small juvenile *T. niloticus* reared in the laboratory. Cube-root transformation applied to weight data. Y = 0.1081 + 0.5247*X; r = 0.996; n = 68. (From Heslinga, G. A., *Proc. Fourth Int. Coral Reef Symp.*, 1, 39, 1981. With permission.)

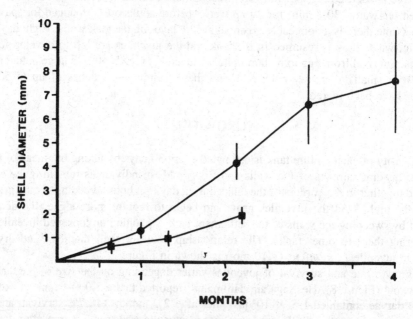

FIGURE 8.   Growth of two juvenile *T. niloticus* cohorts in different culture systems. (●) 5000-ℓ tank; eight water exchanges per day. (■) 100,000-ℓ tank; 0.4 exchanges per day. Data points represent the mean shell diameter with ±1 SD. (From Heslinga, G. A. and Hillmann, A., *Aquaculture*, 22, 35, 1981. With permission.)

Growth during the first 12 months is exponential with ad libitum feeding (Figure 9). Specimens average 6.2 cm in shell diameter after 12 months, with increased variability in size with age. The relationship between shell diameter and dry weight are shown in Figure 10. Growth rates in the laboratory are considerably faster than those reported in nature. Variations in habitat, water temperature, food, and method used to determine age make comparisons between the different field studies difficult.[6]

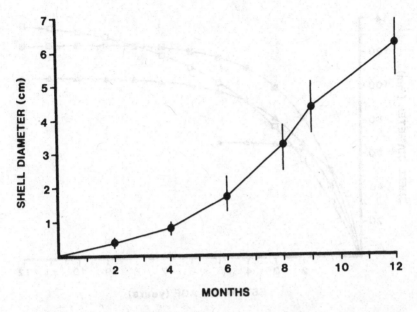

FIGURE 9. Laboratory growth of *T. niloticus* from fertilized egg to 12 months post fertilization. *Ad libitum* feeding. Data points represent mean shell diameters with ±1 SD. (From Heslinga, G. A., *Proc. Fourth Int. Coral Reef Symp.*, 1, 39, 1981. With permission.)

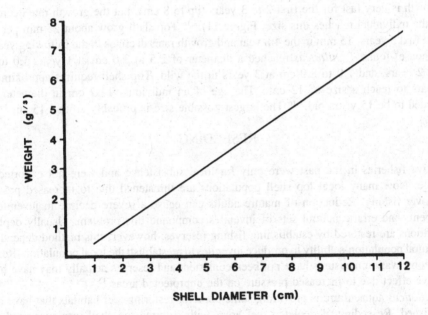

FIGURE 10. Relationship between shell diameter and shell dry weight in large juvenile and adult *T. niloticus*. Cube root transformation applied to weight data. $Y = -0.1802 + 0.6461*X$; $r = 0.997$; $n = 87$. (From Heslinga, G. A., *Proc. Fourth Int. Coral Reef Symp.*, 1, 39, 1981. With permission.)

Laboratory animals maintain nocturnal feeding behavior even with unlimited food. Feeding could possibly be increased by reducing the length of the day photoperiod or reducing light intensity, but this would have serious ramifications on the production of algae on tank surfaces. At present, juveniles only eat naturally occurring algae in the tank and are not supplied with any supplemental algae. If light was reduced, algae would have to be grown separately and introduced into the tanks to maintain a high animal density.[6]

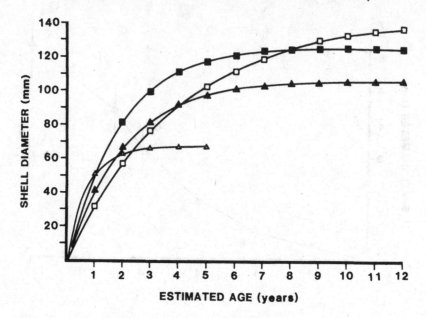

FIGURE 11.   Growth curves calculated from data in previous studies of *T. niloticus*. (■) Rao,[17] (□) Smith,[16] (▲) Asano,[25] (△) Moorhouse,[9] (From Smith, B. D., Master's thesis, University of Guam, 1979.)

Growth is very fast for the first 2 to 3 years (up to 8 cm), but the growth rate decreases after the individual reaches this size (Figure 11).[3,16] Top shell grow about 25 mm per year for the first 3 years, 15 mm in the 4th year and growth rates declines in the following years.[16] Moorhouse[9] found *T. niloticus* reached a diameter of 2.5 to 3.0 cm at 1 year, 5.0 to 6.0 cm at 2 years, and 7.0 to 8.0 cm at 3 years in the wild. Top shell requires approximately 10 years to reach a size of 12 cm.[3] The age of an individual 14.2 cm in diameter was estimated to be 15 years old.[7,16] The largest possible size is probably around 15 cm.[16]

RESEEDING

Native fisheries in the past were only for local subsistence and were not exported for revenue. Now many local top shell populations are threatened due to increased pressure from over fishing. Reduction of mature adults can cause a severe decline in juvenile recruitment, and erratic natural sets of juveniles compound this problem. Usually depleted populations are restored by establishing fishing reserves; however, this method depends on the natural population's ability to produce juveniles to reestablish the local population. Results from conservation measures have not been consistent and reserves actually may have had a negative effect due to increased pressure on the unprotected areas.[12]

T. niloticus aquaculture is providing some hope for restoring reef habitats that have been over fished. Reseeding of depleted reef areas with cultured top shell was suggested as a conservation method in 1930 by Risbec,[20] even though the technology to do so was not available at the time.[15] Now that the technology is available, juvenile reseeding has become a feasible management strategy. The use of reseeded juveniles to increase the natural population and maintain the fishery must be carefully studied before it is professed to be the solution to the problem. Reseeding of abalone in California has not been successful due to predation of newly released juveniles.[21] In addition, reseeding may not be the most cost-effective method for maintaining natural populations.

At present there is no data on mortality rates, sources of mortality, and movement of juvenile *T. niloticus* of different sizes and in different habitat types. These data would be valuable in gaining a basic understanding of top shell ecology and in designing successful reseeding strategies. The needed information could be obtained by using tagged, hatchery-produced juveniles in release and recapture studies. Important variables to analyze include microhabitat, initial size at time of release, reseeding density, and methods used to place juveniles (e.g., by hand, on tank substrata) in the wild.

Evaluating seeded juvenile survival is difficult because individuals may migrate large distances in nature and inhabit inaccessible areas for recapture. Frequent (initially weekly) sampling of 1-m² quadrants, radiating in all directions from the release point, will allow estimation of emigration from the planted area, and will identify the directionality of movement. The size of the area sampled will depend on the rate of movement by the juveniles. Losses not explained by emigration or sampling inefficiency will give an estimate of mortality. Empty shells should also be collected and examined for damage caused by crushing or drilling predators. If a predator species is implicated as an important source of mortality, methods should be developed to exclude the predator from reseeded sites. The true determination of the success of reseeding will be the number of subadults (5 to 6 cm) found in the reseeded area 1 year after planting juveniles. A survival rate (number of animals recovered in the reseeded area after 1 year) greater than 10% should be considered a success.

As discussed in the Ecology section, the main habitat for juvenile top shell is shallow or intertidal reef-flat cobble and rubble, and generally, this is probably the best habitat for release of hatchery-reared seed. A suitable habitat should provide enough food for maximum growth and refuge from predators. The size at which the seed is released is unavoidably a trade off between hatchery costs and field survival. If the relationship is known between these factors, then an optimum size can be determined for release. The first 4 months of juvenile growth are relatively slow and the juveniles are probably extremely vulnerable to predation during this time. For instance, Tegner and Butler[21] found that for abalone, if the juveniles were too small at the time of release, they would be preyed upon almost immediately. It is probably best to retain juveniles in the laboratory during this slow growth period to maintain a high survival. Growth rate and food consumption increases rapidly after the first 4 months, and with increased size, probably some protection from predation. Also limited grow-out space and use of naturally occurring algae for food means it is progressively more expensive to raise the juveniles as the individuals grow larger. For these reasons, it has been suggested that release of seed should be at 10 mm.[6]

If reseeding is going to be successful, methods must be developed that are appropriate to the level of technology and development available on most Pacific islands (e.g., availability of electricity and equipment, and cost of materials.).

FEASIBILITY OF DEVELOPING TOP SHELL AQUACULTURE

The culture of juvenile top shell for reseeding depleted reefs offers an opportunity to maintain the *Trochus* fishery at a high level. Currently, the major restriction to successful culture of this animal on a hatchery scale is the limited time of spontaneous spawning in nature. Spawning occurs only during the new moon lunar phase, and spontaneous spawning does not occur every month and populations probably spawn on different months.[22] Dependence on natural spontaneous spawning is both inefficient and inconvenient for successful hatchery development. Since natural spawning occurs at night, controlled fertilization and following procedures during larval development would have to be done at night. Also, if there are any problems with the larval culture, it will be at least 1 month (probably more) before the next spawning. Development of methods to artificially induce spawning allows control over the time of spawning, number of adults needed for brood stock, controlled fertilization, synchronized larval development, and continuous production of larvae. These qualities enable the best use of the equipment with the minimum of inefficiency.

The development of hatchery methods to produce juveniles for reseeding will help preserve the natural environment in two ways. First, aquaculture is able to supply juveniles of economically important animals for restocking without putting added pressure on the natural populations. The development of hatchery methods will help maintain the natural ecological balance on the reef and keep populations of top shell at high levels capable of sustaining reliable harvests. Second, and more significantly, these methods will contribute to preservation of the islanders' way of life and culture. It is very difficult for islanders to make money and still remain living on a remote island. Maintaining the top shell fishery will help islanders continue to make money by traditional means (fishing) rather than by culturally disruptive means, such as, tourism. The aim of aquaculture in less developed countries is to maintain the level of the desired food source and at the same time keep the local fishermen employed. Aquaculture allows the fisherman to go from being a "hunter/gatherer" to a "farmer" and have more control over his livelihood.

SUMMARY

T. niloticus is an ideal species to culture for the following reasons.

Spawning
1.   Adults are easy to collect and maintain.
2.   Gravid brood stock is available year round.
3.   Spawning occurs spontaneously on a predictable monthly schedule.
4.   Females are very fecund.

Larvae
1.   Lecithotrophic (i.e., feed on yolk) and do not require any supplemental feeding.
2.   Short planktonic period and high survival rate with minimal care.
3.   Resistant to bacterial diseases.

Juveniles
1.   Herbivorous and grow well on algae occurring naturally in grow-out tanks.
2.   Raised in unfiltered water and without antibiotics.
3.   Survive up to 24 hr out of water, which will enable transportation by air freight to locations distant from the hatchery.[15,19]

Adults
1.   Herbivorous and detrivorous, which is a favorable trophic level for culture.
2.   Individuals reach market size in 2 to 3 years after being placed on the reef.
3.   Only the dry shell is exported, therefore the usual health and shipping constraints associated with perishable products does not apply to this fishery.
4.   The skill, investment, and technology required for harvesting top shell is minimal, making this resource easily accessible to islanders.

The combination of biological, economical, and sociological factors make top shell an appropriate species for aquaculture.

REFERENCES

1. **Heslinga, G. A.**, Larval development, settlement and metamorphosis of the tropical gastropod *Trochus niloticus, Malacologia*, 20(2), 349, 1981.
2. **Wells, S. M.**, International trade in ornamental corals and shells, in *Proc. Fourth Int. Coral Reef Symp.*, 1, 323, 1981.
3. **Bouchet, P. and Bour, W.**, The trochus fishery in New Caledonia, *So. Pac. Comm. Fish. Newsl.*, 20, 9, 1980.
4. **Glucksman, J. and Lindholm, R.**, A study of the commercial shell industry in Papua New Guinea since World War Two, with particular reference to village production of trochus *(Trochus niloticus)* and green snail *(Turbo marmoratus), Sci. New Guinea*, 9, 1, 1982.
5. **Van Pel, H.**, *The fishery industry in the Cook Islands*, South Pacific Comm., Noumea, 1955, 1.
6. **Heslinga, G. A.**, Growth and maturity of *Trochus niloticus* in the laboratory, in *Proc. Fourth Int. Coral Reef Symp.*, 1, 39, 1981.
7. **Salvat, B. and Rives, C.**, *Coquillages de Polynésie*, Les Editions du Pacifique, Papeete, 1983, 239.
8. **Anon.** Fishery resources profile no. 3. Trochus (Sici), Ministry of Primary Industries, unpublished report, 1985, 13.
9. **Moorhouse, F. W.**, Notes on the *Trochus niloticus, Sci. Rep. Great Barrier Reef Exped. 1928—1929*, 3, 145, 1932.
10. **Udai, E. S. and van den Andel, W.**, Palau trochus products: An investment opportunity, *Rep. High Comm., U.S. Trust Terr. Pac. Isl.*, 1, 1, 1981.
11. **McGowan, J. A.**, The current status of the *Trochus* industry in Micronesia, *Rep. High Comm., U.S. Trust Terr. Pac. Isl.*, 1, 1, 1956.
12. **Heslinga, G. A., Orak, O., and Ngiramengior, M.**, Coral reef sanctuaries for trochus shells, *Mar. Fish. Rev.*, 46(4), 73, 1984.
13. **Geller, J. B.**, Microstructure of shell repair materials in *Tegula funebralis* (A. Adams, 1855), *The Veliger*, 25(2), 155, 1982.
14. **Maoate, T.**, personal communication, 1984.
15. **Heslinga, G. A. and Hillmann, A.**, Hatchery culture of the commercial top snail *Trochus niloticus* in Palau, Caroline Islands, *Aquaculture*, 22, 35, 1981.
16. **Smith, B. D.**, Growth Rate, Abundance, and Distribution of the Topshell *Trochus niloticus* on Guam, Master's thesis, University of Guam, 1979.
17. **Rao, H. S.**, On the habitat and habits of *Trochus niloticus* Linn. in the Andaman Seas, *Rec. Indian Mus.*, Calcutta, 39, 47, 1937.
18. **Trevelyan, G.**, personal observation, 1984.
19. **Heslinga, G. A.**, Report on Palau's trochus hatchery project, *So. Pac. Comm. Fish. Newsl.*, 20, 4, 1980.
20. **Risbec, J.**, Etude d'un Mollusque Nacier le Troque *(Trochus niloticus* L.), in *Faune de colonies francaises*, Vo. 4, Gruvel, A., Ed., Societe d'Editions Geographiques, Maritimes et Coloniales, Paris, 1930, 149.
21. **Tegner, M. J. and Butler, R. A.**, The survival and mortality of seeded and native Red abalones, *Haliotis rufescens*, on the Palos Verdes peninsula, *Calif. Fish Game*, 71(3), 150, 1985.
22. **Hahn, K. O.**, personal observation, 1984.
23. **Amirthalingam, C.**, Correlation of sex and shell structure in molluscs, *Trochus niloticus* Linn., *Curr. Sci. (Bangalore)*, 1, 72, 1932.
24. **Gail, R.**, Contribution a l'Etude due Troca en Nouvelle-Caledonie, Office de la Recherche Scientifique et Technique Outre-Mer, Noumea, 1, 1958.
25. **Asano, N.**, On the spawning season of top shell, *J. Fish.*, 35, 36, 1939.

CULTURE OF QUEEN CONCH, *STROMBUS GIGAS*, IN THE CARIBBEAN

Kirk O. Hahn

INTRODUCTION

Strombus gigas Linnaeus 1758 has several common names: queen conch, pink conch, and pink-lip conch (U.S.); botuto (Venezuela); cambombia (Panama); carrucho (Puerto Rico); cobo (Cuba); guarura (Los Roques); and lambie (Hispaniola).[1,2] Queen conch is distinguished from the five other species of *Strombus* in the Caribbean by its large size, lack of pronounced spiral grooves, and pink-colored aperture. The juvenile has spines on the spire almost to the apex; shell color is a grayish white with reddish-brown markings; peduncle is white with green patches; antennae are light green with white patches; foot is white to beige with green, purple, and brown patches; and proboscis is greenish-black with white patches.[2,3]

Queen conch is distributed from Bermuda, the Bahamas, southern Florida, the entire Caribbean area, Central America, to northern South America.[4,5] It is the second most valuable fishery in the Caribbean and has been a source of protein for local people for at least the last 100 years.[1,4] A 2.5-year-old queen conch yields up to 1 kg of meat (75% protein by dry weight).[5] Individuals are harvested for food throughout its range but the importance of the fishery depends on local abundance and local food preference.[2]

Queen conch is almost 100% usable. In addition to meat, the shell has considerable ornamental value.[2,6,7] The shell is sold as a curio, carved into cameos, or pulverized for manufacturing porcelain. Pink pearls are occasionally found in the mantle tissue.[2] Queen conch has a total commercial value of less than $5 million (U.S.) per year.[8] Approximately 12% of the total body weight is marketable meat (Figure 1) and the retail price is $2.20 to $3.30/kg.[1,7]

The density of queen conch has been reduced in areas readily accessible in Puerto Rico, Venezuela, and the Grenadines.[9] Presently, there are only three areas that still have commercially exploitable stocks of conch: the Bahamas, Turks and Caicos, and Belize.[5]

The fishery has proceeded from shallow to deeper habitats to maintain the level of catch. The resource is presently in jeopardy of declining to levels incapable of sustaining continued harvest. Queen conch has already been placed on the list of "commerically threatened invertebrates".[10]

Mariculture has been suggested as a method of increasing populations depleted from overfishing. Hatchery-raised juveniles could be released to reseed natural populations in depleted areas. However, it has not been determined what effect natural predators will have on the reseeding efforts.[11]

As of 1985, there were at least eight active research and development programs; Foundation for PRIDE, Department of Marine Science of the University of Puerto Rico, Mexico's Centro de Aquacultura in Quintana Roo, USAID programs in Belize and Haiti, a facility in Martinique operated by the French development organization (IFREMER), Institute for Applied Research in the U.S. Florida Keys, and Bonaire's Carco Project.[8]

ECOLOGY

Habitat

Queen conchs are most abundant in grass beds (*Thalassia testudinum, Syringodium filiforme,* and *Cymodocea manatorum*) at intermediate depths (4 to 8 m).[2,12] They are also found on sand flats, and occasionally on gravel, coral rubble, and smooth, hard coral rock.[2] Animals are found from intertidal down to 60 m, but individuals are rarely found below 30

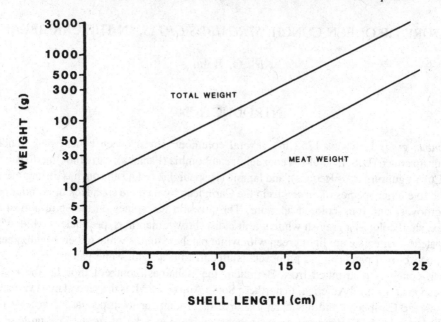

FIGURE 1.   Correlation between maximum shell length of juvenile *Strombus gigas* and both total weight and marketable meat weight. (Berg, C. *Mar. Biol.*, 34, 191, 1976. With permission.)

m.[1,2] The lower depth is probably restricted by the depth of the sea grass.[2] Mean densities in unfished areas are approximately 0.42 individuals per square meter. Juveniles are most commonly found in very shallow grass beds, less than 1.0 m depth, that have abundant food supply, low-velocity currents, and little wave action. Recruitment occurs in shallow sea grass beds during the summer months. Natural recruitment of conchs to depopulated areas is very high and populations are capable of re-establishing natural densities in very short time periods.[12]

Adults are capable of moving large distances (1 to 4 km) when the grass beds are very large. Queen conch migrate seasonally, moving inshore during spring and then moving offshore again in fall. The home range of animals (the area habitually traveled) varies depending on their size. Animals 10 to 13 cm long use 1000 m² and animals 13 to 16 cm long use 2500 to 5000 m².[13] Adult conch travel over a range of several kilometers and have been found 2 km from their release site after 2 months.[1] However, juveniles move very little and remain in small areas for long periods (>1 year).[12]

Growth

The mean natural growth rate of juvenile queen conch is approximately 0.9 cm/month, which is relatively fast.[12] Individuals can grow to 260 mm and weigh up to 3.3 kg. At sexual maturity, conchs stop adding shell length and devote energy instead to the formation and thickening of a flaring lip.[7] Conch become progressively heavier because of shell thickening and lip formation without any major increase in length.[2]

The mean life span of *Strombus gigas* is 5.7 to 6.0 years. There is a high correlation between meat weight, total weight, and shell length. Large juvenile conch (immature) have the largest percentage of marketable meat. An animal should be at least 18.8 cm long and weigh 845 g (approximately 2.5 years old) before being harvested. This size animal yields about 100 g of usable meat.[7] Queen conchs reach market size before reaching sexual maturity which is detrimental to establishing a sustainable fishery.[1,14] The average size is 10.8 cm at 1 year old, 17 cm at 2 years old, 20.5 cm at 3 years old, and flared lip stage after 3 years old.[7] Generally, adult female conch are larger than male conchs.[2]

Table 1
KNOWN PREDATORS OF QUEEN CONCH

Common name	Scientific name
Octopus	*Octopus vulgaris*
Lamp shell	*Xancus angulatus*
Tulip snail	*Fasciolaria tulipa*
Apple murex	*Murex pomum*
Horse conch	*Pleuroploca gigantea*
Hermit crab	*Paguristes grayi*
Giant hermit crab	*Petrochirus diogenes*
Coral crab	*Carpilius corallinus*
Stone crab	*Menippe mercenaria*
Yellow box crab	*Calappa gallus*
Blue crab	*Callenectes sapidus*
Spiny lobster	*Panulirus argus*
Tiger shark	*Galeocerdo cuvieri*
Spotted eagle ray	*Aetobatis narinari*
Southern stingray	*Dasyatis americana*
Massau grouper	*Epinephelus striatus*
Mutton snapper	*Lutjanus analis*
Dog snapper	*L. joco*
Gray snapper	*L. griseus*
Graysby	*Petrometodon cruentata*
Yellowtail snapper	*Ocyurus chrysurus*
White grunt	*Haemulon plumieri*
Bluestriped grunt	*H. sciurus*
Permit	*Trachinotus falcatus*
Hogfish	*Lachnolaimus maximus*
Queen triggerfish	*Balistes vetula*
Porcupinefish	*Diodon hystrix*
Loggerhead turtle	*Caretta caretta*

(From Jory, D. E. and Iversen, E. S., *Proc. Gulf Caribb. Fish. Inst.*, 35, 108, 1983. With permission.)

Predators

There are many animals that prey upon queen conch (gastropods, cephalopods, crustaceans, fish, and reptiles). A total of 28 predatory species have been identified (Table 1).[2,11] There is a 50% decrease in mortality rate in juveniles larger than 10 mm and the number of possible predators reduces after the conch reaches a size range of 10 to 15 cm.[11,15] Once conchs reach the flared-lip stage, they are rarely subject to predation.[7] The level of predation is seasonal and probably related to the water temperature. The highest predation rate is during the summer and the lowest during the winter. Thus, from a mariculture point of view, releasing seed is probably best during the winter when there is reduced predation.[11]

REPRODUCTION

Spawning

Queen conchs have internal fertilization. Copulation and spawning is controlled by the water temperature and usually occurs during the warmer months of the year, although some areas have reproductive activity year round.[1-3,16,17] The main spawning season is from April to November.[2,3,12]

Female queen conch prefer clean sand with low organic content as substratum for spawning of the egg mass.[16] A female is capable of laying 3 to 4 egg masses per month, about 25 egg masses per female per breeding season.[12] The egg mass is crescent shaped and covered with sand grains. The sand grains probably camouflage the egg mass and physically discourage predation.[2] The egg capsules are contained inside a single continuous tube coiled inside a thread running perpendicular to the long axis of the egg mass.[2,16] The average rate of laying a strand of eggs is 1.55 m/hr. Production of the egg mass takes 24 to 36 hr.[16] The total length of the egg tube is approximately 24 to 37 m with 12 to 15 eggs per millimeter and contains approximately 300,000 to 750,000 eggs.[2,4,16]

Larval Development

The embryo develops within the egg capsule. At 25°C, the two- and four-cell stage occurs 9 hr after extrusion of the egg mass, the gastrula develops after 16 hr and the trochophore larva develops after 58 hr.[2] The veliger larva hatches out around 5 days after extrusion. The larva pushes its way out of the capsule through a "door" in the egg capsule and immediately swims toward the water surface.[3,16] The average time of hatch out is 108 hr after spawning with a range of 104 to 114 hr.[3] Torsion and hatch out occur during the same time period.[16] The emerged veliger has two minute lobes, which it immediately uses for swimming. The two lobes divide into four lobes by the fifth day after hatch out, and are used for food gathering and as swimming appendages. These lobes comprise the velum. By the 8th day, the anterior pair of velar lobes divide again.[3]

Veliger larvae eat a narrow range of phytoplankton and cultured larvae cannot survive past day 9 without additional phytoplankton being supplied to the culture tank. *Platymonas tetraselmis* is a suitable phytoplankton for use during larval culture.[16] The veligers continue swimming and feeding near the surface, although they often spend brief periods near the bottom, before assuming a totally benthic existence.[3]

Settlement normally occurs at 18 to 21 days; however, the pelagic period is extended when the veligers are under nourished. Veligers crawl on the sides and bottoms of the tanks as soon as they settle but continue planktonic feeding. The first 10 days after settlement (up to metamorphosis) are transitional for the larvae. Feeding changes from gathering plankton by the velar lobes to grazing benthic diatoms with the proboscis. The veliger stops swimming with the velar lobes and begins crawling with its foot, and the velar lobes cease being used for respiration once the ctenidia develop. The juvenile shell also begins to harden after settlement.[3]

Metamorphosis

Metamorphosis is a critical stage in the life of the queen conch.[16] Metamorphosis takes approximately 12 hr for the transformation from a swimming veliger to a benthic juvenile. There are certain morphological characteristics that indicate when a veliger is approaching metamorphosis. Metamorphosis is complete after development of the proboscis, outward migration of eyes, and disappearance of velar lobes. The time to metamorphosis is highly variable and is dependent upon the water temperature and feeding level. Metamorphosis occurs naturally between 12 to 35 days (usually 18 to 21 days) after hatch out.[3-5,14] Minimum size at metamorphosis is 1.1 mm (not including velar lobes), but most are 1.4 to 1.5 mm. The food source changes from planktonic algae to benthic diatoms after metamorphosis. The most common foods eaten after metamorphosis are *Chaetoceros* spp., *Nitzschia* spp., and *Skeletonema costatum*.[3] After metamorphosis, growth increases to 0.2 mm/day and remains constant for the next 2 weeks. The growth rate then increases to 0.3 to 0.4 mm/day for the next 200 days (Figure 2).[14]

CULTURE METHODS

Collection of Egg Masses

At this point, induction of spawning out of season is impossible. All egg masses are

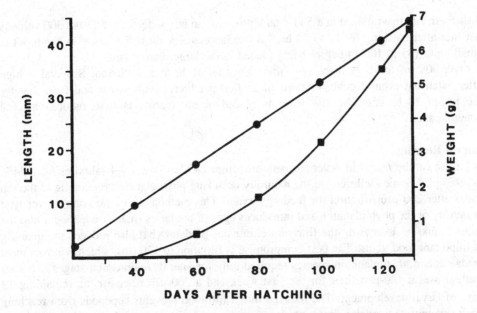

FIGURE 2. Shell length and weight of cultured *Strombus gigas* juveniles. (●) shell length; (■) weight. (Ballantine, D. L. and Appeldoorn, R. S., *Proc. Gulf Caribb. Fish. Inst.*, 35, 57, 1983. With permission.)

collected from the wild.[4] Freshness and predictable hatch out is ensured by collecting egg masses from beneath animals.[18] Egg masses which are in the process of being extruded are preferred.[19] After returning to the lab, the egg mass is gently teased apart (removing large sediments and algae), dipped in a 0.5% solution of Clorox for 45 to 60 sec to remove protozoans and bacteria, and then thoroughly rinsed several times in fresh sea water.[4,18,19] This treatment disinfects the egg mass and gives a 99% hatch out.[4,20] The egg masses are then placed in a mesh bag in a highly aerated aquarium filled with freshly filtered sea water. The bag keeps the egg masses off the bottom, allows good circulation, and facilitates removal of the egg masses for daily water changes and for transfer to the rearing tank for hatching.[18]

An alternate method is used at PRIDE. Egg masses are teased apart but are not treated with Clorox. The separated egg strands are placed in a screen enclosure and hung in flowing raw sea water. A newly laid egg mass or one collected underneath the female begins hatching between 85 to 109 hr after collection at a temperature of 28.5°C. The egg mass is brought into the lab the day before the predicted hatching, rinsed well, and then placed in the rearing tank for hatch out. This method produces healthy larvae, and saves labor, equipment, and handling of the egg mass.[18]

An enclosed, natural breeding habitat was established by PRIDE for production of a reliable supply of egg masses. A 1600-m² enclosure is located in 6 m of water on sand and coral rubble which is covered with algae, and is 75% enclosed by a natural coral reef. The brood stock (4- to 6-year-old adults) are stocked (1:1 sex ratio) at a density of one per 10.3 to 16 m². Mean number of egg masses produced per female per month is 1.2 to 1.7. The brood stock produces an average of 16.6 egg masses per week during the breeding season. It is estimated that a supply of ten egg masses per week will adequately support a 40-tank, commercial-scale hatchery.[21]

Hatch Out

The day before hatching, the egg mass is placed inside a 15-cm diameter PVC dish placed in the upper portion of a 200- to 500-ℓ rearing tank.[18,19] This separates the healthy swimming veligers from the egg mass debris, and weak and dead veligers which remain inside the dish.[18] An egg mass placed in a 500-ℓ tank produces an initial density of 250 to 300 veligers per liter after hatching for 20 to 24 hr.[18] Some laboratories allow the larvae to hatch out in small tanks (35 to 100 ℓ) before being placed in the large rearing tanks.[5]

Over 90% of the eggs hatch out during the first 36 hr after initiation. Survival is high after hatch out even at concentrations of 20,000 per liter.[3] Daily water and tank changes are begun 24 hr after the egg mass is placed in the rearing tank to reduce bacterial contamination.[18]

Larval Rearing

Larvae can be reared in water temperatures from 24°C to 30°C, and salinities from 34‰ to 39‰.[3,5,18] Some facilities use the naturally occurring phytoplankton remaining in the sea water after 200 μm filtration for feeding larvae.[5] This method allows no control over type or density of the phytoplankton and introduces eggs of predators (mainly copepods) into the culture tanks.[3,5] Increasing the filtration eliminates predators but also reduces the quantity of important food items. The best compromise is filtration to 120 μm, which removes most predators and phytoplankton species required only for later developmental stages. Another method uses a 100-μm filter for the first week and a 200-μm filter for the remaining 12 days of larval development. The reduced initial filtration prevents copepods from reaching adult size until the conch veligers are larger and more resistant to predation. Aeration is not necessary as long as there is frequent or continuous introduction of clean sea water, a supply of an appropriate phytoplankton species in the tank, and removal of water from the tank bottom.[3] The optimum density is ten larvae per liter with nonintensive methods (no aeration, low water flow, and use of naturally occurring phytoplankton as the food source).[1,3]

Larval density can be significantly increased with intensive culture techniques (aeration, water filtration to approximately 10 μm, and feeding cultured phytoplankton).[5] The large larvae, 2 to 3 mm across the velar lobes, are difficult to raise in high densities due to their size.[4] Larval densities of 250 to 300 veliger per liter are obtained with aeration and the introduction of cultured phytoplankton.[18] Hensen states the best initial larval concentration is about 3500 larvae per liter after hatch out, which is reduced by mortality to 200 to 600 larvae per liter by day 28, just before metamorphosis.[19]

Feeding begins 2 days after hatch out with 1,000 cells per milliliter and increases until it is 25,000 to 30,000 cells per milliliter 10 days after hatch out.[4] Phytoplankton is maintained at 30,000 cells per milliliter because the ingestion rate of food declines at food densities above 50,000 cells per milliliter or larval densities above 1/mℓ.[4] Strong lighting from above the culture tanks encourages growth and reproduction of the phytoplankton cells supplied for food.[4,5] Phytoplankton cultures must be of high quality when fed to the larvae. Giving no food is preferable to given poor quality food.[18]

The hatchery at PRIDE was designed for efficient larval culture and ease of feeding the developing larvae. Each tank in a module recives approximately 10 to 12 ℓ of phytoplankton, dripped into the tank over 12 hr. The phytoplankton is continuously diluted by water flow through (2 ℓ/min) and evenly distributed by gentle aeration. The larvae are reared at a density of 20 to 39/ℓ in 43 egg-shaped tanks (1000 ℓ) arranged in five modules with 8 to 9 tanks. Feeding efficiency is increased by having the same age class of veligers in each module. Veligers are examined daily to ensure that they ate enough food the previous night. Algae in the gut is visible through the clear shell. The larvae are fed a local phytoplankton, C-*Isochrysis* (C-Iso), (8,500 to 10,000 cells per milliliter) from day 1 (after hatching) to day 14. *Chaetoceros gracile* is supplemented after day 14 at 15% of the total volume of algae given at a density of 12,000 to 15,000 cells per milliliter.[22]

The critical periods during larval culture occur at 1 to 4, 10 to 12, and 18 to 21 days. The mortality at 10 to 12 days is usually massive and occurs within a few hours. The mortality at 18 to 21 days is not as massive and occurs over 2 to 3 days. This mortality is probably caused by high densities of predators in the culture tank. The mortality during the first 4 days of larval development could not be explained.[5]

Most facilities use a static water system since continuous water flow causes fouling of the larval shell with microalgae.[4] The water is changed by stopping the aeration, allowing settlement of the dead and weak veligers, and debris to the tank bottom. PRIDE uses tanks with a valve midway down the side for quick removal of water. The valve is opened and all swimming veligers are collected on a tall cylindrical sieve sitting inside a 20-ℓ bucket. The proper size sieve must be determined before the water change. The minimum size shell length is determined by measuring a sample from the rearing tank. Occasionally the slower growing and thus smaller individuals are removed by using the appropriate size sieve. This procedure allows uniform growth and development for the batch.[18]

The sieves are made of PVC cylinders, 15 cm in diameter. Nitex or polyethylene screen mesh of 157 to 300 μm are used on the bottom and sides of the sieve. Water is changed quickly but carefully so the veligers are not over crowded, do not lose growing time, and are not stressed. As the veligers get larger and hardier, water changes become faster. Water is changed in the morning when the temperature is the lowest (27 to 29°C) as opposed to the afternoon (30 to 31.5°C). The water at PRIDE is filtered to 15 μm to remove predators, macroalgae, and debris but is not sterilized with UV light.[18]

Laughlin and Weil[5] replace only half the water in the rearing tank every 2 days and a dense phytoplankton culture (10% of total volume of rearing tank) is introduced as food for the larvae after each water change. Siddall[4] recommends not changing the water until 4 days after hatch out since the larval heart stops beating when the larva is disturbed (i.e., changing the water). The adult heart is formed by day 4 and does not stop like the larval heart. There is a 100% exchange of water every day after the larvae are 4 days old from hatch out.[4]

The rate of veliger development depends on handling techniques, water temperature, larval density and, most importantly, quality, quantity, and type of food. Veligers develop faster when given high concentrations of food, but at the critical stage at 10 to 12 days, 50 to 70% of the larvae died.[18] Feeding larvae smaller amounts of food slows development and enables the veligers to pass safely through this critical period.[18]

Induction Of Metamorphosis

Once larvae are competent to settle, 12 to 35 days after hatch out (fast development 14 to 19 days and slow development 28 to 35 days) and 1.2 to 1.9 mm in length, metamorphosis can be artificially induced by red algae extracts.[3-5,14] Competent larvae are collected on a nitex screen and exposed to a dilute solution of extracts from red macroalgae (usually *Laurencia obtusa*) for 2 to 3 hr. The larvae are then placed in a shallow, submerged tray filled with fine, filamentous macroalgae (often *Ceramium* sp.). Within 12 hr, 90 to 95% of the larvae complete metamorphosis with nearly 99% survival.[4]

PRIDE does not artificially induce metamorphosis. As soon as the first one or two veligers naturally metamorphose, all veligers are transferred to a screen-enclosed basket on a wet table. Veligers in the basket are exposed to unfiltered sea water (100% water changes occurs every day), diatom and epiphytic algae growth on the screen, macroalgae (*Laurencia* and *Batophora*), sunlight and aeration. Phytoplankton is supplied until no swimming veligers remain. After 3 to 4 weeks, the juveniles are transferred to a larger screen enclosure and freshly collected *Laurencia* and *Batophora* are introduced into the screen enclosure every 1 or 2 weeks.[18]

Food
Phytoplankton Culture

The phytoplankton species usually cultured for food are Tahitian *Isochrysis, Nannochloris oculata, Dunaliella tertiolecta, Thalassiosire weissflogil, Chaetoceros gracile,* and *Tetra-*

selmis chuii.[4,19] Conch larvae are occasionally fed *Emiliania huxleyi, Prorocentrum minimum,* and *Heterocapsa pygmacea.*[23]

A phytoplankton species with good food value produces larvae with a siphonal length of 0.9 to 1.2 mm in 20 days and greater than 80% survival.[23] Using these criteria, Tahitian *Isochrysis, E. huxleyi, P. minimum,* and *H. pygmacea* are classified as phytoplankton species with good food value for conch larvae. There is a positive correlation between food value, and protein content, lipid content, and fatty-acid composition of the phytoplankton.[23,24] A lipid content of approximately 15% (dry weight) in the phytoplankton is optimal for larval conchs.[24] Tahitian *Isochrysis* has a cell content of 41% protein and 20 to 23% lipid, and is the most stable food source for conch larvae.[18,23,25] *E. huxleyi* has 23% lipid, *P. minimum* has 32% protein and 12% lipid, and *H. pygmacea* has 13% lipid. The four phytoplankton species with good food value contained the fatty acids 20:5ω3 and/or 22:6ω3. Fastest growth in conch larvae occurred when both fatty acids were present in large quantities in the phytoplankton. *Dunaliella tertiolecta* has a poor food value. It has a lipid content of only 7% and lacks the fatty acids 20:5ω3 and 22:6ω3.[23]

The phytoplankton culture room is the most expensive part of a hatchery due to the high cost of electricity for lights and air conditioning. Reducing these costs is possible by raising the phytoplankton cultures under ambient conditions (e.g., partially shaded side of a building). Variable air temperatures and sunlight cause unpredictable phytoplankton cultures, although they are of good quality.[18]

The culture method at the PRIDE laboratory originally used 20-ℓ carboys to raise the phytoplankton. Monocultures of Tahitian *Isochrysis, Nanocloris, D. tertiolecta,* and *Thalassiorsire weissflogil* were raised in an insulated, clean room with artificial lighting and a constant air temperature of 27.5°C. The carboys were first chemically sterilized with Clorox (3 mℓ/ℓ of seawater) and dechlorinated after 4 hr with 10 mℓ of sodium thiosulfate, which removes the residual chlorine. A treatment of chloramphenicol, 24 to 36 hr before using the phytoplankton in a carboy, killed any bacteria and increased the quality of the food. Each carboy was used for only 6 days after antibiotic treatment, which protected against bacteria build up. Hatchery production required 4 to 5 carboys in continuous cycle through the phytoplankton culture process with usually 2 to 3 carboys being harvested at one time.[18]

The techniques for culturing phytoplankton have recently been modified at PRIDE. A two-story geodesic dome building was built to incorporate the newly developed methods. Twenty-four tanks are arranged along the outer perimeter of the second floor of the geodesic dome. Sunlight enters through the transparent fiberglass roof. Greenhouse shade cloth (50%), a ceiling extractor fan, and an oscillating fan keep water temperatures between 28 and 34°C. Also, the cultured phytoplankton species were changed for better larval growth. A local phytoplankton, C-*Isochrysis* (C-Iso) is cultured instead of Tahitian *Isochrysis.* C-Iso is the main food cultured because it has proven nutritionally valuable to veligers at all stages, was isolated from local waters, is probably a food source for conch in the wild, and is highly resistant to *Vibrio* bacteria. *Chaetoceros gracile* is grown and fed in smaller quantities as a supplementary food. It grows well in a tropical environment and adds nutritional value to the veliger. The new facility produces 800 ℓ of C-Iso and 40 ℓ of *C. gracile* daily.[22]

The phytoplankton are cultured in F/1.5 media, modified version of Guillard's F/2 media, using batch culture techniques. All salt water is filtered to 10 μm and UV sterilized. From flask transfer day to feeding day, the growth cycle is 14 days for C-Iso and 12 days for *C. gracile.* The algae are first cultured in 175 mℓ flasks for 5 days in a temperature-controlled cool room. A 20-ℓ carboy is inoculated with three flasks of phytoplankton culture. After 4 days, a 200-ℓ clear fiberglass tank (46 cm × 152 cm) is inoculated with the phytoplankton in half a carboy.[22]

C-Iso and *C. gracile* tanks are used after 4 days of growth when the cell density is

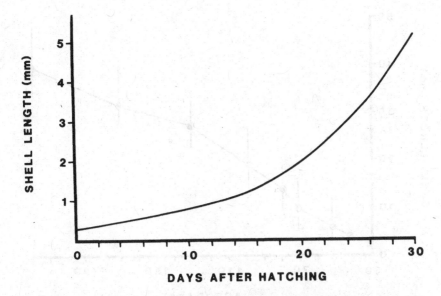

FIGURE 3. Larval and early juvenile growth of *Strombus gigas*. (From Ballantine, D. L. and Appeldoorn, R. S., *Proc. Gulf Caribb. Fish. Inst.*, 35, 57, 1983. With permission.)

approximately 1.5×10^6 cells per milliliter determined by a colorimeter. The quality of the phytoplankton is determined daily. Healthy cultures have few clumps or protozoa, and 24-hr TCBS plates are negative for *Vibrio* spp. C-Iso cultures should have 70 to 100% motility and *C. gracile* cultures should be actively dividing. During feeding, the phytoplankton is siphoned from the culture tanks into five food tubes set in the floor which correspond to specific modules of larval rearing tanks on the first floor of the geodesic dome. At the maximum level of larval production (3×10^6 veligers), 400 ℓ of C-Iso are needed daily.[22]

Macroalgae

Enteromorpha prolifera Muller and *Spyridia filamentosa* (Wulfen) are readily eaten by juvenile queen conch.[26] Under ideal conditions juveniles feed actively during the day and night.[3] Feeding rate is 11.3% of body meat weight per day for *S. filamentosa*, and 10.3% for *E. prolifera;* however, the feeding rate decreases with increased meat weight. *E. prolifera* produces a 8.2% increase and *S. filamentosa* produces a 9.9% increase in meat weight after 21 days.[26]

Food preference in nature changes over the year due to availability and desirability. Queen conch eat all plants found in the habitat and the dominant alga in the area is the dominant food item. Queen conch will ingest pieces of grass up to 35 mm long, although most of the food eaten is algae. Large amounts of sand are also ingested during feeding.[2]

Growth

The growth rate of *Strombus gigas* larvae is dependent on the phytoplankton diet.[24] Growth for larvae and early juveniles averages 53 μm/day (Figures 3 and 4).[14,27] Average length of animals is 2.2 mm (range 1.8 to 2.9 mm) after 27 days and 31.7 mm after 171 days.[3] Larvae fed Tahitian *Isochrysis galbana* Parke grow better than with any other unialgal diet.[23,25] The daily growth rates with different foods are: Tahitian *Isochrysis* — 64 μm/day; *E. huxleyi* — 58 μm/day; *P. minimum* — 52 μm/day; *H. pygmacea* — 33 μm/day; and *D. tertiolecta* — 22 μm/day. Larvae fed Tahitian *Isochrysis* reach 800 μm in 13 days, as opposed to 22 to 25 days with other diets.[25]

Brownell[3] reported a growth of 9 mm/month for the first 9 months, Laughlin and Weil[5]

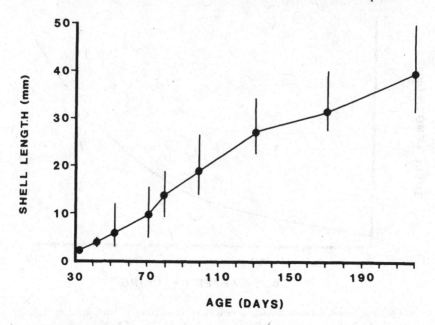

FIGURE 4.   Growth of juvenile *Strombus gigas* reared in the laboratory. Error bars represent size range at each sample period. (Brownell, W. N., Berg, C. J., and Haines, K. C., *F.A.O. Fish. Rep.*, 200, 59, 1977. With permission.)

reported 8 mm/month for the first 5 months and then decreasing to 3 mm/month, and Ballantine and Appeldoorn[13] reported 12 mm/month. Generally, juvenile conch grow an average of 5 to 9 mm/month in the field.[5] Therefore, 2 to 3 years is required for the conch to reach a market size of 200 mm.

The estimated survival from 2 mm to between 20 and 30 mm is as high as 90%.[4] Growth rate is fastest (2 to 3 times) during the first 2 to 3 months and then slowly decreases.[5] This effect is most noticeable in tanks with relatively low conch densities.[5] Growth is greatly reduced at higher rearing densities (Figures 5 and 6).[5,28] Growth rate varies from 0.21 cm/month at 50 juveniles per square meter up to 0.42 cm/month at 25 juveniles per square meter.[5] Growth is inversely proportional to density and follows the equation: Growth (mm) $= -1.08 + 1398 * (\text{individuals/m}^2)^{-1}$.[5,28] Increased locomotor activity and mucus production by juveniles reared at high densities may explain in the lower growth, even though food is plentiful.[29] In addition, there is considerable variation in growth between different tanks. The important factors for growth are light intensity, bottom type, concentrations of metabolic wastes, and availability of appropriate algae for food.[3]

Mortality is not affected by the rearing density but is almost double (20%) when sand is on the bottom of the tank. Ideally juveniles should be raised at densities of 25 to 50 conch per square meter and sand should not be put on the bottom of the tank due to high mortality. Metamorphosed juveniles are left in the rearing tanks until they reach an average size of 10 mm (about 2 months) and are then transferred to grow-out tanks for 1 year (about 6 cm) before being released into the wild.[5]

HATCHERY LOCATION AND COST

A desirable location for a hatchery is in Bahamian or Caribbean waters. A local adult conch population ensures a reliable source of egg masses and grow-out habitats for reseeding or stocking juvenile conch. A productive facility must be large enough to start a batch of conch larvae every day while another batch is undergoing metamorphosis. The facility should

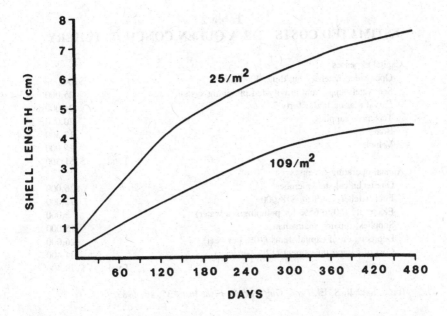

FIGURE 5. Comparison of size/age relationships of two densities of juvenile *Strombus gigas* reared in 200-ℓ tanks. (Laughlin, R. A. and Weil, M. E., *Proc. Gulf Caribb. Fish. Inst.*, 35, 64, 1983. With permission.)

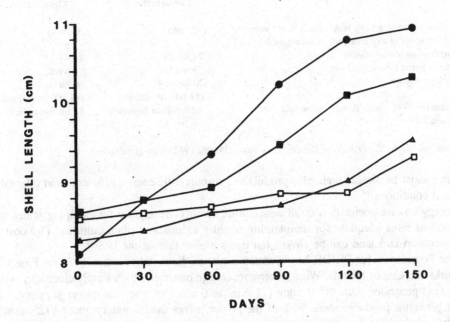

FIGURE 6. Comparison of size/age relationships of juvenile *Strombus gigas* reared at three densities in the laboratory and growth in nature. (●) nature; (■) 17/m²; (△) 31/m²; (□) 54/m². (Laughlin, R. A. and Weil, M. E., *Proc. Gulf Caribb. Fish. Inst.*, 35, 64, 1983. With permission.)

be simple, use wind and solar energy, and use local materials. Routine techniques and specialized equipment guarantee good performance and facilitate training local technicians. Economics, engineering, management techniques, marketing, and biology are highly interdependent when planning a commercial hatchery. A successful facility in a remote developing country must be started with the understanding that its construction, operation, and main-

Table 2
ESTIMATED COSTS FOR A QUEEN CONCH HATCHERY

Capital expenses
Open-sided hatchery building: 4000 m^2	$160,000
Sea water supply and gravity-fed plumbing system	25,000
Larval rearing tanks: forty 730-ℓ	24,000
Technical supplies	20,000
Boat	17,500
Vehicle	15,000
Total	$261,500

Annual operating expenses
One technical staff member	$26,000
Four hatchery staff @ $18,000	72,000
Electricity (50 to 65% for pumping sea water)	5,000
Supplies, repairs, maintenance	15,000
Depreciation of capital items (10% per year)	26,000
Cost of capital investment (12% per year)	34,400
Total	$178,400

(From Siddall, S. E., *Proc. Gulf Caribb. Fish. Inst.*, 35, 46, 1983.

Table 3
ESTIMATED PRODUCTION FOR A QUEEN CONCH HATCHERY

	Conservative	Optimistic
Number of larval rearing tanks which hold "successful" cultures at any time (out of 40 available)	30 tanks	35 tanks
Length of spawning season	24 weeks	24 weeks
Time required for each culture	3 weeks	3 weeks
Stocking density	20 larvae/ℓ (14,600 larvae/tank)	60 larvae/ℓ (43,800 larvae/tank)
Production (90% survival from larvae to 2-cm juveniles)	3.15 million juveniles	11 million juveniles

(From Siddall, S. E., *Proc. Gulf Caribb. Fish. Inst.*, 35, 46, 1983. With permission.)

tenance must be simple, reliable, productive, resourceful, energy efficient, and adaptable to local conditions.[18]

Energy costs associated with an aquaculture facility are mainly for pumping water and aeration at rates adequate for maintaining healthy animal and plant cultures. The cost of electricity on an island can be almost ten times higher than in the U.S.[18]

The Foundation for PRIDE has operated a marine field station since 1976 on Pine Cay in Turks and Caicos Islands. Wind generators charge batteries which supply electricity when the island generator shuts off at night (11 p.m. to 6 a.m.) or when the power goes out. The wind generator produces about 50% of the power, saves the laboratory about $125/month, and has worked for 3 years without requiring any maintenance.[18]

Table 2 gives the estimated cost of building a full production hatchery and Table 3 gives the estimated production levels of the hatchery. The cost of hatchery produced juveniles is between 1.62 to 5.67 cents (U.S.) per animal for 2-cm juveniles (~5 months old). These production costs require a survival rate of 4% (@1.62 cents per juvenile) to 14% (@5.67 cents per juvenile) for 19 months in the wild to break even.[4]

GROW OUT

Both intensive and extensive culture methods for grow out and reseeding of depleted areas

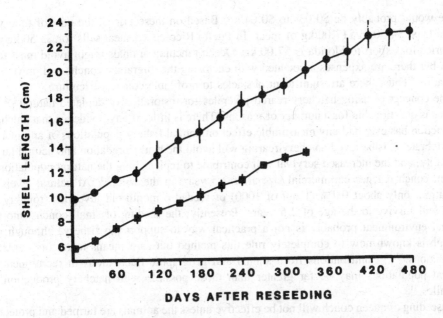

FIGURE 7. Size/age relationships of two groups of juvenile *Strombus gigas* with different initial shell lengths reared in the wild. (■) initial size — 5.8 cm; (●) initial size — 9.9 cm. Error bars represent ± 1 SD. (Laughlin, R. A. and Weil, M. E., *Proc. Gulf Caribb. Fish. Inst.*, 35, 64, 1983. With permission.)

are being investigated by researchers. Several methods have been proposed for grow out: establishing protected areas for ocean ranching by reseeding sandy grass flats, stocking shallow channel-like raceways or ponds along the shore with water circulation by the tidal fluctuation, and penning or embaying stocked juveniles.[18] Intensive culture of conch in pens does not seem practical due to the slow growth associated with this method. Penning is only practical if the structures are built with inexpensive materials, monitored on a regular basis, placed in protected locations with good water exchange, and food is added to ensure good growth. Growth and survival of small conchs (<3 cm) in floating cages indicates this technique can play a important method for over-wintering hatchery-reared conchs. Growth is rapid in cages, the method is inexpensive, and the conchs eat the algae growing on the cage.[30]

The success of reseeding depends on the relationship between growth and mortality of juveniles after release. The important factors are: size of conchs at release, time of release, and the characteristics of the release site.[13] The growth rate of reseeded animals is higher than that of animals kept in the lab (Figure 7).[5] The growth rate in the field is 5.7 to 9.0 mm/month for individuals 9 cm long.[5,13] Appeldoorn and Ballantine[13] found 2- to 5-cm juveniles have a mortality rate of 4 to 5%/day after reseeding. Survival is greatly increased if juveniles are released at a size larger than 4 cm.[7] The optimal nursery habitat must be determined for further progress in reseeding.[13] Reseeding may be possible if combined with strong management mearues for the restoration of conch stocks to their former level.[31]

EVALUATION OF RESEEDING FOR FISHERY MANAGEMENT

Chanley[31] believes the use of hatchery-produced juveniles for reseeding is impractical and cites several reasons for this conclusion. Queen conch are voracious herbivores and require a 20:1 food conversion ratio for satisfactory growth. Since only 12% of conch live weight is edible meat, about 77 kg of algae are required to produce 0.5 kg of conch meat.[7] The present cost in the U.S. of commercially harvested algae is $0.10/kg. The cost of cultivated algae would probably be $0.03, to $0.04/kg. Based on these prices, the cost of food will range from $4.50 to $17.00/kg of meat. In Puerto Rico conch meat sells for $3.50/kg and the price for export to Florida is $7.00/kg. Ocean ranching animals might avoid most food costs but there are expenses associated with enclosing the migratory conchs and preventing predation. Thus, there are significant obstacles to profitable conch mariculture.[31]

The concept of using hatchery-reared juveniles for restoration of depleted natural populations is questionable for a number of reasons. There is little, if any, evidence that hatchery production has ever had any measurable effect on natural fishery populations of any marine invertebrate. It is believed hatchery rearing will avoid the heavy predation and natural larval mortality, and the increased survival will contribute to replenishing the natural population.[31] Queen conch reaches commercial size after 2.5 years in the wild.[7,13] At natural levels of predation, only about 0.1% (1 out of 1000) of seeded 6-month-old juveniles (roughly 60 mm) will survive to the age of 2.5 years. Presently, the seeding of small conchs into the natural environment probably is not a practical way to support the fishery, although not enough is known now to completely rule this method out as a means to restore depleted populations.[13,32] Contributions from annual and geographic fluctuations in recruitment of natural populations may be far greater than those possible with hatchery production of juveniles.[31]

Reseeding of queen conch will not be effective unless the animals are farmed and protected from predation, used for a put and take fishery, released under unusual or special circumstances, or are introduced to an area where they do not exist and thereafter protected so the population can maintain itself by natural reproduction.[31]

The cause of the decline in the conch populations is not from recruitment failure but from overfishing. Reseeding, even if it is cost effective, does not solve the problem. In Chanley's opinion, conch populations can only be restored by effective fishery management, which is not possible unless more is known about the biology of the animal. Hatchery-reared juveniles are useful for conducting research into the behavior, movement, mortality, causes of mortality, desired habitat, food preference, growth, etc. Based upon this information, fishery management programs which have a realistic chance of restoring depleted conch populations can be developed. This is where the technology of conch hatchery production may be most effective.[31]

REFERENCES

1. **Brownell, W. N. and Stevely, J. M.,** The biology, fisheries, and management of the queen conch, *Strombus gigas, Mar. Fish. Rev.,* 43 (7), 1, 1981.
2. **Randall, J. E.,** Contributions to the biology of the Queen conch, *Strombus gigas, Bull. Mar. Sci.,* 14, 246, 1964.
3. **Brownell, W. N.,** Reproduction, laboratory culture, and growth of *Strombus gigas, S. costatus,* and *S. pugilus* in Los Roques, Venezuela, *Bull. Mar. Sci.,* 27(4), 668, 1977.
4. **Siddall, S. E.,** Biological and economic outlook for hatchery production of juvenile Queen conch, *Proc. Gulf Caribb. Fish. Inst.,* 35, 46, 1983.
5. **Laughlin, R. A. and Weil, M. E.,** Queen conch mariculture and restoration in the archipielago de Los Roques: preliminary results, *Proc. Gulf Caribb. Fish. Inst.,* 35, 64, 1983.
6. **Menzel, R. W.,** Possibilities of molluscan cultivation in the Caribbean, *F.A.O. Fish. Rep.,* 1971(2), 183, 1971.
7. **Berg, C. J.,** Growth of the Queen conch *Strombus gigas,* with a discussion of the practicality of its mariculture, *Mar. Biol.,* 34, 191, 1976.
8. **Siddall, S. E.,** Synopsis of recent research on the queen conch *Strombus gigas* Linné, *J. Shellfish Res.,* 4(1), 1, 1984.
9. **Adams, J. E.,** Conch fishing industry of Union Island, Grenadines, West Indies, *J. Trop. Sci.,* 12, 279, 1970.

10. **Goodwin, M. H.,** Overview of conch fisheries and culture, *Proc. Gulf Caribb. Fish. Inst.,* 35, 43, 1983.
11. **Jory, D. E. and Iversen, E. S.,** Queen conch predators: not a roadblock to mariculture, *Proc. Gulf Caribb. Fish. Inst.,* 35, 108, 1983.
12. **Weil, M. E. and Laughlin, G. R.,** Biology, population dynamics, and reproduction of the queen conch *Strombus gigas* Linné in the archipielago de Los Roques National Park, *J. Shellfish Res.,* 4(1), 45, 1984.
13. **Appeldoorn, R. S. and Ballantine, D. L.,** Field release of cultured Queen conch in Puerto Rico: implications for stock restoration, *Proc. Gulf Caribb. Fish. Inst.,* 35, 89, 1983.
14. **Ballantine, D. L. and Appeldoorn, R. S.,** Queen conch culture and future prospects in Puerto Rico, *Proc. Gulf Caribb. Fish. Inst.,* 35, 57, 1983.
15. **Appeldoorn, R. S.,** The effect of size on mortality of small juvenile conchs *(Strombus gigas* Linné and *S. costatus* Gmelin), *J. Shellfish Res.,* 4(1), 37, 1984.
16. **D'Asaro, C. N.,** Organogenesis, development and metamorphosis in the Queen conch *Strombus gigas,* with notes on the breeding habits, *Bull. Mar. Sci.,* 15, 359, 1965.
17. **Blakesley, H. L.,** A contribution of the fisheries and biology of the queen conch, *Strombus gigas* L. in Belize, Abstract, Am. Fish. Soc., Ann. Meet., Vancouver, Canada, 1977.
18. **Davis, M. and Hesse, C.,** Third world level conch mariculture in the Turks and Caicos Islands, *Proc. Gulf Caribb. Fish. Inst.,* 35, 73, 1983.
19. **Hensen, R. R.,** Queen conch management and culture in the Netherlands Antilles, *Proc. Gulf Caribb. Fish. Inst.,* 35, 53, 1983.
20. **Siddall, S. E.,** Temporal changes in the salinity and temperature requirement of tropical mussel larvae, *Proc. World Maricult. Soc.,* 9, 549, 1979.
21. **Davis, M., Mitchell, B. A., and Brown, J. L.,** Breeding behavior of the queen conch *Strombus gigas* Linné held in a natural enclosed habitat, *J. Shellfish Res.,* 4(1), 17, 1984.
22. **Ray, M. and Davis, M.,** Algae production for commercially grown queen conch *(Strombus gigas), Proc. Gulf Caribb. Fish. Inst.,* 38, (in press), 1987.
23. **Pillsbury, K. S.,** The relative food value and biochemical composition of five phytoplankton diets for queen conch, *Strombus gigas* (Linne) larvae, *J. Exp. Mar. Biol. Ecol.,* 90, 221, 1985.
24. **Pillsbury, K.,** Lipid requirements for larvae of *Strombus gigas* (Linné), *J. Shellfish Res.,* 4(1), 111, 1984.
25. **Pillsbury, K.,** Nutritional value of three species of algae to larvae of the queen conch *Strombus gigas* (Linné), *J. Shellfish Res.,* 4(1), 98, 1984.
26. **Creswell, L.,** Ingestion, assimilation, and growth of juveniles of the queen conch *Strombus gigas* Linné fed experimental diets, *J. Shellfish Res.,* 4(1), 23, 1984.
27. **Brownell, W. N., Berg, C. J., and Haines, K. C.,** Fisheries and aquaculture of the conch, *Strombus gigas,* in the Caribbean, *F.A.O. Fish. Rep.,* 200, 59, 1977.
28. **Appeldoorn, R. S. and Sanders, I. M.,** Quantification of the density-growth relationship in hatchery-reared juvenile conchs *(Strombus gigas* Linné and *S. costatus* Gmelin), *J. Shellfish Res.,* 4(1), 63, 1984.
29. **Siddall, S. E.,** Density-dependent levels of activity of juveniles of the queen conch *Strombus gigas* Linné, *J. Shellfish Res.,* 4(1), 67, 1984.
30. **Iversen, E. S.,** Feasibility of increasing Bahamian conch production by mariculture, *Proc. Gulf Caribb. Fish. Inst.,* 35, 83, 1983.
31. **Chanley, P.,** Queen Conch culture; is it practical?, *Catch '82, Shellfish. Newsl.,* 16, 12, 1982.
32. **Appeldoorn, R. S.,** personal communication, 1986.

Index

INDEX

Printed in the United States
by Baker & Taylor Publisher Services